Category Theory for the Sciences

Category Theory for the Sciences

David I. Spivak

The MIT Press
Cambridge, Massachusetts
London, England

Library of Congress Cataloging-in-Publication Data

Spivak, David I., 1978– author.
Category theory for the sciences / David I. Spivak.
    pages cm
Includes bibliographical references and index.
ISBN 978-0-262-02813-4 (hardcover : alk. paper) 1.
Science—Mathematical models. 2. Categories (Mathematics) I. Title.
Q175.32.M38S65 2014
512'.62—dc23
                            2014007215

10   9   8   7   6

Contents

Acknowledgments

I would like to express my deep appreciation to the many scientists with whom I have worked over the past six years. It all started with Paea LePendu, who first taught me about databases when I was naively knocking on doors in the University of Oregon computer science department. This book would never have been written if Tristan Nguyen and Dave Balaban had not noticed my work and encouraged me to continue. Dave Balaban and Peter Gates have been my scientific partners since the beginning, working hard to understand what I am offering and working just as hard to help me understand all that I am missing. Peter Gates has deepened my understanding of data in profound ways.

I have also been tremendously lucky to know Haynes Miller, who made it possible for me to settle at MIT, with the help of Clark Barwick and Jacob Lurie. I knew that MIT would be the best place in the world for me to pursue this type of research, and it consistently lives up to expectation. Researchers like Markus Buehler and his graduate students Tristan Giesa and Dieter Brommer have been a pleasure to work with, and the many materials science examples scattered throughout this book are a testament to how much our work together has influenced my thinking.

I would also like to thank the collaborators and conversation partners with whom I have discussed subjects written about in this book. Besides the people mentioned previously, these include Steve Awodey, Allen Brown, Adam Chlipala, Carlo Curino, Dan Dugger, Henrik Forssell, David Gepner, Jason Gross, Bob Harper, Ralph Hutchison, Robert Kent, Jack Morava, Scott Morrison, David Platt, Joey Perricone, Dylan Rupel, Guarav Singh, Sam Shames, Nat Stapleton, Patrick Schultz, Ka Yu Tam, Ryan Wisnesky, Jesse Wolfson, and Elizabeth Wood.

I would like to thank Peter Kleinhenz and Peter Gates for reading an earlier version of this book and providing invaluable feedback before I began teaching the 18-S996 class at MIT in spring 2013. In particular, the first figure of the book, Figure 1.1, is a slight alteration of a diagram Gates sent me to help motivate the book for scientists. I would also like to greatly thank the 18-S996 course grader Darij Grinberg, who not only was the best grader I have had in my 14 years of teaching, but gave me more comments than anyone else on the book itself. I would like to thank the students from the 18-S996 class at MIT who found typos, pointed out unclear explanations, and generally helped improve the book in many ways: Aaron Brookner, Leon Dimas, Dylan Erb, Deokhwan Kim, Taesoo Kim, Owen Lewis, Yair Shenfeld, and Adam Strandberg, among others. People outside the class, V. Galchin, K. Hofmeyr, D. McAdams, D. Holmes, C. McNally, P. O'Neill, and R. Harper, also contributed to finding errata and making improvements.

I'd also like to thank Marie Lufkin Lee, Marc Lowenthal, Katherine Almeida, and everyone else at MIT Press who helped get this book ready for publication. And thanks to Laura Baldwin, who helped me work through some painful LaTeX issues. The book is certainly far better than when I originally submitted it. I also appreciate the willingness of the Press to work with me in making a copy of this book publicly available.

Thanks also to my teacher Peter Ralston, who taught me to repeatedly question the obvious. My ability to commit to a project like this one and to see it to fruition has certainly been enhanced since I studied with him.

Finally, I acknowledge my appreciation for support from the Office of Naval Research and Air Force Office of Scientific Research[1] without which this book would not have been remotely possible. I believe that the funding of basic research is an excellent way of ensuring that the United States remains a global leader in the years to come.

---

[1]Grant numbers: N000140910466, N000141010841, N000141310260, FA9550-14-1-0031.

Chapter 1

Introduction

The diagram in Figure 1.1 is intended to evoke thoughts of the scientific method.

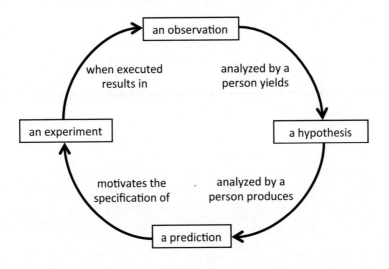

Figure 1.1

An observation analyzed by a person yields a hypothesis, which analyzed by a person produces a prediction, which motivates the specification of an experiment, which when executed results in an observation.

Its statements look valid, and a good graphic can be very useful for leading a reader through a story that the author wishes to tell.

But a graphic has the power to evoke feelings of understanding without really meaning much. The same is true for text: it is possible to use a language like English to express ideas that are never made rigorous or clear. When someone says, "I believe in free will," what does she believe in? We may all have some concept of what she's saying—something we can conceptually work with and discuss or argue about. But to what extent are we all discussing the same thing, the thing she intended to convey?

Science is about agreement. When we supply a convincing argument, the result of this convincing is agreement. When, in an experiment, the observation matches the hypothesis—success!—that is agreement. When my methods make sense to you, that is agreement. When practice does not agree with theory, that is disagreement. Agreement is the good stuff in science; it is the celebratory moment.

But it is easy to think we are in agreement, when we really are not. Modeling our thoughts on heuristics and graphics may be convenient for quick travel down the road, but we are liable to miss our turnoff at the first mile. The danger is in mistaking convenient conceptualizations for what is actually there. It is imperative that we have the ability at any time to ground in reality. What does that mean?

Data. Hard evidence. The physical world. It is here that science is grounded and heuristics evaporate. So let's look again at Figure 1.1. It is intended to evoke an idea of how science is performed. Do hard evidence and data back up this theory? Can we set up an experiment to find out whether science is actually performed according to such a protocol? To do so we have to shake off the impressions evoked by the diagram and ask, What does this diagram intend to communicate?

In this book I will use a mathematical tool called *ologs*, or ontology logs, to give some structure to the kinds of ideas that are often communicated in graphics. Each olog inherently offers a framework in which to record data about the subject. More precisely, it encompasses a *database schema*, which means a system of interconnected tables that are initially empty but into which data can be entered. For example, consider the following olog:

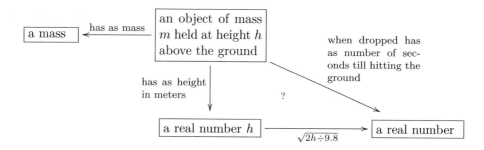

This olog represents a framework in which to record data about objects held above the ground, their mass, their height, and a comparison (the question mark) between the number of seconds till they hit the ground and a certain real-valued function of their height. Ologs are discussed in detail throughout this book.

Figure 1.1 looks like an olog, but it does not conform to the rules laid out for ologs (see Section 2.3). In an olog, every arrow is intended to represent a mathematical function. It is difficult to imagine a function that takes in predictions and outputs experiments, but such a function is necessary in order for the arrow

$$\boxed{\text{a prediction}} \xrightarrow{\text{motivates the specification of}} \boxed{\text{an experiment}}$$

in Figure 1.1 to make sense. To produce an experiment design from a prediction probably requires an expert, and even then the expert may be motivated to specify a different experiment on Tuesday than he is on Monday. But perhaps this criticism leads to a way forward. If we say that every arrow represents a function *when in the context of a specific expert who is actually doing the science at a specific time*, then Figure 1.1 begins to make sense. In fact, the figure is reconsidered in Section 7.3 (Example 7.3.3.10), where background methodological context is discussed.

This book extols the virtues of a new branch of mathematics, *category theory*, which was invented for powerful communication of ideas between different fields and subfields within mathematics. By powerful communication of ideas I mean something precise. Different branches of mathematics can be formalized into categories. These categories can then be connected by functors. And the sense in which these functors provide powerful communication of ideas is that facts and theorems proven in one category can be transferred through a connecting functor to yield proofs of analogous theorems in another category. A functor is like a conductor of mathematical truth.

I believe that the language and tool set of category theory can be useful throughout science. We build scientific understanding by developing models, and category theory is the study of basic conceptual building blocks and how they cleanly fit together to make such models. Certain structures and conceptual frameworks show up again and again in our understanding of reality. No one would dispute that vector spaces are ubiquitous throughout the sciences. But so are hierarchies, symmetries, actions of agents on objects, data models, global behavior emerging as the aggregate of local behavior, self-similarity, and the effect of methodological context.

Some ideas are so common that our use of them goes virtually undetected, such as set-theoretic intersections. For example, when we speak of a material that is both lightweight and ductile, we are intersecting two sets. But what is the use of even mentioning this set-theoretic fact? The answer is that when we formalize our ideas, our understanding is clarified. Our ability to communicate with others is enhanced, and the possibility for developing new insights expands. And if we are ever to get to the point that we can input our ideas into computers, we will need to be able to formalize these ideas first.

It is my hope that this book will offer scientists a new vocabulary in which to think and communicate, and a new pipeline to the vast array of theorems that exist and are considered immensely powerful within mathematics. These theorems have not made their way into the world of science, but they are directly applicable there. Hierarchies are partial orders, symmetries are group elements, data models are categories, agent actions are monoid actions, local-to-global principles are sheaves, self-similarity is modeled by operads, context can be modeled by monads. All of these will be discussed in the book.

1.1   A brief history of category theory

The paradigm shift brought on by Einstein's theory of relativity led to a widespread realization that there is no single perspective from which to view the world. There is no background framework that we need to find; there are infinitely many different frameworks and perspectives, and the real power lies in being able to translate between them. It is in this historical context that category theory got its start.[1]

Category theory was invented in the early 1940s by Samuel Eilenberg and Saunders Mac Lane. It was specifically designed to bridge what may appear to be two quite different fields: topology and algebra. Topology is the study of abstract shapes such as 7-dimensional spheres; algebra is the study of abstract equations such as $y^2z = x^3 - xz^2$. People had already created important and useful links (e.g., cohomology theory) between these fields, but Eilenberg and Mac Lane needed to precisely compare different links with one another. To do so they first needed to boil down and extract the fundamental nature of these two fields. But in doing so, the ideas they worked out amounted to a framework that fit not only topology and algebra, but many other mathematical disciplines as well.

At first category theory was little more than a deeply clarifying language for existing difficult mathematical ideas. However, in 1957 Alexander Grothendieck used category theory to build new mathematical machinery (new cohomology theories) that granted unprecedented insight into the behavior of algebraic equations. Since that time, categories have been built specifically to zoom in on particular features of mathematical subjects and study them with a level of acuity that is unavailable elsewhere.

Bill Lawvere saw category theory as a new foundation for all mathematical thought. Mathematicians had been searching for foundations in the nineteenth century and were reasonably satisfied with set theory as *the foundation*. But Lawvere showed that the category of sets is simply one category with certain nice properties, not necessarily the

---

[1] The following history of category theory is far too brief and perhaps reflects more of the author's aesthetic than any kind of objective truth. References are Kromer [19], Marquis [30], and Landry and Marquis [22].

center of the mathematical universe. He explained how whole algebraic theories can be viewed as examples of a single system. He and others went on to show that higher-order logic was beautifully captured in the setting of category theory (more specifically toposes). It is here also that Grothendieck and his school worked out major results in algebraic geometry.

In 1980, Joachim Lambek showed that the types and programs used in computer science form a specific kind of category. This provided a new semantics for talking about programs, allowing people to investigate how programs combine and compose to create other programs, without caring about the specifics of implementation. Eugenio Moggi brought the category-theoretic notion of monads into computer science to encapsulate ideas that up to that point were considered outside the realm of such theory.

It is difficult to explain the clarity and beauty brought to category theory by people like Daniel Kan and André Joyal. They have each repeatedly extracted the essence of a whole mathematical subject to reveal and formalize a stunningly simple yet extremely powerful pattern of thinking, revolutionizing how mathematics is done.

All this time, however, category theory was consistently seen by much of the mathematical community as ridiculously abstract. But in the twenty-first century it has finally come to find healthy respect within the larger community of pure mathematics. It is the language of choice for graduate-level algebra and topology courses, and in my opinion will continue to establish itself as the basic framework in which to think about and express mathematical ideas.

As mentioned, category theory has branched out into certain areas of science as well. Baez and Dolan [6] have shown its value in making sense of quantum physics, it is well established in computer science, and it has found proponents in several other fields as well. But to my mind, we are at the very beginning of its venture into scientific methodology. Category theory was invented as a bridge, and it will continue to serve in that role.

1.2 Intention of this book

The world of *applied mathematics* is much smaller than the world of *applicable mathematics*. As mentioned, this book is intended to create a bridge between the vast array of mathematical concepts that are used daily by mathematicians to describe all manner of phenomena that arise in our studies and the models and frameworks of scientific disciplines such as physics, computation, and neuroscience.

For the pure mathematician I try to prove that concepts such as categories, functors, natural transformations, limits, colimits, functor categories, sheaves, monads, and operads—concepts that are often considered too abstract even for math majors—can be communicated to scientists with no math background beyond linear algebra. If this material is as teachable as I think, it means that category theory is not esoteric but well aligned with ideas that already make sense to the scientific mind. Note, however, that this book is example-based rather than proof-based, so it may not be suitable as a reference for students of pure mathematics.

For the scientist I try to prove the claim that category theory includes a formal treatment of conceptual structures that the scientist sees often, perhaps without realizing that there is well-oiled mathematical machinery to be employed. A major topics is the structure of information itself: how data is made meaningful by its connections, both internal and outreaching, to other data.[2] Note, however, that this book should certainly not be taken as a reference on scientific matters themselves. One should assume that any account of physics, materials science, chemistry, and so on, has been oversimplified. The intention is to give a flavor of how category theory may help model scientific ideas, not to explain those ideas in a serious way.

Data gathering is ubiquitous in science. Giant databases are currently being mined for unknown patterns, but in fact there are many (many) known patterns that simply have not been catalogued. Consider the well-known case of medical records. In the early twenty-first century, it is often the case that a patient's medical history is known by various doctor's offices but quite inadequately shared among them. Sharing medical records often means faxing a handwritten note or a filled-in house-created form from one office to another.

Similarly, in science there exists substantial expertise making brilliant connections between concepts, but this expertise is conveyed in silos of English prose known as journal articles. Every scientific journal article has a methods section, but it is almost impossible to read a methods section and subsequently repeat the experiment—the English language is inadequate to precisely and concisely convey what is being done.

The first thought I wish to convey in this book is that reusable methodologies can be formalized and that doing so is inherently valuable. Consider the following analogy. Suppose one wants to add up the area of a region in space (or the area under a curve). One breaks the region down into small squares, each with area A, and then counts the number of squares, say n. One multiplies these numbers together and says that the region has an area of about nA. To obtain a more precise and accurate result, one repeats the process with half-size squares. This methodology can be used for any area-finding problem (of which there are more than a first-year calculus student generally realizes) and thus it deserves to be formalized. But once we have formalized this methodology, it can be taken to its limit, resulting in integration by Riemann sums. Formalizing the problem can lead

---

[2]The word *data* is generally considered to be the plural form of the word *datum*. However, individual datum elements are only useful when they are organized into structures (e.g., if one were to shuffle the cells in a spreadsheet, most would consider the data to be destroyed). It is the whole organized structure that really houses the information; the data must be in formation in order to be useful. Thus I use the word *data* as a collective noun (akin to *sand*); it bridges the divide between the *individual datum elements* (akin to grains of sand) and the *data set* (akin to a sand pile).

to powerful techniques that were unanticipated at the outset.

I intend to show that category theory is incredibly efficient as a language for experimental design patterns, introducing formality while remaining flexible. It forms a rich and tightly woven conceptual fabric that allows the scientist to maneuver between different perspectives whenever the need arises. Once she weaves that fabric into her own line of research, she has an ability to think about models in a way that simply would not occur without it. Moreover, putting ideas into the language of category theory forces a person to clarify her assumptions. This is highly valuable both for the researcher and for her audience.

What must be recognized in order to find value in this book is that conceptual chaos is a major problem. Creativity demands clarity of thinking, and to think clearly about a subject requires an organized understanding of how its pieces fit together. Organization and clarity also lead to better communication with others. Academics often say they are paid to think and understand, but that is not the whole truth. They are paid to think, understand, and *communicate their findings*. Universal languages for science, such as calculus and differential equations, matrices, or simply graphs and pie charts, already exist, and they grant us a cultural cohesiveness that makes scientific research worthwhile. In this book I attempt to show that category theory can be similarly useful in describing complex scientific understandings.

1.3 What is requested from the student

The only way to learn mathematics is by doing exercises. One does not get fit by merely looking at a treadmill or become a chef by merely reading cookbooks, and one does not learn math by watching someone else do it. There are about 300 exercises in this book. Some of them have solutions in the text, others have solutions that can only be accessed by professors teaching the class.

A good student can also make up his own exercises or simply play around with the material. This book often uses databases as an entry to category theory. If one wishes to explore categorical database software, FQL (functorial query language) is a great place to start. It may also be useful in solving some of the exercises.

1.4 Category theory references

I wrote this book because the available books on category theory are almost all written for mathematicians (the rest are written for computer scientists). One book, *Conceptual Mathematics* by Lawvere and Schanuel [24], offers category theory to a wider audience, but its style is not appropriate for a course or as a reference. Still, it is very well written and clear.

The bible of category theory is *Categories for the Working Mathematician* by Mac Lane [29]. But as the title suggests, it was written for working mathematicians and would be opaque to my target audience. However, once a person has read the present book, Mac Lane's book may become a valuable reference.

Other good books include Awodey's *Category theory* [4], a recent gentle introduction by Simmons [37], and Barr and Wells's *Category Theory for Computing Science*, [11]. A paper by Brown and Porter, "Category Theory: an abstract setting for analogy and comparison" [9] is more in line with the style of this book, only much shorter. Online, I find Wikipedia [46] and a site called *nLab* [34] to be quite useful.

This book attempts to explain category theory by examples and exercises rather than by theorems and proofs. I hope this approach will be valuable to the working scientist.

Chapter 2

The Category of Sets

The theory of sets was invented as a foundation for all of mathematics. The notion of sets and functions serves as a basis on which to build intuition about categories in general. This chapter gives examples of sets and functions and then discusses commutative diagrams. Ologs are then introduced, allowing us to use the language of category theory to speak about real world concepts. All this material is basic set theory, but it can also be taken as an investigation of the *category of sets*, which is denoted **Set**.

2.1 Sets and functions

People have always found it useful to put things into bins.

$$\boxed{\text{a thing}} \xrightarrow{\text{ is put into }} \boxed{\text{a bin}}$$

The study of sets is the study of things in bins.

2.1.1 Sets

You probably have an innate understanding of what a set is. We can think of a set X as a collection of *elements* $x \in X$, each of which is recognizable as being in X and such that for each pair of named elements $x, x' \in X$ we can tell if $x = x'$ or not.[1] The set of pendulums is the collection of things we agree to call pendulums, each of which is

---

[1]Note that the symbol x', read "x-prime," has nothing to do with calculus or derivatives. It is simply notation used to name a symbol that is somehow like x. This suggestion of kinship between x and x' is meant only as an aid for human cognition, not as part of the mathematics.

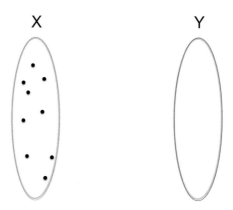

Figure 2.1 A set X with nine elements, and a set Y with no elements, $Y = \varnothing$.

recognizable as being a pendulum, and for any two people pointing at pendulums we can tell if they're pointing at the same pendulum or not.

Notation 2.1.1.1. The symbol $\varnothing$ denotes the set with no elements (see Figure 2.1), which can also be written as { }. The symbol $\mathbb{N}$ denotes the set of natural numbers:

$$\mathbb{N} := \{0, 1, 2, 3, 4, \ldots, 877, \ldots\}. \tag{2.1}$$

The symbol $\mathbb{Z}$ denotes the set of integers, which contains both the natural numbers and their negatives,

$$\mathbb{Z} := \{\ldots, -551, \ldots, -2, -1, 0, 1, 2, \ldots\}. \tag{2.2}$$

If A and B are sets, we say that A is a *subset* of B, and write $A \subseteq B$, if every element of A is an element of B. So we have $\mathbb{N} \subseteq \mathbb{Z}$. Checking the definition, one sees that for any set A, we have (perhaps uninteresting) subsets $\varnothing \subseteq A$ and $A \subseteq A$. We can use *set-builder notation* to denote subsets. For example, the set of even integers can be written $\{n \in \mathbb{Z} \mid n \text{ is even}\}$. The set of integers greater than 2 can be written in many ways, such as

$$\{n \in \mathbb{Z} \mid n > 2\} \qquad \text{or} \qquad \{n \in \mathbb{N} \mid n > 2\} \qquad \text{or} \qquad \{n \in \mathbb{N} \mid n \geqslant 3\}.$$

The symbol $\exists$ means "there exists." So we could write the set of even integers as

$$\{n \in \mathbb{Z} \mid n \text{ is even}\} = \{n \in \mathbb{Z} \mid \exists m \in \mathbb{Z} \text{ such that } 2m = n\}.$$

The symbol $\exists!$ means "there exists a unique." So the statement "$\exists! x \in \mathbb{R}$ such that $x^2 = 0$" means that there is one and only one number whose square is 0. Finally, the symbol $\forall$

means "for all." So the statement "$\forall m \in \mathbb{N} \; \exists n \in \mathbb{N}$ such that $m < n$" means that for every number there is a bigger one.

As you may have noticed in defining $\mathbb{N}$ and $\mathbb{Z}$ in (2.1) and (2.2), we use the colon-equals notation "$A := XYZ$" to mean something like "define A to be XYZ." That is, a colon-equals declaration does not denote a fact of nature (like $2 + 2 = 4$) but a choice of the writer.

We also often discuss a certain set with one element, denoted $\{\odot\}$, as well as the familiar set of real numbers, $\mathbb{R}$, and some variants such as $\mathbb{R}_{\geqslant 0} := \{x \in \mathbb{R} \mid x \geqslant 0\}$.

Exercise 2.1.1.2.

Let $A := \{1, 2, 3\}$. What are all the subsets of A? Hint: There are eight. $\diamond$

Solution 2.1.1.2.

The most obvious ones are these six:

$$\{1\}, \quad \{2\}, \quad \{3\}, \quad \{1, 2\}, \quad \{1, 3\}, \quad \{2, 3\}.$$

But the empty set and the whole set are always subsets too:

$$\varnothing, \quad \{1, 2, 3\}.$$

$\blacklozenge$

A set can have other sets as elements. For example, the set

$$X := \{\{1, 2\}, \{4\}, \{1, 3, 6\}\}$$

has three elements, each of which is a set.

2.1.2   Functions

If X and Y are sets, then a *function f from X to Y*, denoted $f \colon X \to Y$, is a mapping that sends each element $x \in X$ to an element of Y, denoted $f(x) \in Y$. We call X the *domain* of the function f, and we call Y the *codomain* of f.

Note that for every element $x \in X$, there is exactly one arrow emanating from x, but for an element $y \in Y$, there can be several arrows pointing to y, or there can be no arrows pointing to y (see Figure 2.2).

Slogan 2.1.2.1.

Given a function $f \colon X \to Y$, we think of X as a set of things, and Y as a set of bins. The function tells us in which bin to put each thing.

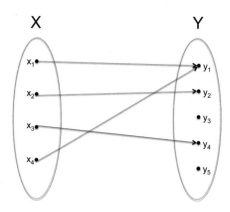

Figure 2.2 A function from a set X to a set Y.

Application 2.1.2.2. In studying the mechanics of materials, one wishes to know how a material responds to tension. For example, a rubber band responds to tension differently than a spring does. To each material we can associate a force-extension curve, recording how much force the material carries when extended to various lengths. Once we fix a methodology for performing experiments, finding a material's force-extension curve would ideally constitute a function from the set of materials to the set of curves.

◊◊

Exercise 2.1.2.3.

Here is a simplified account of how the brain receives light. The eye contains about 100 million photoreceptor (PR) cells. Each connects to a retinal ganglion (RG) cell. No PR cell connects to two different RG cells, but usually many PR cells can attach to a single RG cell.

Let PR denote the set of photoreceptor cells, and let RG denote the set of retinal ganglion cells.

a. According to the above account, does the connection pattern constitute a function $RG \to PR$, a function $PR \to RG$, or neither one?

b. Would you guess that the connection pattern that exists between other areas of the brain are function-like? Justify your answer.

◊

Solution 2.1.2.3.

a. To every element of PR we associate an element of RG, so this is a function $PR \to RG$.

b. (Any justified guess is legitimate.) With no background in the subject, I might guess this happens in any case of immediate perception being translated to neural impulses.

$\blacklozenge$

Example 2.1.2.4. Suppose that X is a set and $X' \subseteq X$ is a subset. Then we can consider the function $X' \to X$ given by sending every element of X' to "itself" as an element of X. For example, if $X = \{a, b, c, d, e, f\}$ and $X' = \{b, d, e\}$, then $X' \subseteq X$. We turn that into the function $X' \to X$ given by $b \mapsto b, d \mapsto d, e \mapsto e$.[2]

As a matter of notation, we may sometimes say the following: Let X be a set, and let $i: X' \subseteq X$ be a subset. Here we are making clear that X' is a subset of X, but that i is the name of the associated function.

Exercise 2.1.2.5.

Let $f: \mathbb{N} \to \mathbb{N}$ be the function that sends every natural number to its square, e.g., $f(6) = 36$. First fill in the blanks, then answer a question.

a. $2 \mapsto$ _____

b. $0 \mapsto$ _____

c. $-2 \mapsto$ _____

d. $5 \mapsto$ _____

e. Consider the symbol $\to$ and the symbol $\mapsto$. What is the difference between how these two symbols are used so far in this book?

$\diamond$

Solution 2.1.2.5.

a. 4

b. 0

c. The function does not apply to -2 because -2 is not an element of $\mathbb{N}$.

---

[2]This kind of arrow, $\mapsto$, is read "maps to." A function $f: X \to Y$ means a rule for assigning to each element $x \in X$ an element $f(x) \in Y$. We say that "x maps to $f(x)$" and write $x \mapsto f(x)$.

d. 25

e. The symbol $\to$ is used to denote a function from one set to another. For example, the arrow in $g\colon X \to Y$ is a symbol that tells us that g is the name of a function from set X to set Y. The symbol $\mapsto$ is used to tell us where the function sends a specific element of the domain. So in our squaring function $f\colon \mathbb{N} \to \mathbb{N}$, we write $5 \mapsto 25$ because the function f sends 5 to 25.

♦

Given a function $f\colon X \to Y$, the elements of Y that have at least one arrow pointing to them are said to be *in the image* of f; that is, we have

$$\mathrm{im}(f) := \{y \in Y \mid \exists x \in X \text{ such that } f(x) = y\}. \tag{2.3}$$

The image of a function f is always a subset of its codomain, $\mathrm{im}(f) \subseteq Y$.

Exercise 2.1.2.6.

If $f\colon X \to Y$ is depicted by Figure 2.2, write its image, $\mathrm{im}(f)$ as a set.   ◇

Solution 2.1.2.6.

The image is $\mathrm{im}(f) = \{y_1, y_2, y_4\}$.   ♦

Given a function $f\colon X \to Y$ and a function $g\colon Y \to Z$, where the codomain of f is the same set as the domain of g (namely, Y), we say that f and g are *composable*

$$X \xrightarrow{\ f\ } Y \xrightarrow{\ g\ } Z.$$

The *composition of f and g* is denoted by $g \circ f\colon X \to Z$. See Figure 2.3.

Slogan 2.1.2.7.

Given composable functions $X \xrightarrow{f} Y \xrightarrow{g} Z$, we have a way of putting every thing in X into a bin in Y, and we have a way of putting each bin from Y into a larger bin in Z. The composite, $g \circ f\colon X \to Z$, is the resulting way that every thing in X is put into a bin in Z.

Exercise 2.1.2.8.

If $A \subseteq X$ is a subset, Example 2.1.2.4 showed how to think of it as a function $i\colon A \to X$. Given a function $f\colon X \to Y$, we can compose $A \xrightarrow{i} X \xrightarrow{f} Y$ and get a function $f \circ i\colon A \to Y$. The image of this function is denoted

$$f(A) := \mathrm{im}(f \circ i),$$

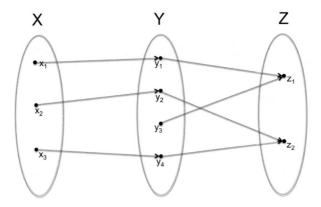

Figure 2.3 Functions $f\colon X \to Y$ and $g\colon Y \to Z$ compose to a function $g \circ f\colon X \to Z$ (follow the arrows).

see (2.3) for the definition of image.

Let $X = Y := \mathbb{Z}$, let $A := \{-1, 0, 1, 2, 3\} \subseteq X$, and let $f\colon X \to Y$ be given by $f(x) = x^2$. What is the image set $f(A)$?   ◊

Solution 2.1.2.8.

By definition of image (see (2.3), we have

$$f(A) = \{y \in \mathbb{Z} \mid \exists a \in A \text{ such that } f \circ i(a) = y\}.$$

Since $A = \{-1, 0, 1, 2, 3\}$ and since $i(a) = a$ for all $a \in A$, we have $f(A) = \{0, 1, 4, 9\}$. Note that an element of a set can only be in the set once; even though $f(-1) = f(1) = 1$, we need only mention 1 once in $f(A)$. In other words, if a student has an answer such as $\{1, 0, 1, 4, 9\}$, this suggests a minor confusion.   ♦

Notation 2.1.2.9. Let X be a set and $x \in X$ an element. There is a function $\{\odot\} \to X$ that sends $\odot \mapsto x$. We say that this function *represents* $x \in X$. We may denote it $x\colon \{\odot\} \to X$.

Exercise 2.1.2.10.

Let X be a set, let $x \in X$ be an element, and let $x\colon \{\odot\} \to X$ be the function representing it. Given a function $f\colon X \to Y$, what is $f \circ x$?   ◊

Solution 2.1.2.10.

It is the function $\{\odot\} \to Y$ that sends $\odot$ to $f(x)$. In other words, it represents the element $f(x)$.   ♦

Remark 2.1.2.11. Suppose given sets A, B, C and functions $A \xrightarrow{f} B \xrightarrow{g} C$. The *classical order* for writing their composition has been used so far, namely, $g \circ f\colon A \to C$. For any element $a \in A$, we write $g \circ f(a)$ to mean $g(f(a))$. This means "do g to whatever results from doing f to a."

However, there is another way to write this composition, called *diagrammatic order*. Instead of $g \circ f$, we would write $f; g\colon A \to C$, meaning "do f, then do g." Given an element $a \in A$, represented by $a\colon \{\odot\} \to A$, we have an element $a; f; g$.

Let X and Y be sets. We write $\mathrm{Hom}_{\mathbf{Set}}(X, Y)$ to denote the set of functions $X \to Y$.[3] Note that two functions $f, g\colon X \to Y$ are equal if and only if for every element $x \in X$, we have $f(x) = g(x)$.

Exercise 2.1.2.12.

Let $A = \{1, 2, 3, 4, 5\}$ and $B = \{x, y\}$.

a. How many elements does $\mathrm{Hom}_{\mathbf{Set}}(A, B)$ have?

b. How many elements does $\mathrm{Hom}_{\mathbf{Set}}(B, A)$ have?

                                                                                                        ◊

Solution 2.1.2.12.

a. 32. For example, $1 \mapsto x, 2 \mapsto x, 3 \mapsto x, 4 \mapsto y, 5 \mapsto x$.

b. 25. For example, $x \mapsto 1, y \mapsto 4$.

                                                                                                        ♦

---

[3]The notation $\mathrm{Hom}_{\mathbf{Set}}(-, -)$ will make more sense later, when it is seen in a larger context. See especially Section 5.1.

Exercise 2.1.2.13.

a. Find a set A such that for all sets X there is exactly one element in $\text{Hom}_{\textbf{Set}}(X, A)$. Hint: Draw a picture of proposed A's and X's. How many dots should be in A?

b. Find a set B such that for all sets X there is exactly one element in $\text{Hom}_{\textbf{Set}}(B, X)$.

◊

Solution 2.1.2.13.

a. Here is one: $A := \{☺\}$. (Here is another, $A := \{48\}$, and another, $A := \{a_1\}$).

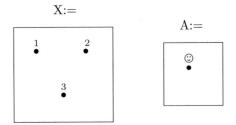

Why? We are trying to count the number of functions $X \to A$. Regardless of X and A, in order to give a function $X \to A$ one must answer the question, Where do I send x? several times, once for each element $x \in X$. Each element of X is sent to an element in A. For example, if $X = \{1, 2, 3\}$, then one asks three questions: Where do I send 1? Where do I send 2? Where do I send 3? When A has only one element, there is only one place to send each x. A function $X \to \{☺\}$ would be written $1 \mapsto ☺, 2 \mapsto ☺, 3 \mapsto ☺$. There is only one such function, so $\text{Hom}_{\textbf{Set}}(X, \{☺\})$ has one element.

b. $B = \varnothing$ is the only possibility.

$$B := \boxed{}$$

To give a function $B \to X$ one must answer the question, Where do I send b? for each $b \in B$. Because B has no elements, no questions must be answered in order to provide such a function. There is one way to answer all the necessary questions, because doing so is immediate ("vacuously satisfied"). It is like commanding John to "assign a letter grade to every person who is over 14 feet tall." John is finished with his job the moment the command is given, and there is only one way for him to finish the job. So $\text{Hom}_{\textbf{Set}}(\varnothing, X)$ has one element.

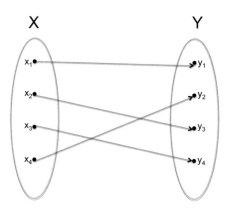

Figure 2.4 An isomorphism $X \xrightarrow{\cong} Y$.

For any set X, we define the *identity function on X*, denoted

$$\text{id}_X \colon X \to X,$$

to be the function such that for all $x \in X$, we have $\text{id}_X(x) = x$.

Definition 2.1.2.14 (Isomorphism). Let X and Y be sets. A function $f \colon X \to Y$ is called an *isomorphism*, denoted $f \colon X \xrightarrow{\cong} Y$, if there exists a function $g \colon Y \to X$ such that $g \circ f = \text{id}_X$ and $f \circ g = \text{id}_Y$.

$$X \overset{f}{\underset{g}{\rightleftarrows}} Y$$

In this case we also say that f is *invertible* and that g is *the inverse* of f. If there exists an isomorphism $X \xrightarrow{\cong} Y$, we say that X and Y are *isomorphic* sets and may write $X \cong Y$.

Example 2.1.2.15. If X and Y are sets and $f \colon X \to Y$ is an isomorphism, then the analogue of Figure 2.2 will look like a perfect matching, more often called a *one-to-one correspondence*. That means that no two arrows will hit the same element of Y, and every element of Y will be in the image. For example, Figure 2.4 depicts an isomorphism $X \xrightarrow{\cong} Y$ between four element sets.

Application 2.1.2.16. There is an isomorphism between the set Nuc_{DNA} of nucleotides found in DNA and the set Nuc_{RNA} of nucleotides found in RNA. Indeed, both sets have four elements, so there are 24 different isomorphisms. But only one is useful in biology. Before we say which one it is, let us say there is also an isomorphism $\text{Nuc}_{\text{DNA}} \cong \{A, C, G, T\}$ and an isomorphism $\text{Nuc}_{\text{RNA}} \cong \{A, C, G, U\}$, and we will use the letters as abbreviations for the nucleotides.

The convenient isomorphism $\text{Nuc}_{\text{DNA}} \xrightarrow{\cong} \text{Nuc}_{\text{RNA}}$ is that given by RNA transcription; it sends

$$A \mapsto U, \quad C \mapsto G, \quad G \mapsto C, \quad T \mapsto A.$$

(See also Application 5.1.2.21.) There is also an isomorphism $\text{Nuc}_{\text{DNA}} \xrightarrow{\cong} \text{Nuc}_{\text{DNA}}$ (the matching in the double helix), given by

$$A \mapsto T, \quad C \mapsto G, \quad G \mapsto C, \quad T \mapsto A.$$

Protein production can be modeled as a function from the set of 3-nucleotide sequences to the set of eukaryotic amino acids. However, it cannot be an isomorphism because there are $4^3 = 64$ triplets of RNA nucleotides but only 21 eukaryotic amino acids.

◊◊

Exercise 2.1.2.17.

Let $n \in \mathbb{N}$ be a natural number, and let X be a set with exactly n elements.

a. How many isomorphisms are there from X to itself?

b. Does your formula from part (a) hold when $n = 0$?

◊

Solution 2.1.2.17.

a. There are $n!$, pronounced "n factorial." For example, if $X = \{a, b, c, d\}$, then we have $4! = 4 * 3 * 2 * 1 = 24$ isomorphisms $X \xrightarrow{\cong} X$. One such isomorphism is $a \mapsto a, b \mapsto d, c \mapsto b, d \mapsto b$. The heuristic reason that the answer is 4! is that there are four ways to pick where a goes, but then only three remaining ways to pick where b goes, then only two remaining ways to pick were c goes, and then only one remaining way to pick where d goes. To really understand this answer, list all the isomorphisms $\{1, 2, 3, 4\} \xrightarrow{\cong} \{1, 2, 3, 4\}$ for yourself.

b. Yes, there is one function $\varnothing \to \varnothing$ and it is an isomorphism.

♦

Proposition 2.1.2.18. *The following facts hold about isomorphism.*

1. *Any set A is isomorphic to itself; i.e., there exists an isomorphism $A \xrightarrow{\cong} A$.*

2. *For any sets A and B, if A is isomorphic to B, then B is isomorphic to A.*

3. *For any sets A, B, and C, if A is isomorphic to B, and B is isomorphic to C, then A is isomorphic to C.*

Proof.   1. The identity function $\mathrm{id}_A \colon A \to A$ is invertible; its inverse is id_A because $\mathrm{id}_A \circ \mathrm{id}_A = \mathrm{id}_A$.

2. If $f \colon A \to B$ is invertible with inverse $g \colon B \to A$, then g is an isomorphism with inverse f.

3. If $f \colon A \to B$ and $f' \colon B \to C$ are each invertible with inverses $g \colon B \to A$ and $g' \colon C \to B$, then the following calculations show that $f' \circ f$ is invertible with inverse $g \circ g'$:

$$(f' \circ f) \circ (g \circ g') = f' \circ (f \circ g) \circ g' = f' \circ \mathrm{id}_B \circ g' = f' \circ g' = \mathrm{id}_C$$
$$(g \circ g') \circ (f' \circ f) = g \circ (g' \circ f') \circ f = g \circ \mathrm{id}_B \circ f = g \circ f = \mathrm{id}_A$$

$\square$

Exercise 2.1.2.19.

Let A and B be these sets:

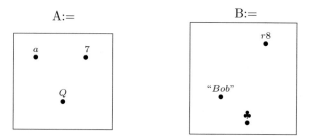

Note that the sets A and B are isomorphic. Suppose that $f \colon B \to \{1, 2, 3, 4, 5\}$ sends "Bob" to 1, sends ♣ to 3, and sends $r8$ to 4. Is there a canonical function $A \to \{1, 2, 3, 4, 5\}$ corresponding to f?[4]   ◊

---

[4]Canonical, as used here, means something like "best choice," a choice that stands out as the only reasonable one.

Solution 2.1.2.19.

No. There are a lot of choices, and none is any more reasonable than any other, i.e., none are canonical. (In fact, there are six choices; do you see why?)

The point of this exercise is to illustrate that even if one knows that two sets are isomorphic, one cannot necessarily treat them as the same. To treat them as the same, one should have in hand a specified *isomorphism* $g\colon A \xrightarrow{\cong} B$, such as $a \mapsto r8$, $7 \mapsto$ "*Bob*", $Q \mapsto$ ♣. Now, given $f\colon B \to \{1,2,3,4,5\}$, there is a canonical function $A \to \{1,2,3,4,5\}$ corresponding to f, namely, $f \circ g$. ♦

Exercise 2.1.2.20.

Find a set A such that for any set X, there is an isomorphism of sets

$$X \cong \mathrm{Hom}_{\mathbf{Set}}(A, X).$$

Hint: A function $A \to X$ points each element of A to an element of X. When would there be the same number of ways to do that as there are elements of of X? ◊

Solution 2.1.2.20.

Let $A = \{\odot\}$. Then to point each element of A to an element of X, one must simply point $\odot$ to an element of X. The set of ways to do that can be put in one-to-one correspondence with the set of elements of X. For example, if $X = \{1,2,3\}$, then $\odot \mapsto 3$ is a function $A \to X$ representing the element $3 \in X$. See Notation 2.1.2.9. ♦

Notation 2.1.2.21. For any natural number $n \in \mathbb{N}$, define a set

$$\underline{n} := \{1,2,3,\ldots,n\}. \tag{2.4}$$

We call $\underline{n}$ the *numeral set* of size n. So, in particular, $\underline{2} = \{1,2\}, \underline{1} = \{1\}$, and $\underline{0} = \varnothing$.

Let A be any set. A function $f\colon \underline{n} \to A$ can be written as a length n sequence

$$f = (f(1), f(2), \ldots, f(n)). \tag{2.5}$$

We call this the *sequence notation* for f.

Exercise 2.1.2.22.

a. Let $A = \{a,b,c,d\}$. If $f\colon \underline{10} \to A$ is given in sequence notation by (a,b,c,c,b,a,d,d,a,b), what is $f(4)$?

b. Let $s\colon \underline{7} \to \mathbb{N}$ be given by $s(i) = i^2$. Write s in sequence notation.

◊

Solution 2.1.2.22.

a. *c*

b. $(1, 4, 9, 16, 25, 36, 49)$

◆

Definition 2.1.2.23 (Cardinality of finite sets). Let A be a set and $n \in \mathbb{N}$ a natural number. We say that A *has cardinality* n, denoted

$$|A| = n,$$

if there exists an isomorphism of sets $A \cong \underline{n}$. If there exists some $n \in \mathbb{N}$ such that A has cardinality n, then we say that A is *finite*. Otherwise, we say that A is *infinite* and write $|A| \geqslant \infty$.

Exercise 2.1.2.24.

a. Let $A = \{5, 6, 7\}$. What is $|A|$?

b. What is $|\{1, 1, 2, 3, 5\}|$?

c. What is $|\mathbb{N}|$?

d. What is $|\{n \in \mathbb{N} \mid n \leqslant 5\}|$?

◇

Solution 2.1.2.24.

a. $|5, 6, 7| = 3$.

b. $|\{1, 1, 2, 3, 5\}| = 4$. As explained in Solution 2.1.2.8, a set contains each of its elements only once. So we have $\{1, 1, 2, 3, 5\} = \{1, 2, 3, 5\}$, which has cardinality 4.

c. $|\mathbb{N}| \geqslant \infty$.

d. $|\{n \in \mathbb{N} \mid n \leqslant 5\}| = |\{0, 1, 2, 3, 4, 5\}| = 6$.

◆

We will see in Corollary 3.4.5.6 that for any $m, n \in \mathbb{N}$, there is an isomorphism $\underline{m} \cong \underline{n}$ if and only if $m = n$. So if we find that A has cardinality m and that A has cardinality n, then $m = n$.

Proposition 2.1.2.25. *Let A and B be finite sets. If there is an isomorphism of sets* $f\colon A \to B$, *then the two sets have the same cardinality,* $|A| = |B|$.

Proof. If $f\colon A \to B$ is an isomorphism and $B \cong \underline{n}$, then $A \cong \underline{n}$ because the composition of two isomorphisms is an isomorphism.

$\square$

2.2 Commutative diagrams

At this point it is difficult to precisely define diagrams or commutative diagrams in general, but we can get a heuristic idea.[5] Consider the following picture:

$$
\begin{array}{ccc}
A & \xrightarrow{f} & B \\
& {\scriptstyle h}\searrow & \downarrow{\scriptstyle g} \\
& & C
\end{array}
\tag{2.6}
$$

We say this is a *diagram of sets* if each of A, B, C is a set and each of f, g, h is a function. We say this diagram *commutes* if $g \circ f = h$. In this case we refer to it as a commutative triangle of sets, or, more generally, as a *commutative diagram* of sets.

Application 2.2.1.1. In its most basic form, the central dogma of molecular biology is that DNA codes for RNA codes for protein. That is, there is a function from DNA triplets to RNA triplets and a function from RNA triplets to amino acids. But sometimes we just want to discuss the translation from DNA to amino acids, and this is the composite of the other two. The following commutative diagram is a picture of this fact

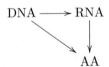

$\diamond\diamond$

Consider the following picture:

$$
\begin{array}{ccc}
A & \xrightarrow{f} & B \\
{\scriptstyle h}\downarrow & & \downarrow{\scriptstyle g} \\
C & \xrightarrow{i} & D
\end{array}
$$

---

[5]Commutative diagrams are precisely defined in Section 6.1.2.

We say this is a *diagram of sets* if each of A, B, C, D is a set and each of f, g, h, i is a function. We say this diagram *commutes* if $g \circ f = i \circ h$. In this case we refer to it as a commutative square of sets. More generally, it is a commutative diagram of sets.

Application 2.2.1.2. Given a physical system S, there may be two mathematical approaches $f \colon S \to A$ and $g \colon S \to B$ that can be applied to it. Either of those results in a prediction of the same sort, $f' \colon A \to P$ and $g' \colon B \to P$. For example, in mechanics we can use either the Lagrangian approach or the Hamiltonian approach to predict future states. To say that the diagram

$$
\begin{array}{ccc}
S & \longrightarrow & A \\
\downarrow & & \downarrow \\
B & \longrightarrow & P
\end{array}
$$

commutes would say that these approaches give the same result.

◇◇

Note that diagram (2.6) is considered to be the same diagram as each of the following:

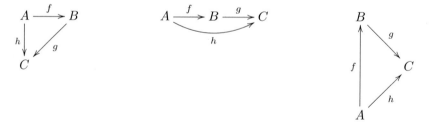

In all these we have $h = g \circ f$, or in diagrammatic order, $h = f; g$.

2.3   Ologs

In this book I ground the mathematical ideas in applications whenever possible. To that end I introduce ologs, which serve as a bridge between mathematics and various conceptual landscapes. The following material is taken from Spivak and Kent [43], an introduction to ologs.

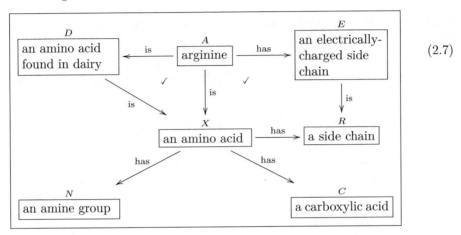

$$\text{(2.7)}$$

2.3.1 Types

A type is an abstract concept, a distinction the author has made. Each type is represented as a box containing a *singular indefinite noun phrase*. Each of the following four boxes is a type:

$$\boxed{\text{a man}} \qquad\qquad \boxed{\text{an automobile}} \qquad\qquad \text{(2.8)}$$

> a pair (a, w), where w is a woman and a is an automobile

> a pair (a, w), where w is a woman and a is a blue automobile owned by w

Each of the four boxes in (2.8) represents a type of thing, a whole class of things, and the label on that box is what one should call *each example* of that class. Thus ⌜a man⌝ does not represent a single man but the set of men, each example of which is called "a man." Similarly, the bottom right box represents an abstract type of thing, which probably has more than a million examples, but the label on the box indicates the common name for each such example.

Typographical problems emerge when writing a text box in a line of text, e.g., the text box $\boxed{\text{a man}}$ seems out of place, and the more in-line text boxes there are, the worse it gets. To remedy this, I denote types that occur in a line of text with corner symbols; e.g., I write ⌜a man⌝ instead of $\boxed{\text{a man}}$.

2.3.1.1   Types with compound structures

Many types have compound structures, i.e., they are composed of smaller units. Examples include

$$
\boxed{\begin{array}{l}\text{a man and}\\ \text{a woman}\end{array}}
\qquad
\boxed{\begin{array}{l}\text{a food portion } f \text{ and}\\ \text{a child } c \text{ such that } c\\ \text{ate all of } f\end{array}}
\qquad
\boxed{\begin{array}{l}\text{a triple } (p,a,j), \text{ where } p\\ \text{is a paper, } a \text{ is an author}\\ \text{of } p, \text{ and } j \text{ is a journal in}\\ \text{which } p \text{ was published}\end{array}}
\qquad (2.9)
$$

It is good practice to declare the variables in a compound type, as in the last two cases of (2.9). In other words, it is preferable to replace the first box in (2.9) with something like

$$
\boxed{\begin{array}{l}\text{a man } m \text{ and}\\ \text{a woman } w\end{array}}
\qquad \text{or} \qquad
\boxed{\begin{array}{l}\text{a pair } (m,w),\\ \text{where } m \text{ is a man}\\ \text{and } w \text{ is a woman}\end{array}}
$$

so that the variables (m, w) are clear.

Rules of good practice 2.3.1.2. A type is presented as a text box. The text in that box should

(i) begin with the word *a* or *an*;

(ii) refer to a distinction made and recognizable by the olog's author;

(iii) refer to a distinction for which instances can be documented;

(iv) be the common name that each instance of that distinction can be called; and

(v) declare all variables in a compound structure.

The first, second, third, and fourth rules ensure that the class of things represented by each box appears to the author to be a well defined set, and that the class is appropriately named. The fifth rule encourages good readability of arrows (see Section 2.3.2).

I do not always follow the rules of good practice throughout this book. I think of these rules being as followed "in the background," but I have nicknamed various boxes. So ⌜Steve⌝ may stand as a nickname for ⌜a thing classified as Steve⌝ and ⌜arginine⌝ as a nickname for ⌜a molecule of arginine⌝. However, one should always be able to rename each type according to the rules of good practice.

2.3.2 Aspects

An aspect of a thing x is a way of viewing it, a particular way in which x can be regarded or measured. For example, a woman can be regarded as a person; hence "being a person" is an aspect of a woman. A molecule has a molecular mass (say in daltons), so "having a molecular mass" is an aspect of a molecule. In other words, when it comes to ologs, the word *aspect* simply means function. The domain A of the function $f\colon A \to B$ is the thing we are measuring, and the codomain is the set of possible answers or results of the measurement.

$$\boxed{\text{a woman}} \xrightarrow{\text{is}} \boxed{\text{a person}} \qquad (2.10)$$

$$\boxed{\text{a molecule}} \xrightarrow{\text{has as molecular mass (Da)}} \boxed{\text{a positive real number}} \qquad (2.11)$$

So for the arrow in (2.10), the domain is the set of women (a set with perhaps 3 billion elements); the codomain is the set of persons (a set with perhaps 6 billion elements). We can imagine drawing an arrow from each dot in the "woman" set to a unique dot in the "person" set, just as in Figure 2.2. No woman points to two different people nor to zero people—each woman is exactly one person—so the rules for a function are satisfied. Let us now concentrate briefly on the arrow in (2.11). The domain is the set of molecules, the codomain is the set $\mathbb{R}_{>0}$ of positive real numbers. We can imagine drawing an arrow from each dot in the "molecule" set to a single dot in the "positive real number" set. No molecule points to two different masses, nor can a molecule have no mass: each molecule has exactly one mass. Note, however, that two different molecules can point to the same mass.

2.3.2.1 Invalid aspects

To be valid an aspect must be a functional relationship. Arrows may on their face appear to be aspects, but on closer inspection they are not functional (and hence not valid as aspects).

Consider the following two arrows:

$$\boxed{\text{a person}} \xrightarrow{\text{has}} \boxed{\text{a child}} \qquad (2.12^*)$$

$$\boxed{\text{a mechanical pencil}} \xrightarrow{\text{uses}} \boxed{\text{a piece of lead}} \qquad (2.13^*)$$

A person may have no children or may have more than one child, so the first arrow is invalid: it is not a function. Similarly, if one drew an arrow from each mechanical pencil to each piece of lead it uses, one would not have a function.

Warning 2.3.2.2. The author of an olog has a worldview, some fragment of which is captured in the olog. When person A examines the olog of person B, person A may or may not agree with it. For example, person B may have the following olog

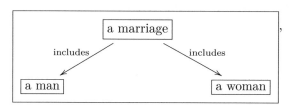

,

which associates to each marriage a man and a woman. Person A may take the position that some marriages involve two men or two women and thus see B's olog as wrong. Such disputes are not "problems" with either A's olog or B's olog; they are discrepancies between worldviews. Hence, a reader R may see an olog in this book and notice a discrepancy between R's worldview and my own, but this is not a problem with the olog. Rules are enforced to ensure that an olog is structurally sound, not to ensure that it "correctly reflects reality," since worldviews can differ.

Consider the aspect $\ulcorner$an object$\urcorner$ $\xrightarrow{\text{has}}$ $\ulcorner$a weight$\urcorner$. At some point in history, this would have been considered a valid function. Now we know that the same object would have a different weight on the moon than it has on earth. Thus, as worldviews change, we often need to add more information to an olog. Even the validity of $\ulcorner$an object on earth$\urcorner$ $\xrightarrow{\text{has}}$ $\ulcorner$a weight$\urcorner$ is questionable, e.g., if I am considered to be the same object on earth before and after I eat Thanksgiving dinner. However, to build a model we need to choose a level of granularity and try to stay within it, or the whole model would evaporate into the nothingness of truth. Any level of granularity is called *a stereotype*; e.g., we stereotype objects on earth by saying they each have a weight. A stereotype is a lie, more politely a conceptual simplification, that is convenient for the way we want to do business.

Remark 2.3.2.3. In keeping with Warning 2.3.2.2, the arrows in (2.12*) and (2.13*) may not be wrong but simply reflect that the author has an idiosyncratic worldview or vocabulary. Maybe the author believes that every mechanical pencil uses exactly one piece of lead. If this is so, then $\ulcorner$a mechanical pencil$\urcorner$ $\xrightarrow{\text{uses}}$ $\ulcorner$a piece of lead$\urcorner$ is indeed a valid aspect. Similarly, suppose the author meant to say that each person *was once* a child, or that a person has an inner child. Since every person has one and only one inner child (according to the author), the map $\ulcorner$a person$\urcorner$ $\xrightarrow{\text{has as inner child}}$ $\ulcorner$a child$\urcorner$ is a valid aspect. We cannot fault the olog for its author's view, but note that we have changed the name of the label to make the intention more explicit.

2.3.2.4   Reading aspects and paths as English phrases

Each arrow (aspect) $X \xrightarrow{f} Y$ can be read by first reading the label on its source box X, then the label on the arrow f, and finally the label on its target box Y. For example, the arrow

$$\boxed{\text{a book}} \xrightarrow{\text{has as first author}} \boxed{\text{a person}} \qquad (2.14)$$

is read "a book has as first author a person."

Remark 2.3.2.5. Note that the map in (2.14) is a valid aspect, but a similarly benign-looking map $\ulcorner$a book$\urcorner \xrightarrow{\text{has as author}} \ulcorner$a person$\urcorner$ would not be valid, because it is not functional. When creating an olog, one must be vigilant about this type of mistake because it is easy to miss, and it can corrupt the olog.

Sometimes the label on an arrow can be shortened or dropped altogether if it is obvious from context (see Section 2.3.3). Here is a common example from the way I write ologs.

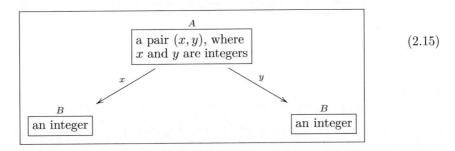

$$(2.15)$$

Neither arrow is readable by the preceding protocol (e.g., "a pair (x,y), where x and y are integers x an integer" is not an English sentence), and yet it is clear what each map means. For example, given $(8, 11)$ in A, arrow x would yield 8 and arrow y would yield 11. The label x can be thought of as a nickname for the full name "yields as the value of x," and similarly for y. I do not generally use the full name, so as not to clutter the olog.

One can also read paths through an olog by inserting the word *which* (or *who*) after each intermediate box. For example, olog (2.16) has two paths of length 3 (counting arrows in a chain):

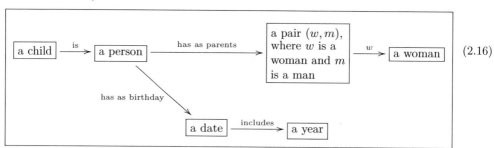

The top path is read "a child is a person, who has as parents a pair (w, m), where w is a woman and m is a man, which yields, as the value of w, a woman." The reader should read and understand the content of the bottom path, which associates to every child a year.

2.3.2.6   Converting nonfunctional relationships to aspects

There are many relationships that are not functional, and these cannot be considered aspects. Often the word *has* indicates a relationship—sometimes it is functional, as in ⌜a person⌝ $\xrightarrow{\text{has}}$ ⌜a stomach⌝, and sometimes it is not, as in ⌜a father⌝ $\xrightarrow{\text{has}}$ ⌜a child⌝. Clearly, a father may have more than one child. This one is easily fixed by realizing that the arrow should go the other way: there is a function ⌜a child⌝ $\xrightarrow{\text{has}}$ ⌜a father⌝.

What about ⌜a person⌝ $\xrightarrow{\text{owns}}$ ⌜a car⌝. Again, a person may own no cars or more than one car, but this time a car can be owned by more than one person too. A quick fix would be to replace it by ⌜a person⌝ $\xrightarrow{\text{owns}}$ ⌜a set of cars⌝. This is okay, but the relationship between ⌜a car⌝ and ⌜a set of cars⌝ then becomes an issue to deal with later. There is another way to indicate such nonfunctional relationships. In this case it would look like this:

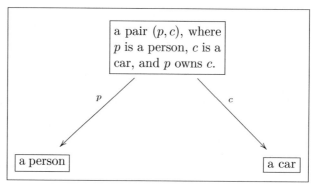

This setup will ensure that everything is properly organized. In general, relationships can involve more than two types, and in olog form looks like this:

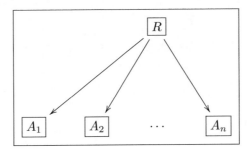

For example,

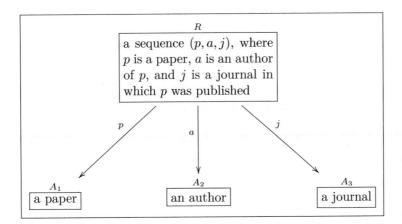

Exercise 2.3.2.7.

On page 27, the arrow in (2.12*) was indicated as an invalid aspect:

$$\boxed{\text{a person}} \xrightarrow{\text{has}} \boxed{\text{a child}} \qquad\qquad (2.12^*)$$

Create a valid olog that captures the parent-child relationship; your olog should still have boxes ⌜a person⌝ and ⌜a child⌝ but may have an additional box.   ◊

Solution 2.3.2.7.

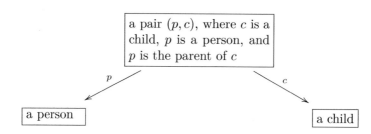

♦

Rules of good practice 2.3.2.8. An aspect is presented as a labeled arrow pointing from a source box to a target box. The arrow label text should

(i) begin with a verb;

(ii) yield an English sentence, when the source box text followed by the arrow text followed by the target box text is read;

(iii) refer to a functional relationship: each instance of the source type should give rise to a specific instance of the target type;

(iv) constitute a useful description of that functional relationship.

2.3.3 Facts

In this section I discuss facts, by which I mean path equivalences in an olog. It is the notion of path equivalences that makes category theory so powerful.

A *path* in an olog is a head-to-tail sequence of arrows. That is, any path starts at some box B_0, then follows an arrow emanating from B_0 (moving in the appropriate direction), at which point it lands at another box B_1, then follows any arrow emanating from B_1, and so on, eventually landing at a box B_n and stopping there. The number of arrows is the *length* of the path. So a path of length 1 is just an arrow, and a path of length 0 is just a box. We call B_0 the *source* and B_n the *target* of the path.

Given an olog, its author may want to declare that two paths are equivalent. For example, consider the two paths from A to C in the olog

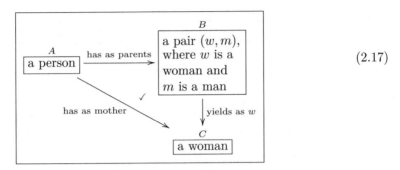

(2.17)

We know as English speakers that a woman parent is called a mother, so these two paths $A \to C$ should be equivalent. A mathematical way to say this is that the triangle in olog (2.17) *commutes*. That is, path equivalences are simply commutative diagrams, as in Section 2.2. In the preceding example we concisely say "a woman parent is equivalent to a mother." We declare this by defining the diagonal map in (2.17) to be *the composition* of the horizontal map and the vertical map.

I generally prefer to indicate a commutative diagram by drawing a check mark, $\checkmark$, in the region bounded by the two paths, as in olog (2.17). Sometimes, however, one cannot do this unambiguously on the two-dimensional page. In such a case I indicate the commutative diagram (fact) by writing an equation. For example, to say that the diagram

$$A \xrightarrow{\ f\ } B$$
$$h \downarrow \qquad \downarrow g$$
$$C \xrightarrow{\ i\ } D$$

commutes, we could either draw a check mark inside the square or write the equation

$$_A[f, g] \simeq {}_A[h, i]$$

above it.[6] Either way, it means that starting from A, "doing f, then g" is equivalent to "doing h, then i."

---

[6] We defined function composition in Section 2.1.2, but here we are using a different notation. There we used *classical order*, and our path equivalence would be written $g \circ f = i \circ h$. As discussed in Remark 2.1.2.11, category theorists and others often prefer the *diagrammatic order* for writing compositions, which is $f; g = h; i$. For ologs, we roughly follow the latter because it makes for better English sentences, and for the same reason, we add the source object to the equation, writing $_A[f, g] \simeq {}_A[h, i]$.

Here is another example:

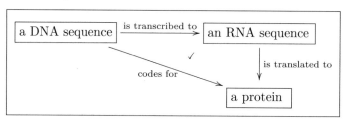

Note how this diagram gives us the established terminology for the various ways in which DNA, RNA, and protein are related in this context.

Exercise 2.3.3.1.

Create an olog for human nuclear biological families that includes the concepts of person, man, woman, parent, father, mother, and child. Make sure to label all the arrows and that each arrow indicates a valid aspect in the sense of Section 2.3.2.1. Indicate with check marks ($\checkmark$) the diagrams that are intended to commute. If the 2-dimensionality of the page prevents a check mark from being unambiguous, indicate the intended commutativity with an equation. $\diamond$

Solution 2.3.3.1.

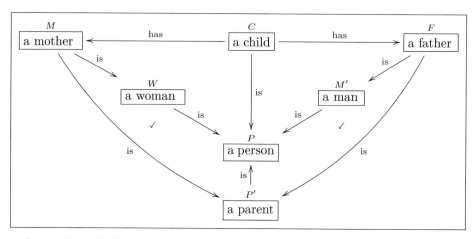

Note that neither of the two triangles from child to person commute. To say that they did commute would be to say that "a child and its mother are the same person" and that "a child and its father are the same person." $\blacklozenge$

Example 2.3.3.2 (Noncommuting diagram). In my conception of the world, the following diagram does not commute:

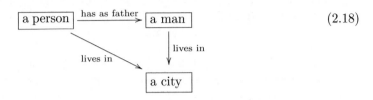

$$(2.18)$$

The noncommutativity of diagram (2.18) does not imply that no person lives in the same city as his or her father. Rather it implies that it is not the case that *every* person lives in the same city as his or her father.

Exercise 2.3.3.3.

Create an olog about a scientific subject, preferably one you think about often. The olog should have at least five boxes, five arrows, and one commutative diagram. ◊

Solution 2.3.3.3.

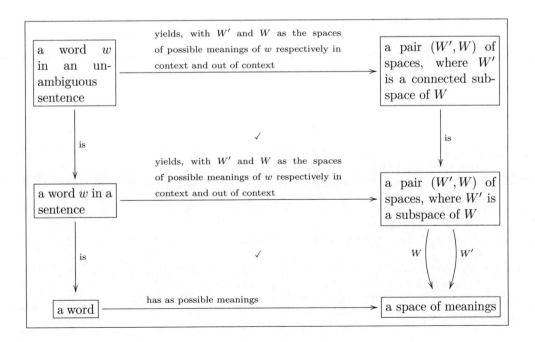

To be clear, the check mark in the lower half of the diagram indicates that the square including W commutes; the square that includes W' does not. The reason is that the space of possible definitions for a word includes the meanings for that word in all contexts. The space of definitions for a word in context is smaller than the space of definitions for a word taken out of context, because out of context the meaning of a word is more ambiguous.

In my conception, a word has a space rather than simply a set of meanings. For example, consider the sentence, "He wore a large hat." Here, the word *large* has a space of meanings, though I might say that the space is connected in that the meaning of *large* is fluid but not ambiguous. On the other hand, in the ambiguous sentence, "Kids make nutritious snacks," the word *make* has two disconnected spaces of meanings: either the kids assemble snacks or they are themselves considered to be snacks. ♦

2.3.3.4 A formula for writing facts as English

Every fact consists of two paths, say, P and Q, that are to be declared equivalent. The paths P and Q will necessarily have the same source, say, s, and target, say, t, but their lengths may be different, say, m and n respectively.[7] We draw these paths as

$$P: \quad \overset{a_0=s}{\bullet} \xrightarrow{f_1} \overset{a_1}{\bullet} \xrightarrow{f_2} \overset{a_2}{\bullet} \xrightarrow{f_3} \cdots \xrightarrow{f_{m-1}} \overset{a_{m-1}}{\bullet} \xrightarrow{f_m} \overset{a_m=t}{\bullet} \qquad (2.19)$$

$$Q: \quad \overset{b_0=s}{\bullet} \xrightarrow{g_1} \overset{b_1}{\bullet} \xrightarrow{g_2} \overset{b_2}{\bullet} \xrightarrow{g_3} \cdots \xrightarrow{g_{n-1}} \overset{b_{n-1}}{\bullet} \xrightarrow{g_n} \overset{b_n=t}{\bullet}$$

Every part ℓ of an olog (i.e., every box and every arrow) has an associated English phrase, which we write as $\langle\!\langle \ell \rangle\!\rangle$. Using a dummy variable x, we can convert a fact into English too. The following general formula may be a bit difficult to understand (see Example 2.3.3.5). The fact $P \simeq Q$ from (2.19) can be Englished as follows:

> Given $x, \langle\!\langle s \rangle\!\rangle$ consider the following. $\qquad (2.20)$
> We know that x is $\langle\!\langle s \rangle\!\rangle$,
> which $\langle\!\langle f_1 \rangle\!\rangle$ $\langle\!\langle a_1 \rangle\!\rangle$, which $\langle\!\langle f_2 \rangle\!\rangle$ $\langle\!\langle a_2 \rangle\!\rangle$, which ... $\langle\!\langle f_{m-1} \rangle\!\rangle$ $\langle\!\langle a_{m-1} \rangle\!\rangle$, which $\langle\!\langle f_m \rangle\!\rangle$ $\langle\!\langle t \rangle\!\rangle$,
> that we call $P(x)$.
> We also know that x is $\langle\!\langle s \rangle\!\rangle$,
> which $\langle\!\langle g_1 \rangle\!\rangle$ $\langle\!\langle b_1 \rangle\!\rangle$, which $\langle\!\langle g_2 \rangle\!\rangle$ $\langle\!\langle b_2 \rangle\!\rangle$, which ... $\langle\!\langle g_{n-1} \rangle\!\rangle$ $\langle\!\langle b_{n-1} \rangle\!\rangle$, which $\langle\!\langle g_n \rangle\!\rangle$ $\langle\!\langle t \rangle\!\rangle$,
> that we call $Q(x)$.
> Fact: Whenever x is $\langle\!\langle s \rangle\!\rangle$, we will have $P(x) = Q(x)$.

---

[7]If the source equals the target, $s = t$, then it is possible to have $m = 0$ or $n = 0$, and the ideas that follow still make sense.

Example 2.3.3.5. Consider the olog

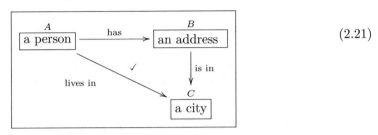

(2.21)

To put the fact that diagram (2.21) commutes into English, we first English the two paths: F="a person has an address which is in a city" and G="a person lives in a city." The source of both is s = "a person" and the target of both is t = "a city." Write:

Given x, a person, consider the following.
We know that x is a person,
who has an address, which is in a city,
that we call $P(x)$.
We also know that x is a person,
who lives in a city
that we call $Q(x)$.
Fact: Whenever x is a person, we will have $P(x) = Q(x)$.

More concisely, one reads olog 2.21 as

A person x has an address, which is in a city, and this is the city x lives in.

Exercise 2.3.3.6.

This olog was taken from Spivak [38].

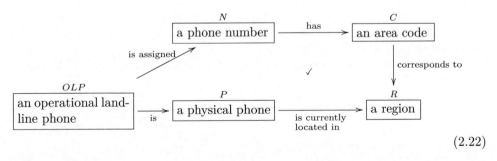

(2.22)

It says that a landline phone is physically located in the region to which its phone number is assigned. Translate this fact into English using the formula from (2.20). ◊

Solution 2.3.3.6.

Given x, an operational landline phone, consider the following.
We know that x is an operational landline phone,
which is assigned a phone number, which has an area code,
which corresponds to a region
that we call $P(x)$.
We also know that x is an operational landline phone,
which is a physical phone, which is currently located in a region
that we call $Q(x)$.
Fact: Whenever x is an operational landline phone, we will have
$P(x) = Q(x)$.

♦

Exercise 2.3.3.7.

In olog (2.22), suppose that the box ⌜an operational landline phone⌝ is replaced with the box ⌜an operational cell phone⌝. Would the diagram still commute?   ◇

Solution 2.3.3.7.

No, it would not commute. A cell phone is assigned a phone number, which has an area code, which corresponds to a region. However, the phone as a physical object can be operational even if it is not currently located in that region.   ♦

2.3.3.8 Images

This section discusses a specific kind of fact, generated by any aspect. Recall that every function has an image (2.3), meaning the subset of elements in the codomain that are "hit" by the function. For example, the function $f \colon \mathbb{Z} \to \mathbb{Z}$ given by $f(x) = 2 * x \colon \mathbb{Z} \to \mathbb{Z}$ has as image the set of all even numbers.

Similarly, the set of mothers arises as the image of the "has as mother" function:

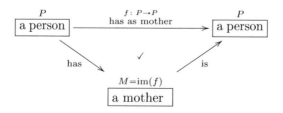

Exercise 2.3.3.9.

For each of the following types, write a function for which it is the image, or write "not clearly useful as an image type."

a. ⌜a book⌝

b. ⌜a material that has been fabricated by a working process of type T⌝

c. ⌜a bicycle owner⌝

d. ⌜a child⌝

e. ⌜a used book⌝

f. ⌜a primary residence⌝

◊

Solution 2.3.3.9.

Every set X is the image of the identity function $\mathrm{id}_X \colon X \to X$, but this does not fulfill the purpose of this exercise, which is to help the reader understand images of functions and how they are useful in ologs. The following solutions reflect only my aesthetic about what is useful—readers' answers may be different and yet correct.

a. Not clearly useful as an image type.

b. This is the image of the aspect

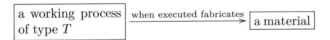

c. This is the image of the aspect

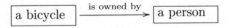

assuming every bicycle is owned by one person.

d. Not clearly useful as an image type.

e. Not clearly useful as an image type.

f. This is the image of the aspect

assuming each person has a primary residence.

The point is that the notion of image creates new types out of existing aspects, or functions. This connection puts the function first and derives the type from it as its image. A bicycle owner is not a type of person until we have the function that assigns ownership.

◆

Chapter 3

Fundamental Considerations in Set

In this chapter we continue to pursue an understanding of sets. We begin by examining how to combine sets in various ways to get new sets. To that end, products and coproducts are introduced, and then more complex limits and colimits, with the aim of conveying a sense of their *universal properties*. The chapter ends with some additional interesting constructions in **Set**.

3.1 Products and coproducts

This section introduces two concepts that are likely to be familiar, although perhaps not by their category-theoretic names: product and coproduct. Each is an example of a large class of ideas that exist far beyond the realm of sets (see Section 6.1.1).

3.1.1 Products

Definition 3.1.1.1. Let X and Y be sets. The *product of X and Y*, denoted $X \times Y$, is defined as the set of ordered pairs (x, y), where $x \in X$ and $y \in Y$. Symbolically,

$$X \times Y = \{(x, y) \mid x \in X, \ y \in Y\}.$$

There are two natural *projection functions*, $\pi_1 \colon X \times Y \to X$ and $\pi_2 \colon X \times Y \to Y$.

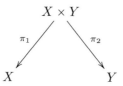

Example 3.1.1.2 (Grid of dots). Let $X = \{1, 2, 3, 4, 5, 6\}$ and $Y = \{\clubsuit, \diamondsuit, \heartsuit, \spadesuit\}$. Then we can draw $X \times Y$ as a 6 by 4 grid of dots, and the projections as projections

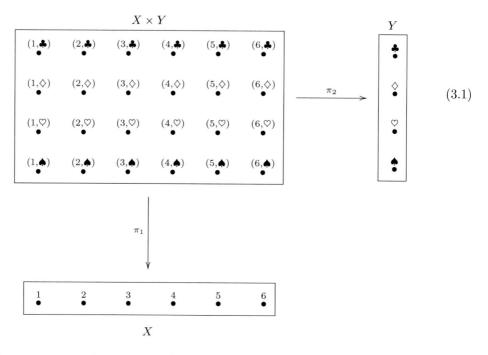

$$(3.1)$$

Application 3.1.1.3. A traditional (Mendelian) way to predict the genotype of offspring based on the genotype of its parents is by the use of Punnett squares. If F is the set of possible genotypes for the female parent, and M is the set of possible genotypes of the male parent, then $F \times M$ is drawn as a square, called a Punnett square, in which every combination is drawn.   ◊◊

Exercise 3.1.1.4.

How many elements does the set $\{a, b, c, d\} \times \{1, 2, 3\}$ have?   ◊

Solution 3.1.1.4.

$4 \times 3 = 12$. ♦

Application 3.1.1.5. Suppose we are conducting experiments about the mechanical properties of materials, as in Application 2.1.2.2. For each material sample we will produce multiple data points in the set ⌜extension⌝ × ⌜force⌝ ≅ $\mathbb{R} \times \mathbb{R}$.

◊◊

Remark 3.1.1.6. It is possible to take the product of more than two sets as well. For example, if A, B, and C are sets, then $A \times B \times C$ is the set of triples

$$A \times B \times C := \{(a, b, c) \mid a \in A, b \in B, c \in C\}.$$

This kind of generality is useful in understanding multiple dimensions, e.g., what physicists mean by ten-dimensional space. It comes under the heading of *limits* (see Section 6.1.3).

Example 3.1.1.7. Let $\mathbb{R}$ be the set of real numbers. By $\mathbb{R}^2$ we mean $\mathbb{R} \times \mathbb{R}$. Similarly, for any $n \in \mathbb{N}$, we define $\mathbb{R}^n$ to be the product of n copies of $\mathbb{R}$.

According to Penrose [35], Aristotle seems to have conceived of space as something like $S := \mathbb{R}^3$ and of time as something like $T := \mathbb{R}$. Space-time, had he conceived of it, would probably have been $S \times T \cong \mathbb{R}^4$. He, of course, did not have access to this kind of abstraction, which was probably due to Descartes. (The product $X \times Y$ is often called *Cartesian product*, in his honor.)

Exercise 3.1.1.8.

Let $\mathbb{Z}$ denote the set of integers, and let $+ : \mathbb{Z} \times \mathbb{Z} \to \mathbb{Z}$ denote the addition function and $\cdot : \mathbb{Z} \times \mathbb{Z} \to \mathbb{Z}$ denote the multiplication function. Which of the following diagrams commute?

a.

$$
\begin{array}{ccc}
\mathbb{Z} \times \mathbb{Z} \times \mathbb{Z} & \xrightarrow{(a,b,c) \mapsto (a \cdot b, a \cdot c)} & \mathbb{Z} \times \mathbb{Z} \\
{\scriptstyle (a,b,c) \mapsto (a+b,c)} \downarrow & & \downarrow {\scriptstyle (x,y) \mapsto x+y} \\
\mathbb{Z} \times \mathbb{Z} & \xrightarrow[(x,y) \mapsto xy]{} & \mathbb{Z}
\end{array}
$$

b.

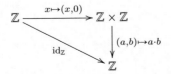

c.

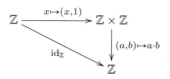

$\diamond$

Solution 3.1.1.8.

a. This diagram does not commute because $a \cdot b + a \cdot c \neq (a + b) \cdot c$, at least for some integers $a, b, c \in \mathbb{Z}$, e.g., $a = 0, b = 1, c = 1$.

b. This diagram does not commute because $x \cdot 0 \neq x$, at least for some integers $x \in \mathbb{Z}$.

c. This diagram commutes. For every integer $x \in \mathbb{Z}$, we have $x \cdot 1 = x$.

$\blacklozenge$

3.1.1.9   Universal property for products

A universal property is an abstract quality that characterizes a given construction. For example, the following proposition says that the product construction is characterized as possessing a certain quality.

Proposition 3.1.1.10 (Universal property for product). *Let X and Y be sets. For any set A and functions $f \colon A \to X$ and $g \colon A \to Y$, there exists a unique function $A \to X \times Y$ such that the following diagram commutes:*

(3.2)

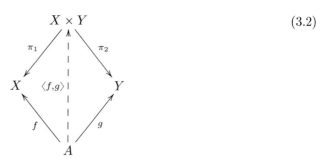

We say this function is induced by f and g, *and we denote it*

$$\langle f, g \rangle \colon A \to X \times Y, \qquad \text{where} \qquad \langle f, g \rangle(a) = (f(a), g(a)).$$

That is, we have $\pi_1 \circ \langle f, g \rangle = f$ *and* $\pi_2 \circ \langle f, g \rangle = g$, *and* $\langle f, g \rangle$ *is the only function for which that is so.*

Proof. Suppose given f, g as in the proposition statement. To provide a function $\ell \colon A \to X \times Y$ is equivalent to providing an element $\ell(a) \in X \times Y$ for each $a \in A$. We need such a function $\ell = \langle f, g \rangle$, for which $\pi_1 \circ \langle f, g \rangle = f$ and $\pi_2 \circ \langle f, g \rangle = g$. An element of $X \times Y$ is an ordered pair (x, y), and we can use $\langle f, g \rangle(a) = (x, y)$ if and only if $x = \pi_1(x, y) = f(a)$ and $y = \pi_2(x, y) = g(a)$. So it is necessary and sufficient to define

$$\langle f, g \rangle(a) := (f(a), g(a))$$

for all $a \in A$.

$\square$

Example 3.1.1.11 (Grid of dots, continued). It is important that the reader sees the universal property for products as completely intuitive.

Recall that if X and Y are sets, say, of cardinalities $|X| = m$ and $|Y| = n$ respectively, then $X \times Y$ is an $m \times n$ grid of dots, and it comes with two canonical projections $X \xleftarrow{\pi_1} X \times Y \xrightarrow{\pi_2} Y$. These allow us to extract from every grid element $z \in X \times Y$ its column $\pi_1(z) \in X$ and its row $\pi_2(z) \in Y$.

Suppose that each person in a classroom picks an element of X and an element of Y. Thus we have functions $f \colon C \to X$ and $g \colon C \to Y$. But is not picking a column and a row the same thing as picking an element in the grid? The two functions f and g induce a unique function $C \to X \times Y$. How does this function $C \to X \times Y$ compare with the original functions f and g? The commutative diagram (3.2) sums up the connection.

Example 3.1.1.12. Let $\mathbb{R}$ be the set of real numbers, and let $0 \in \mathbb{R}$ be the origin. As in Notation 2.1.2.9, it is represented by a function $z \colon \{\odot\} \to \mathbb{R}$, with $z(\odot) = 0$. Thus we can draw functions

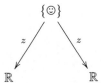

The universal property for products guarantees a function $\langle z, z \rangle \colon \{\odot\} \to \mathbb{R} \times \mathbb{R}$, which represents the origin in $(0, 0) \in \mathbb{R}^2$.

Exercise 3.1.1.13.

For every set A there is some relationship between the following three sets:

$$\mathrm{Hom}_{\mathbf{Set}}(A, X), \qquad \mathrm{Hom}_{\mathbf{Set}}(A, Y), \qquad \text{and} \qquad \mathrm{Hom}_{\mathbf{Set}}(A, X \times Y).$$

What is it?

Hint: This problem is somewhat recursive in that you will use products in your formula.   ◊

Solution 3.1.1.13.

There is an isomorphism

$$\mathrm{Hom}_{\mathbf{Set}}(A, X \times Y) \xrightarrow{\cong} \mathrm{Hom}_{\mathbf{Set}}(A, X) \times \mathrm{Hom}_{\mathbf{Set}}(A, Y).$$

In an attempt to make this more concrete, suppose we have a height function from the set P of people to the set $\mathbb{Z}$ of integers and a name function from the set of people to the set S of strings. That is, we have an element of $\mathrm{Hom}_{\mathbf{Set}}(P, \mathbb{Z})$ and an element of $\mathrm{Hom}_{\mathbf{Set}}(P, S)$. From this we get an element of $\mathrm{Hom}_{\mathbf{Set}}(P, \mathbb{Z}) \times \mathrm{Hom}_{\mathbf{Set}}(P, S)$. That is, the two pieces of information combine into a single piece of information if we pack the height and the name into a single datum, i.e., an element of $\mathbb{Z} \times S$.   ◆

Exercise 3.1.1.14.

a. Let X and Y be sets. Construct the swap map $s \colon X \times Y \to Y \times X$ using only the universal property for products. If $\pi_1 \colon X \times Y \to X$, $\pi_2 \colon X \times Y \to Y$, $p_1 \colon Y \times X \to Y$, and $p_2 \colon Y \times X \to X$ are the projection functions, write s in terms of the symbols $\pi_1, \pi_2, p_1, p_2, \circ$, and $\langle \, , \, \rangle$.

b. Can you prove that s is an isomorphism using only the universal property for products?

                                                                                                    ◊

Solution 3.1.1.14.

a.

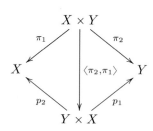

b. Consider the map $\langle p_2, p_1 \rangle \colon Y \times X \to X \times Y$. Let $s \colon X \times Y \to X \times Y$ be the composite $\langle p_2, p_1 \rangle \circ \langle \pi_2, \pi_1 \rangle$. We have $\pi_1 \circ s = \pi_1$ by the following calculation:

$$\pi_1 \circ s = \pi_1 \circ \langle p_2, p_1 \rangle \circ \langle \pi_2, \pi_1 \rangle$$
$$= p_2 \circ \langle \pi_2, \pi_1 \rangle = \pi_1,$$

and by a similar calculation, $\pi_2 \circ s = \pi_2$. But we also have $\pi_1 \circ \mathrm{id}_{X \times Y} = \pi_1$ and $\pi_2 \circ \mathrm{id}_{X \times Y} = \pi_2$. Thus the universal property (Proposition 3.1.1.10) implies that $s = \mathrm{id}_{X \times Y}$.

By similar reasoning, if $t \colon Y \times X \to Y \times X$ is the composite $\langle \pi_2, \pi_1 \rangle \circ \langle p_2, p_1 \rangle$, we can show that $t = \mathrm{id}_{Y \times X}$. By Definition 2.1.2.14, the functions s and t constitute an isomorphism $X \times Y \to Y \times X$.

♦

Example 3.1.1.15. Suppose given sets X, X', Y, Y' and functions $m \colon X \to X'$ and $n \colon Y \to Y'$. We can use the universal property for products to construct a function $s \colon X \times Y \to X' \times Y'$.

The universal property (Proposition 3.1.1.10) says that to get a function from any set A to $X' \times Y'$, we need two functions, namely, some $f \colon A \to X'$ and some $g \colon A \to Y'$. Here we want to use $A := X \times Y$.

What we have readily available are the two projections $\pi_1 \colon X \times Y \to X$ and $\pi_2 \colon X \times Y \to Y$. But we also have $m \colon X \to X'$ and $n \colon Y \to Y'$. Composing, we set $f := m \circ \pi_1$ and $g := n \circ \pi_2$.

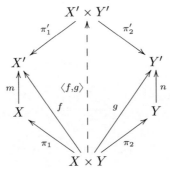

The dotted arrow is often called the *product* of $m \colon X \to X'$ and $n \colon Y \to Y'$. Here it is denoted $\langle f, g \rangle$, but f and g were not given variables. Since writing $\langle m \circ \pi_1, n \circ \pi_2 \rangle$ is clunky notation, we instead denote this function

$$m \times n \colon X \times Y \to X' \times Y'.$$

3.1.1.16   Ologging products

Given two objects c, d in an olog, there is a canonical label $\langle\!\langle c \times d \rangle\!\rangle$ for their product $c \times d$, written in terms of the labels $\langle\!\langle c \rangle\!\rangle$ and $\langle\!\langle d \rangle\!\rangle$. Namely,

$$\langle\!\langle c \times d \rangle\!\rangle := \text{``a pair } (x, y), \text{ where } x \text{ is } \langle\!\langle c \rangle\!\rangle \text{ and } y \text{ is } \langle\!\langle d \rangle\!\rangle.\text{''}$$

The projections $c \leftarrow c \times d \rightarrow d$ can be labeled "yields, as x," and "yields, as y," respectively.

Suppose that e is another object, and $p\colon e \rightarrow c$ and $q\colon e \rightarrow d$ are two arrows. By the universal property for products (Proposition 3.1.1.10), p and q induce a unique arrow $e \rightarrow c \times d$, making the evident diagrams commute. This arrow can be labeled

$$\text{``yields, insofar as it } \langle\!\langle p \rangle\!\rangle \ \langle\!\langle c \rangle\!\rangle \text{ and } \langle\!\langle q \rangle\!\rangle \ \langle\!\langle d \rangle\!\rangle,\text{''}.$$

Example 3.1.1.17. Every car owner owns at least one car, but there is no obvious function $\ulcorner$a car owner$\urcorner \rightarrow \ulcorner$a car$\urcorner$ because he or she may own more than one. One good choice would be the car that the person drives most often, which can be called his or her primary car. Also, given a person and a car, an economist could ask how much utility the person would get out of the car. From all this we can put together the following olog involving products:

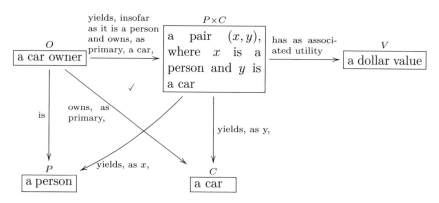

The composite map $O \rightarrow V$ tells us the utility a car owner gets out of their primary car.

3.1.2   Coproducts

We can characterize the coproduct of two sets with its own universal property.

Definition 3.1.2.1. Let X and Y be sets. The *coproduct of X and Y*, denoted $X \sqcup Y$, is defined as the disjoint union of X and Y, i.e., the set for which an element is either an element of X or an element of Y. If something is an element of both X and Y, then we include both copies, and distinguish between them, in $X \sqcup Y$. See Example 3.1.2.2.

There are two natural inclusion functions, $i_1 \colon X \to X \sqcup Y$ and $i_2 \colon Y \to X \sqcup Y$.

$$(3.3)$$

Example 3.1.2.2. The coproduct of $X := \{a, b, c, d\}$ and $Y := \{1, 2, 3\}$ is

$$X \sqcup Y \cong \{a, b, c, d, 1, 2, 3\}.$$

The coproduct of X and itself is

$$X \sqcup X \cong \{a_1, b_1, c_1, d_1, a_2, b_2, c_2, d_2\}.$$

The names of the elements in $X \sqcup Y$ are not so important. What is important are the inclusion maps i_1, i_2 from (3.3), which ensure that we know where each element of $X \sqcup Y$ came from.

Example 3.1.2.3 (Airplane seats).

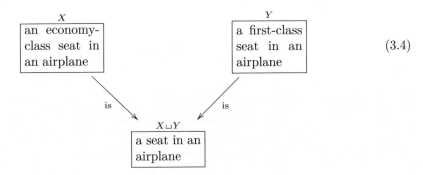

$$(3.4)$$

Exercise 3.1.2.4.

Would you say that ⌜a phone⌝ is the coproduct of ⌜a cell phone⌝ and ⌜a landline phone⌝?   ◊

Solution 3.1.2.4.

Let's make the case that ⌜a phone⌝ is the coproduct ⌜a cell phone⌝⊔⌜a landline phone⌝. First, there is no overlap between cell phones and landline phones (nothing is both). But is it true that every phone is either a cell phone or a landline? There used to be something called car phones, which were mobile in that they worked from any location but were immobile in the sense that the said location had to be within a given car. So, if at the time this solution is being read, there are phones that are neither landlines nor cell phones, then the answer to this question is no. But if every phone is either a cell phone or a landline, then the answer to this question is yes.   ◆

Example 3.1.2.5 (Disjoint union of dots). Below, X and Y are sets, having six and four elements respectively, and $X \sqcup Y$ is their coproduct, which has ten elements.

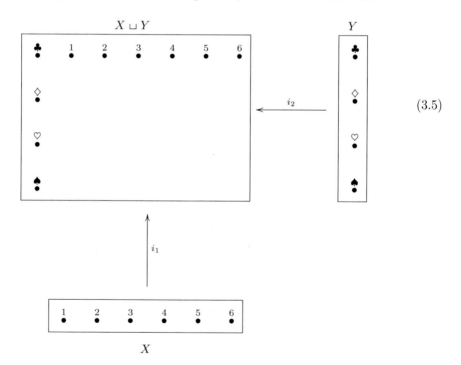

$$(3.5)$$

3.1.2.6   Universal property for coproducts

Proposition 3.1.2.7 (Universal property for coproduct). *Let X and Y be sets. For any set A and functions $f\colon X \to A$ and $g\colon Y \to A$, there exists a unique function $X \sqcup Y \to A$*

such that the following diagram commutes:

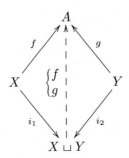

We say this function is induced by f *and* g, *and we denote it* [1]

$$\begin{cases} f \\ g \end{cases} : X \sqcup Y \to A.$$

That is, we have $\begin{cases} f \\ g \end{cases} \circ i_1 = f$ *and* $\begin{cases} f \\ g \end{cases} \circ i_2 = g$, *and* $\begin{cases} f \\ g \end{cases}$ *is the only function for which that is so.*

Proof. Suppose given f, g as in the proposition statement. To provide a function $\ell \colon X \sqcup Y \to A$ is equivalent to providing an element $f(m) \in A$ for each $m \in X \sqcup Y$. We need such a function $\ell = \begin{cases} f \\ g \end{cases}$ such that $\begin{cases} f \\ g \end{cases} \circ i_1 = f$ and $\begin{cases} f \\ g \end{cases} \circ i_2 = g$. But each element $m \in X \sqcup Y$ is either of the form $i_1 x$ or $i_2 y$ and cannot be of both forms. So we assign

$$\begin{cases} f \\ g \end{cases} (m) = \begin{cases} f(x) & \text{if } m = i_1 x, \\ g(y) & \text{if } m = i_2 y. \end{cases} \tag{3.6}$$

This assignment is necessary and sufficient to make all relevant diagrams commute.

$\square$

Slogan 3.1.2.8.

    Any time behavior is determined by cases, there is a coproduct involved.

---

[1]We are using a two-line symbol, which is a bit unusual. A certain function $X \sqcup Y \to A$ is being denoted by the symbol $\begin{cases} f \\ g \end{cases}$, called *case notation*. The reasoning for this will be clear from the proof, especially (3.6).

Exercise 3.1.2.9.

Let $f\colon \mathbb{Z} \to \mathbb{N}$ be the function defined by

$$f(n) = \begin{cases} n & \text{if } n \geqslant 0, \\ -n & \text{if } n < 0. \end{cases}$$

a. What is the standard name for f?

b. In the terminology of Proposition 3.1.2.7, what are A, X, Y, and $X \sqcup Y$?

◇

Solution 3.1.2.9.

a. The standard name for f is absolute value, also written $f(n) = |n|$.

b. Here $A = \mathbb{N}$, $X = \{n \in \mathbb{Z} \mid n \geqslant 0\}$, $Y = \{n \in \mathbb{Z} \mid n < 0\}$, so we have a natural isomorphism $X \sqcup Y \cong \mathbb{Z}$.

◆

Application 3.1.2.10 (Piecewise defined curves). In science, curves are often defined or considered piecewise. For example, in testing the mechanical properties of a material, we might be interested in various regions of deformation, such as elastic, plastic, or post-fracture. These are three intervals on which the material displays different kinds of properties.

For real numbers $a \leqslant b \in \mathbb{R}$, let $[a, b] := \{x \in \mathbb{R} \mid a \leqslant x \leqslant b\}$ denote the closed interval. Given a function $[a, b] \xrightarrow{f} \mathbb{R}$ and a function $[c, d] \xrightarrow{g} \mathbb{R}$, the universal property for coproducts implies that they extend uniquely to a function $[a, b] \sqcup [c, d] \to \mathbb{R}$, which will appear as a piecewise defined curve,

$$\begin{Bmatrix} f \\ g \end{Bmatrix} (x) = \begin{cases} f(x) & \text{if } x \in [a, b], \\ g(x) & \text{if } x \in [c, d]. \end{cases}$$

Often we are given a curve on $[a, b]$ and another on $[b, c]$, where the two curves agree at the point b. This situation is described by pushouts, which are mild generalizations of coproducts (see Section 3.3.2).

◇◇

Example 3.1.2.11 (Airplane seats, continued). The universal property for coproducts says the following. Any time we have a function $X \to A$ and a function $Y \to A$, we get a unique function $X \sqcup Y \to A$. For example, every economy-class seat in an airplane and every first-class seat in an airplane is actually *in a particular airplane*. Every economy-class seat has a price, as does every first-class seat.

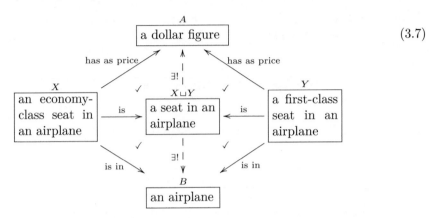

(3.7)

The universal property for coproducts formalizes the following intuitively obvious fact:

> If we know how economy-class seats are priced and we know how first-class seats are priced, and if we know that every seat is either economy class or first class, then we automatically know how all seats are priced.

To say it another way (and using the other induced map),

> If we keep track of which airplane every economy-class seat is in and we keep track of which airplane every first-class seat is in, and if we know that every seat is either economy class or first class, then we require no additional tracking for any airplane seat whatsoever.

Exercise 3.1.2.12.

Write the universal property for coproduct, in terms of a relationship between the following three sets:

$$\mathrm{Hom}_{\mathbf{Set}}(X, A), \qquad \mathrm{Hom}_{\mathbf{Set}}(Y, A), \qquad \text{and} \qquad \mathrm{Hom}_{\mathbf{Set}}(X \sqcup Y, A).$$

◊

Solution 3.1.2.12.

$$\mathrm{Hom}_{\mathbf{Set}}(X \sqcup Y, A) \xrightarrow{\cong} \mathrm{Hom}_{\mathbf{Set}}(X, A) \times \mathrm{Hom}_{\mathbf{Set}}(Y, A).$$

To assign an A value to each element of $X \sqcup Y$, you can delegate responsibility: have one person assign an A value to each element of X, and have another person assign an A value to each element of Y. One function is equivalent to two.   ♦

Example 3.1.2.13. In the following olog the types A and B are disjoint, so the coproduct $C = A \sqcup B$ is just the union.

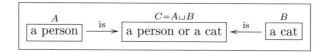

Example 3.1.2.14. In the following olog A and B are not disjoint, so care must be taken to differentiate common elements.

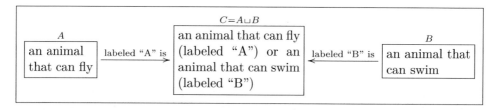

Since ducks can both swim and fly, each duck is found twice in C, once labeled "A", a flyer, and once labeled "B", a swimmer. The types A and B are kept disjoint in C, which justifies the name disjoint union.

Exercise 3.1.2.15.

Following Section 3.1.1.16, devise a naming system for coproducts, the inclusions, and the universal maps. Try it out by making an olog (involving coproducts) that discusses the idea that both a .wav file and an .mp3 file can be played on a modern computer. Be careful that your arrows are valid (see Section 2.3.2.1).   ◊

Solution 3.1.2.15.

Given two objects c, d in an olog, there is a canonical label "$c \sqcup d$" for their coproduct $c \sqcup d$, written in terms of the labels "c" and "d." Namely,

$$\langle\!\langle c \sqcup d \rangle\!\rangle := \text{``} \langle\!\langle c \rangle\!\rangle \text{ (indicated as being ``} \langle\!\langle c \rangle\!\rangle \text{'') or } \langle\!\langle d \rangle\!\rangle \text{ (indicated as being ``} \langle\!\langle d \rangle\!\rangle \text{'')."}$$

The inclusions $c \to c \sqcup d \leftarrow d$ can be labeled "after being tagged "c" is" and "after being tagged "d" is" respectively.

For example,

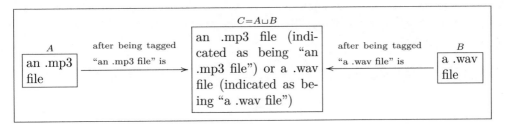

3.2 Finite limits in Set

This section discusses *limits* of variously shaped diagrams of sets. This is made more precise in Section 6.1.3, which discusses arbitrary limits in arbitrary categories.

3.2.1 Pullbacks

Definition 3.2.1.1 (Pullback). Suppose given the following diagram of sets and functions:

$$\begin{array}{c} Y \\ \downarrow g \\ X \xrightarrow{f} Z \end{array} \tag{3.8}$$

Its *fiber product* is the set

$$X \times_Z Y := \{(x, z, y) \mid f(x) = z = g(y)\}.$$

There are obvious projections $\pi_1 \colon X \times_Z Y \to X$ and $\pi_2 \colon X \times_Z Y \to Y$ (e.g., $\pi_2(x, z, y) = y$). The following diagram commutes:

$$\begin{array}{ccc} X \times_Z Y & \xrightarrow{\pi_2} & Y \\ \pi_1 \downarrow & \lrcorner & \downarrow g \\ X & \xrightarrow{f} & Z \end{array} \tag{3.9}$$

Given the setup of diagram (3.8), we define a *pullback of X and Y over Z* to be any set W for which we have an isomorphism $W \xrightarrow{\cong} X \times_Z Y$. The corner symbol ⌐ in diagram (3.9) indicates that $X \times_Z Y$ is a pullback.

Exercise 3.2.1.2.

Let X, Y, Z be as drawn and $f \colon X \to Z$ and $g \colon Y \to Z$ the indicated functions.

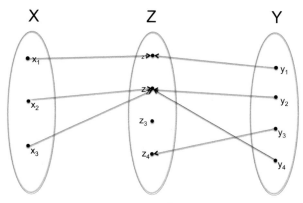

What is the fiber product of the diagram $X \xrightarrow{f} Z \xleftarrow{g} Y$?

Solution 3.2.1.2.

It is the five-element set
$$X \times_Z Y = \{(x_1, z_1, y_1), (x_2, z_2, y_2), (x_2, z_2, y_4), (x_3, z_2, y_2), (x_3, z_2, y_4)\}.$$

Exercise 3.2.1.3.

a. Draw a set X with five elements and a set Y with three elements. Color each element of X and each element of Y red, blue, or green,[2] and do so in a random-looking way. Considering your coloring of X as a function $X \to C$, where $C = \{\text{red, blue, green}\}$, and similarly obtaining a function $Y \to C$, draw the fiber product $X \times_C Y$.

b. The universal property for products guarantees a function $X \times_C Y \to X \times Y$, which will be an injection. This means that the drawing you made of the fiber product can be embedded into the 5×3 grid. Draw the grid and indicate this subset.

---

[2]You may use shadings rather than coloring, if you prefer.

Solution 3.2.1.3.

a. Let $X = \{1, 2, 3, 4, 5\}$ and $Y = \{a, b, c\}$. The fiber product is shown in part (b).

b.

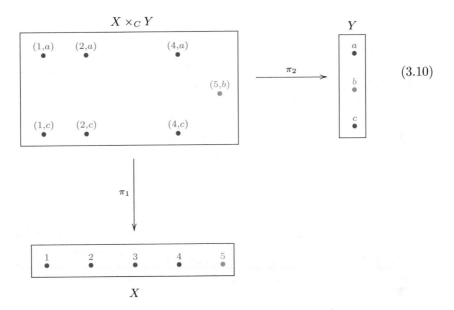

$$(3.10)$$

Note that inside the set of $X \times Y = \underline{15}$ possible (x, y) pairs is the set of pairs that agree on color—this is $X \times_C Y$. The grid $X \times Y$ is not drawn, but it includes the drawn dots, $X \times_C Y \subseteq X \times Y$, as well as eight nondrawn dots such as $(3, a)$, which "couldn't agree on a color."

♦

Remark 3.2.1.4. Some may prefer to denote the fiber product in (3.8) by $f \times_Z g$ rather than $X \times_Z Y$. The former is mathematically better notation, but human-readability is often enhanced by the latter, which is also more common in the literature. We use whichever is more convenient.

Exercise 3.2.1.5.

Let $f \colon X \to Z$ and $g \colon Y \to Z$ be functions.

a. Suppose that $Y = \emptyset$; what can you say about $X \times_Z Y$?

b. Suppose now that Y is any set but that Z has exactly one element; what can you say about $X \times_Z Y$?

◊

Solution 3.2.1.5.

a. If $Y = \emptyset$, then $X \times_Z Y = \emptyset$ regardless of X, Y, Z, f, and g.

b. We always have that $X \times_Z Y$ is the set of all triples (x, z, y), where $x \in X, y \in Y, z \in Z$ and $f(x) = z = g(y)$. If Z has only one element, say, $Z = \{\odot\}$, then for all $x \in X$ and $y \in Y$, we have $f(x) = \odot = g(y)$. So $X \times_{\{\odot\}} Y = \{(x, \odot, z) \mid x \in X, y \in Y\}$. But this set is isomorphic to the set $\{(x, y) \mid x \in X, y \in Y\}$. In other words, if Z has one element, then $X \times_Z Y \cong X \times Y$. One way of seeing this is by looking at Exercise 3.2.1.3 and thinking about what happens when there is only one color.

♦

Exercise 3.2.1.6.

Let $S = \mathbb{R}^3, T = \mathbb{R}$, and think of them as (Aristotelian) space and time, with the origin in $S \times T$ given by the center of mass of MIT at the time of its founding. Let $Y = S \times T$, and let $g_1 \colon Y \to S$ be one projection and $g_2 \colon Y \to T$ the other projection. Let $X = \{\odot\}$ be a set with one element, and let $f_1 \colon X \to S$ and $f_2 \colon X \to T$ be given by the origin in both cases.

a. What are the fiber products W_1 and W_2:

b. Interpret these sets in terms of the center of mass of MIT at the time of its founding.

◊

Solution 3.2.1.6.

Let $(s_0, t_0) \in S \times T$, where $s_0 \in S$ and $t_0 \in T$, be the center of mass of MIT at the time of its founding.

a. By definition $W_1 = \{(\odot, s, (s,t)) \mid s = s_0\}$ and $W_2 = \{(\odot, t, (s,t)) \mid t = t_0\}$.

b. We interpret both W_1 and W_2 as subsets of $Y = S \times T$. There is an isomorphism $W_1 = \{(s,t) \mid s = s_0\}$ and $W_2 = \{(s,t) \mid t = t_0\}$. In other words, we can interpret W_1 as the point in space $s_0 \in S$ throughout all time; this is like the time line for that point. We can interpret W_2 as all of space, as it existed at the time when MIT was founded.

◆

3.2.1.7 Using pullbacks to define new ideas from old

The fiber product of a diagram can serve to define a new concept. For example, olog (3.13) defines what it means for a cell phone to have a bad battery, in terms of the length of time for which it remains charged. Being explicit reduces the chance of misunderstandings between different groups of people. This can be useful in situations like audits and those in which one is trying to reuse or understand data gathered by others.

Example 3.2.1.8. Consider the following two ologs. The one on the right is the pullback of the one on the left.

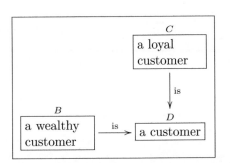 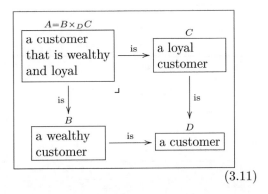

$$(3.11)$$

Check from Definition 3.2.1.1 that the label "a customer that is wealthy and loyal" is fair and straightforward as a label for the fiber product $A = B \times_D C$, given the labels on B, C, and D.

Remark 3.2.1.9. Note that in diagram (3.11) the upper left box in the pullback could have been (noncanonically named) ⌜a good customer⌝. If it were taken to be the fiber product, then the author would be effectively *defining* a good customer to be one that is wealthy and loyal.

Exercise 3.2.1.10.

For each of the following, an author has proposed that the right-hand diagram is a pullback. Do you think their labels are appropriate or misleading; that is, is the label in the upper left box of the pullback reasonable given the rest of the olog, or is it suspect in some way?

a.

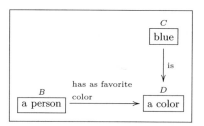 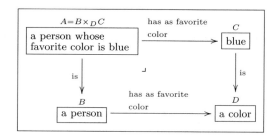

b.

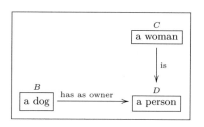 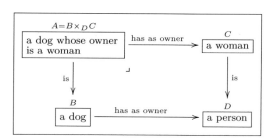

c.

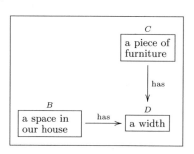 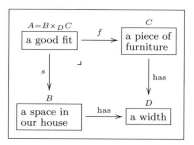

◊

Solution 3.2.1.10.

a. This is appropriate.

b. This is appropriate.

c. This is misleading. If a piece of furniture has precisely the same width as a space in our house, it is not a good fit—it is terribly frustrating but not a fit.

♦

Exercise 3.2.1.11.

Consider your olog from Exercise 2.3.3.1. Are any of the commutative squares in it actually pullback squares? ◊

Solution 3.2.1.11.

Yes, both commutative squares are pullbacks. That is, a mother is a parent who is a woman, and a father is a parent who is a man. ♦

Definition 3.2.1.12 (Preimage). Let $f\colon X \to Y$ be a function and $y \in Y$ an element. The *preimage of y under f*, denoted $f^{-1}(y)$, is the subset $f^{-1}(y) := \{x \in X \mid f(x) = y\}$. If $Y' \subseteq Y$ is any subset, the *preimage of Y' under f*, denoted $f^{-1}(Y')$, is the subset $f^{-1}(Y') = \{x \in X \mid f(x) \in Y'\}$.

Exercise 3.2.1.13.

Let $f\colon X \to Y$ be a function and $y \in Y$ an element. Draw a pullback diagram in which the fiber product is isomorphic to the preimage $f^{-1}(y)$. ◊

Solution 3.2.1.13.

It is often useful to think of an element $y \in Y$ as a function $y\colon \{\odot\} \to Y$, as in Notation 2.1.2.9. Then the following diagram is a pullback:

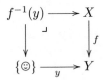

♦

Exercise 3.2.1.14.

Consider the function $f \colon \mathbb{N} \to \mathbb{N}$, where $f(n) = n + 3$. Let $A = \{i \in \mathbb{N} \mid i \geqslant 7\}$, and let $g \colon A \to \mathbb{N}$ be the inclusion, e.g., $g(17) = 17$. What is the pullback of the following diagram?

$$
\begin{array}{c}
A \\
\downarrow g \\
\mathbb{N} \xrightarrow{\ f\ } \mathbb{N}
\end{array}
$$

$\diamond$

Solution 3.2.1.14.

The pullback is isomorphic to the set $\{(n, i) \in \mathbb{N} \times \mathbb{N} \mid n + 3 = i \geqslant 7\} \cong \{n \in \mathbb{N} \mid n \geqslant 4\}$.

$\blacklozenge$

Proposition 3.2.1.15 (Universal property for pullback). *Suppose given the diagram of sets and functions as below:*

$$
\begin{array}{c}
Y \\
\downarrow u \\
X \xrightarrow{\ t\ } Z
\end{array}
$$

For any set A and the following commutative solid-arrow diagram (i.e., functions $f \colon A \to X$ and $g \colon A \to Y$ such that $t \circ f = u \circ g$), there is a unique function $A \to X \times_Z Y$ such that the diagram commutes:

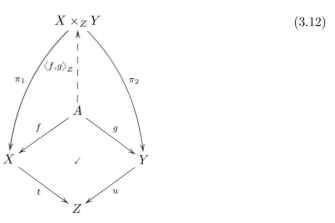

$$(3.12)$$

Exercise 3.2.1.16.

a. Create an olog whose underlying shape is a commutative square. Now add the fiber product so that the shape is the same as that of diagram (3.12).

b. Use your result to devise English labels to the object $X \times_Z Y$, to the projections π_1, π_2, and to the dotted map $A \xrightarrow{\langle f, g \rangle_Z} X \times_Z Y$, such that these labels are as canonical as possible.

◊

Solution 3.2.1.16.

a.

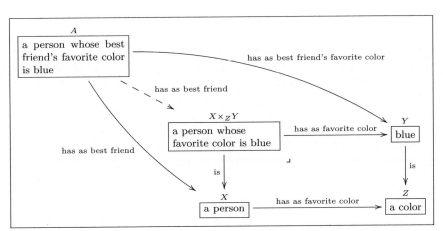

b. The answer to part (a) is not as general as possible, so we proceed in two steps. First we will give a good guess for the answer to part (a), then we will give the general answer.

What makes this example particularly nice is that the function $u \colon Y \to Z$ is labeled "is", suggesting that Y is a subset of Z. In this case, the map π_1 is labeled the same as u (namely, "is"), the map π_2 is labeled the same as t, and the map $\langle f, g \rangle \colon A \to X \times_Z Y$ is labeled the same as f. If $\langle\!\langle X \rangle\!\rangle, \langle\!\langle Y \rangle\!\rangle, \langle\!\langle Z \rangle\!\rangle, \langle\!\langle t \rangle\!\rangle$, and $\langle\!\langle u \rangle\!\rangle$ are the labels for X, Y, Z, t, and u respectively, then the object $X \times_Z Y$ is labeled "$\langle\!\langle X \rangle\!\rangle$, which $\langle\!\langle t \rangle\!\rangle\langle\!\langle Y \rangle\!\rangle$." For the part (a) example, it would be ⌜a person who has as favorite color blue⌝. See Example 3.4.5.10.

But, in general, we cannot expect either t or u to be an "is." In general, $X \times_Z Y$ should be labeled "a pair (x, z), where x is $\langle\!\langle X \rangle\!\rangle$, y is $\langle\!\langle Y \rangle\!\rangle$, and $x \langle\!\langle t \rangle\!\rangle \langle\!\langle Z \rangle\!\rangle$ that is the same as $\langle\!\langle y \rangle\!\rangle\langle\!\langle u \rangle\!\rangle$." The maps π_1 and π_2 should simply be labeled "yields, as x" and "yields, as y." The map $\langle f, g \rangle_Z$ should be labeled "yields, insofar as it $\langle\!\langle f \rangle\!\rangle\langle\!\langle X \rangle\!\rangle$ and $\langle\!\langle g \rangle\!\rangle\langle\!\langle Y \rangle\!\rangle$ and these agree as $\langle\!\langle X \rangle\!\rangle\langle\!\langle t \rangle\!\rangle\langle\!\langle Z \rangle\!\rangle$ and $\langle\!\langle Y \rangle\!\rangle\langle\!\langle u \rangle\!\rangle\langle\!\langle Z \rangle\!\rangle$."

$\blacklozenge$

3.2.1.17 Pasting diagrams for pullback

Consider the following diagram, which includes a left-hand square, a right-hand square, and a big rectangle:

$$
\begin{array}{ccccc}
A' & \xrightarrow{f'} & B' & \xrightarrow{g'} & C' \\
{\scriptstyle i}\big\downarrow & \lrcorner & {\scriptstyle j}\big\downarrow & \lrcorner & \big\downarrow{\scriptstyle k} \\
A & \xrightarrow{f} & B & \xrightarrow{g} & C
\end{array}
$$

The right-hand square has a corner symbol indicating that $B' \cong B \times_C C'$ is a pullback. But the corner symbol in the leftmost corner is ambiguous; it might be indicating that the left-hand square is a pullback, or it might be indicating that the big rectangle is a pullback. It turns out not to be ambiguous because the left-hand square is a pullback if and only if the big rectangle is. This is the content of the following proposition.

Proposition 3.2.1.18. *Consider the diagram:*

$$
\begin{array}{ccccc}
& & B' & \xrightarrow{g'} & C' \\
& & {\scriptstyle j}\big\downarrow & \lrcorner & \big\downarrow{\scriptstyle k} \\
A & \xrightarrow{f} & B & \xrightarrow{g} & C
\end{array}
$$

where $B' \cong B \times_C C'$ is a pullback. Then there is an isomorphism $A \times_B B' \cong A \times_C C'$. In other words, there is an isomorphism

$$ A \times_B (B \times_C C') \cong A \times_C C'. $$

Proof. We first provide a map $\phi \colon A \times_B (B \times_C C') \to A \times_C C'$. An element of $A \times_B (B \times_C C')$ is of the form $(a, b, (b, c, c'))$ such that $f(a) = b$, $g(b) = c$ and $k(c') = c$. But this implies that $g \circ f(a) = c = k(c')$ so we put $\phi(a, b, (b, c, c')) := (a, c, c') \in A \times_C C'$. Now we provide a proposed inverse, $\psi \colon A \times_C C' \to A \times_B (B \times_C C')$. Given (a, c, c') with $g \circ f(a) = c = k(c')$, let $b = f(a)$ and note that (b, c, c') is an element of $B \times_C C'$. So we can define $\psi(a, c, c') = (a, b, (b, c, c'))$. It is easy to see that ϕ and ψ are inverse.

$\square$

Proposition 3.2.1.18 can be useful in authoring ologs. For example, the type ⌜a cell phone that has a bad battery⌝ is vague, but we can lay out precisely what it means using pullbacks:

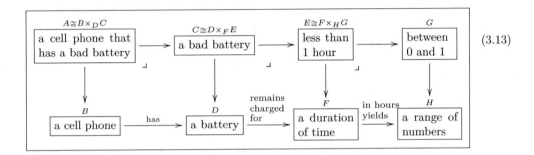

$$(3.13)$$

The category-theoretic fact described here says that since $A \cong B \times_D C$ and $C \cong D \times_F E$, it follows that $A \cong B \times_F E$. That is, we can deduce the definition "a cell phone that has a bad battery is defined as a cell phone that has a battery which remains charged for less than one hour."

Exercise 3.2.1.19.

a. Create an olog that defines two people to be "of approximately the same height" if and only if their height difference is less than half an inch, using a pullback. Your olog can include the box ⌜a real number x such that $-.5 < x < .5$⌝.

b. In the same olog, use pullbacks to make a box for those people whose height is approximately the same as a person named "Mary Quite Contrary."

◇

Solution 3.2.1.19.

Parts (a) and (b) are answered in the same olog (arrow labels are abbreviated).

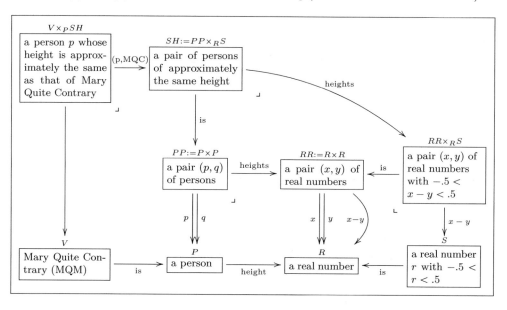

◆

Exercise 3.2.1.20.

Consider the following diagrams. In the left-hand one, both squares commute.

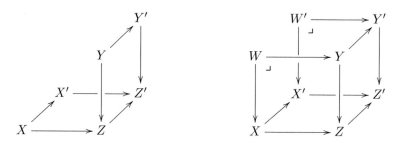

Let $W = X \times_Z Y$ and $W' = X' \times_{Z'} Y'$ be fiber products, and form the right-hand diagram. Use the universal property for fiber products to construct a map $W \to W'$ such that all squares commute.   ◊

Solution 3.2.1.20.

We redraw the right-hand diagram, with arrows labeled and a new dotted arrow:

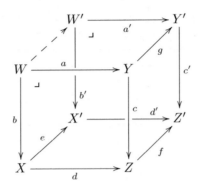

The commutativity of the right, back, and bottom squares can be written equationally as

$$c' \circ g \circ a = f \circ c \circ a = f \circ d \circ b = d' \circ e \circ b.$$

Therefore, the universal property for pullbacks (3.2.1.15) allows us to form the desired map $W \to W'$ as $\langle g \circ a, e \circ b \rangle_{Z'}$ ◆

3.2.2 Spans, experiments, and matrices

Definition 3.2.2.1. Given sets A and B, a *span on A and B* is a set R together with functions $f \colon R \to A$ and $g \colon R \to B$.

$$R \xrightarrow{\ g\ } B$$
$$\downarrow{\scriptstyle f}$$
$$A$$

Application 3.2.2.2. Think of A and B as observables and R as a set of experiments performed on these two variables.

For example, let's rename variables and say that T is the set of possible temperatures of a gas in a fixed container and that P is the set of possible pressures of the gas, so we have the span $T \xleftarrow{f} E \xrightarrow{g} P$. We perform 1,000 experiments in which we change and record the temperature, and we simultaneously record the pressure. The results might look like this:

Experiment E		
ID	**Temperature**	**Pressure**
1	100	72
2	100	73
3	100	72
4	200	140
5	200	138
6	200	141
$\vdots$	$\vdots$	$\vdots$

◇◇

Definition 3.2.2.3. Let A, B, and C be sets, and let $A \xleftarrow{f} R \xrightarrow{g} B$ and $B \xleftarrow{f'} R' \xrightarrow{g'} C$ be spans. Their *composite span* is given by the fiber product $R \times_B R'$ as in this diagram:

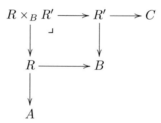

Application 3.2.2.4. Let's look back at the lab's experiment in Application 3.2.2.2, which resulted in a span $T \xleftarrow{f} E \xrightarrow{g} P$. Suppose we notice that something looks a little wrong. The pressure should be linear with the temperature but it does not appear to be. We hypothesize that the volume of the container is increasing under pressure. We look up this container online and see that experiments have been done to measure the volume as the interior pressure changes. That data gives us a span $P \xleftarrow{f'} E' \xrightarrow{g'} V$.

The composite of our lab's span with the online data span yields a span $T \leftarrow E'' \rightarrow V$, where $E'' := E \times_P E'$. What information does this span give us? In explaining it, one might say, "whenever an experiment e in our lab yielded the same pressure as the online experiment e' recorded, we called that a data point e''. Every data point has an associated temperature (from our lab) and an associated volume (from the online experiment). This is the best we can do."

The information we get this way might be seen by some as unscientific, but it certainly is the kind of information people use in business and in everyday life calculation—we get data from multiple sources and put it together. Moreover, it is scientific in the sense that it is reproducible. The way we obtained our T-V data is completely transparent.

◊◊

We can relate spans to matrices of natural numbers, and see a natural categorification of matrix addition and matrix multiplication. If the spans come from experiments, as in Applications 3.2.2.2 and 3.2.2.4, the matrices will look like huge but sparse matrices. Let's go through that.

Let A and B be sets, and let $A \leftarrow R \rightarrow B$ be a span. By the universal property for products, we have a unique map $R \xrightarrow{p} A \times B$.

We make a matrix of natural numbers out of this data as follows. The set of rows is A, the set of columns is B. For elements $a \in A$ and $b \in B$, the (a, b) entry is the cardinality of its preimage, $|p^{-1}(a, b)|$, i.e., the number of elements in R that are sent by p to (a, b).

Suppose we are given two (A, B) spans, i.e., $A \leftarrow R \rightarrow B$ and $A \leftarrow R' \rightarrow B$; we might think of these has having the same *dimensions*, i.e., they are both $|A| \times |B|$ matrices. We can take the disjoint union $R \sqcup R'$, and by the universal property for coproducts we have a unique span $A \leftarrow R \sqcup R' \rightarrow B$ making the requisite diagram commute.[3] The matrix corresponding to this new span will be the sum of the matrices corresponding to the two previous spans out of which it was made.

Given a span $A \leftarrow R \rightarrow B$ and a span $B \leftarrow S \rightarrow C$, the composite span can be formed as in Definition 3.2.2.3. It will correspond to the usual multiplication of an $|A| \times |B|$ matrix by a $|B| \times |C|$ matrix, resulting in a $|A| \times |C|$ matrix.

---

[3]The following diagram commutes:

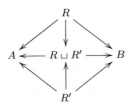

Exercise 3.2.2.5.

Let $A = \{1, 2\}$ and $B = \{1, 2, 3\}$, and consider the span $A \xleftarrow{f} R \xrightarrow{g} B$ given by the table

ID	f : A	g : B
1	1	2
2	1	2
3	1	3
4	2	1
5	2	2
6	2	3
7	2	3
8	2	3

R

So $R = \underline{8}$.

a. What is the matrix corresponding to this span?

b. If $R' \subseteq A \times B$ is a subset, with corresponding span $A \xleftarrow{f'} R' \xrightarrow{g'} B$ given by projections, what can you say about the numbers in the corresponding matrix?

◊

Solution 3.2.2.5.

a. The matrix is

$$\begin{pmatrix} 0 & 2 & 1 \\ 1 & 1 & 3 \end{pmatrix}$$

It is not a coincidence that the sum of the entries is 8, the number of elements in R.

b. Every entry in the matrix would be either 0 or 1.

♦

Construction 3.2.2.6. Given a span $A \xleftarrow{f} R \xrightarrow{g} B$, one can draw a *bipartite graph* with each element of A drawn as a dot on the left, each element of B drawn as a dot on the right, and each element $r \in R$ drawn as an arrow connecting vertex $f(r)$ on the left to vertex $g(r)$ on the right.

Exercise 3.2.2.7.

a. Draw the bipartite graph (as in Construction 3.2.2.6) corresponding to the span $T \xleftarrow{f}$ $E \xrightarrow{g} P$ in Application 3.2.2.2 (assuming the ellipses are vacuous, i.e., assuming that $|E| = 6$).

b. Now make up your own span $P \xleftarrow{f'} E' \xrightarrow{g'} V$ (with $|E'| \geqslant 2$), and write it out in database form as in Application 3.2.2.2 and in bipartite graph form.

c. Draw the composite span $T \leftarrow E \times_P E' \rightarrow V$ as a bipartite graph.

d. Describe in words how the composite span graph (for $T \leftarrow E \times_P E' \rightarrow V$) relates to the graphs of its factors ($T \leftarrow E \rightarrow P$ and $P \leftarrow E' \rightarrow V$).

                                                              ◊

Solution 3.2.2.7.

a. The six rows in that table correspond to the following six lines:

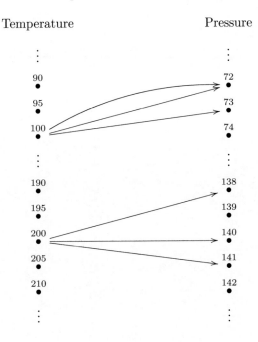

b.

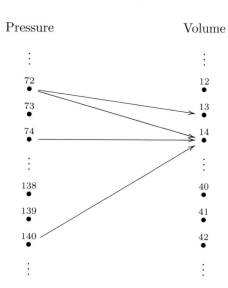

Experiment E′		
ID	**Pressure**	**Volume**
1	72	14
2	140	14
3	72	13
4	74	14

c. Putting these two together, we get

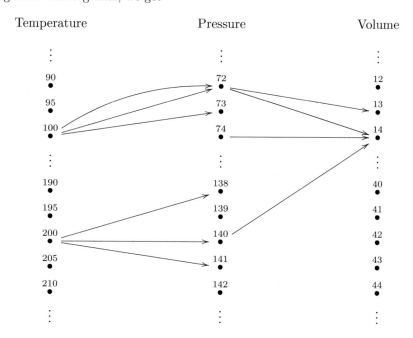

The composite span is

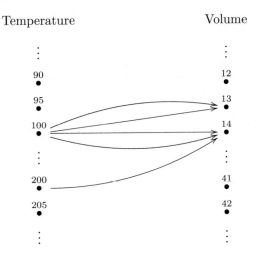

d. A bipartite graph is a set of connectors; it connects dots on the left to dots on the right. Given two composable spans, we have a set of connectors from dots on the left to dots in the middle, and a set of connectors from dots in the middle to dots on the right. We can then count each path from a left-column dot to a right-column dot as a connector and so draw an arrow; this is the graph of the composite span.

◆

3.2.3   Equalizers and terminal objects

Definition 3.2.3.1. Suppose given two parallel arrows

$$X \underset{g}{\overset{f}{\rightrightarrows}} Y. \tag{3.14}$$

The *equalizer of f and g* is the set

$$Eq(f,g) := \{x \in X \mid f(x) = g(x)\}$$

which is a subset of X. Writing $p\colon Eq(f,g) \to X$ for the inclusion, we have a commutative diagram

$$Eq(f,g) \xrightarrow{\ p\ } X \underset{g}{\overset{f}{\rightrightarrows}} Y$$

with $f \circ p = g \circ p$.

Example 3.2.3.2. Suppose one has designed an experiment to test a theoretical prediction. The question is, When does the theory match the experiment? The answer is given by the equalizer of the following diagram:

$$
\boxed{\text{an input}} \quad
\begin{array}{c}
\text{should, according to theory, yield} \\
\xrightarrow{\hspace{6cm}} \\
\text{according to experiment yields}
\end{array}
\quad \boxed{\text{an output}}
$$

The equalizer is the set of all inputs for which the theory and the experiment yield the same output.

Exercise 3.2.3.3.

Create an olog that uses equalizers in a reasonably interesting way. Alternatively, use an equalizer to specify those published authors who have published exactly one paper. Hint: Find a function from authors to papers; then find another.   ◊

Solution 3.2.3.3.

Let A be the set of published authors, let P be the set of published papers, let $F\colon A \to P$ be the function sending each author to the first paper she ever published, and let $L\colon A \to P$ send an author to the last paper she published. Then the equalizer $E \subseteq A$ of F and L is the set of authors for whom the first paper and last paper are the same. ♦

Exercise 3.2.3.4.

Find a universal property enjoyed by the equalizer of two arrows $f\colon X \to Y$ and $g\colon X \to Y$, and present it in the style of Propositions 3.1.1.10, 3.1.2.7, and 3.2.1.15.   ◊

Solution 3.2.3.4.

The equalizer of f and g is a set $Eq(f,g)$ together with a function $p\colon Eq(f,g) \to X$, such that $f \circ p = g \circ p$:

$$
Eq(f,g) \xrightarrow{\ p\ } X \underset{g}{\overset{f}{\rightrightarrows}} Y.
$$

The universal property is that for any other set E and function $q\colon E \to X$ for which $f \circ q = g \circ q$, there exists a unique function $e\colon E \to Eq(f,g)$ such that $p \circ e = q$.   ♦

Exercise 3.2.3.5.

a. A terminal set is a set S such that for every set X, there exists a unique function $X \to S$. Find a terminal set.

b. Do you think that the notion *terminal set* belongs here in Section 3.2, i.e., in the same world as products, pullbacks, and equalizers? Why? Another way to ask this is, If products, pullbacks, and equalizers are all *limits*, then what do limits have in common?

◇

Solution 3.2.3.5.

a. Let $S = \{\odot\}$. Then S is a terminal set. So is $S = \{43\}$. This was the content of Exercise 2.1.2.13, part (a).

b. The notion of a terminal set does fit well into Section 3.2 because it has a similar kind of universal property. Namely, for any other set S' that might fill the position of S, there is a unique map $S' \to S$. See Section 6.1.3.

◆

3.3 Finite colimits in Set

This section parallels Section 3.2. I introduce several types of finite colimits to give the reader some intuition about them without formally defining colimits yet.

3.3.1 Background: equivalence relations

Definition 3.3.1.1 (Equivalence relations and equivalence classes). Let X be a set, and consider the product $X \times X$, as in Definition 3.1.1.1. An *equivalence relation on* X is a subset $R \subseteq X \times X$ satisfying the following properties for all $x, y, z \in X$:

Reflexivity: $(x, x) \in R$;

Symmetry: $(x, y) \in R$ if and only if $(y, x) \in R$;

Transitivity: If $(x, y) \in R$ and $(y, z) \in R$, then $(x, z) \in R$.

If R is an equivalence relation, we often write $x \sim_R y$, or simply $x \sim y$, to mean $(x, y) \in R$. For convenience we may refer to the equivalence relation by the symbol $\sim$, saying that $\sim$ is an equivalence relation on X.

An *equivalence class of* $\sim$ is a subset $A \subseteq X$ such that

- A is nonempty, $A \neq \emptyset$;

- if $x \in A$, then $y \in A$ if and only if $x \sim y$.

Suppose that $\sim$ is an equivalence relation on X. The *quotient of X by* $\sim$, denoted $X/\sim$, is the set of equivalence classes of $\sim$. By definition, for any element $x \in X$, there is exactly one equivalence class A such that $x \in A$. Thus we can define a function $Q: X \to X/\sim$, called the *quotient function*, sending each element $x \in X$ to the equivalence class containing it. Note that for any $y \in X/\sim$, there is some $x \in X$ with $Q(x) = y$; we call x a *representative* of the equivalence class y.

Example 3.3.1.2. Let $\mathbb{Z}$ denote the set of integers. Define a relation $R \subseteq \mathbb{Z} \times \mathbb{Z}$ by

$$R = \{(x, y) \mid \exists n \in \mathbb{Z} \text{ such that } x + 7n = y\}.$$

Then R is an equivalence relation because $x + 7 * 0 = x$ (reflexivity); $x + 7 * n = y$ if and only if $y + 7 * (-n) = x$ (symmetry); and $x + 7n = y$ and $y + 7m = z$ together imply that $x + 7(m + n) = z$ (transitivity).

An example equivalence class $A \subseteq \mathbb{Z}$ for this relation is $A = \{\ldots, -12, -5, 2, 9, \ldots\}$.

Exercise 3.3.1.3.

Let X be the set of people on earth. Define a binary relation $R \subseteq X \times X$ on X as follows. For a pair (x, y) of people, put $(x, y) \in R$ if x cares what happens to y. Justify your answers to the following questions:

a. Is this relation reflexive?

b. Is it symmetric?

c. Is it transitive?

d. What if "cares what happens to" is replaced with "has shaken hands with". Is this relation reflexive, symmetric, transitive?

$\diamond$

Solution 3.3.1.3.

a. Yes; everyone cares what happens to themselves.

b. No; x may care what happens to y without y caring what happens to x.

c. No; I care about you, and you care about John, but I do not give a fig what happens to John—I do not even know the guy.

d. Shaking hands is symmetric but not reflexive and not transitive.

$\blacklozenge$

Example 3.3.1.4 (Partitions). An equivalence relation on a set X can be thought of as a way of partitioning X. A *partition of* X consists of a set I, called *the set of parts*, and for every element $i \in I$, the selection of a subset $X_i \subseteq X$ such that two properties hold:

- Every element $x \in X$ is in some part (i.e., for all $x \in X$, there exists $i \in I$ such that $x \in X_i$).

- No element can be found in two different parts (i.e., if $x \in X_i$ and $x \in X_j$, then $i = j$).

Given a partition of X, we define an equivalence relation $\sim$ on X by putting $x \sim x'$ if x and x' are in the same part (i.e., if there exists $i \in I$ such that $x, x' \in X_i$). The parts become the equivalence classes of this relation. Conversely, given an equivalence relation, one makes a partition on X by taking I to be the set of equivalence classes and, for each $i \in I$, letting X_i be the elements in that equivalence class.

Exercise 3.3.1.5.

Let X and B be sets, and let $f : X \to B$ be a function. Define a subset $R_f \subseteq X \times X$ by

$$R_f = \{(x, y) \mid f(x) = f(y)\}.$$

a. Let $f : \mathbb{R} \to \mathbb{R}$ be given by the cosine function, $f(x) = cos(x)$, and let $R_f \subseteq \mathbb{R} \times \mathbb{R}$ be the relation as defined. Find $x, y \in \mathbb{R}$ such that $x \neq y$, but $(x, y) \in R_f$.

b. Is R_f an equivalence relation, for any f?

c. Are all equivalence relations on X obtainable in this way (as R_f for some function having domain X)?

d. Does this viewpoint on equivalence classes relate to that of Example 3.3.1.4?

◊

Solution 3.3.1.5.

a. Let $x = 0.7$, and let $y = 2 * \pi + 0.7$, where $\pi \approx 3.14159$ denotes the ratio of a circle's circumference to its diameter.

b. Yes; reflexive because $f(x) = f(x)$; symmetric because if $f(x) = f(y)$, then $f(y) = f(x)$; transitive because if $f(x) = f(y)$ and $f(y) = f(z)$, then $f(x) = f(z)$.

c. Let $B = X/\sim$ be the quotient, and let $f : X \to B$ be the quotient function, sending every element in X to its equivalence class. By definition, two elements $x, y \in X$ are equivalent if and only if they are in the same equivalence class, $f(x) = f(y)$.

CHAPTER 3. FUNDAMENTAL CONSIDERATIONS IN SET

d. Yes; the fibers of f form a partition. That is, we can take B to be the set of parts, and for every $b \in B$, define the subset $X_b \subseteq X$ to be the fiber $X_b = f^{-1}(b)$. Then every element of X is in some part (fiber), and no element of X is in two different fibers.

♦

Exercise 3.3.1.6.

Take a set I of sets. That is, suppose I is a set and that for each element $i \in I$, you are given a set X_i. For every two elements $i, j \in I$, say that $i \sim j$ if X_i and X_j are isomorphic. Is this relation an equivalence relation on I?   ◊

Solution 3.3.1.6.

Yes; $\sim$ is an equivalence relation on I. This is precisely the content of Proposition 2.1.2.18.   ♦

Any relation can be enlarged to an equivalence relation with minimal alteration.

Proposition 3.3.1.7 (Generating equivalence relations). *Let X be a set and $R \subseteq X \times X$ any subset. There exists a relation $S \subseteq X \times X$ such that*

- *S is an equivalence relation;*

- *$R \subseteq S$;*

- *for any equivalence relation S' such that $R \subseteq S'$, we have $S \subseteq S'$.*

The relation S' is called the equivalence relation generated by R.

Proof. Let L_R be the set of all equivalence relations on X that contain R. In other words, each element $\ell \in L_R$ is an equivalence relation, so we have $R \subseteq \ell \subseteq X \times X$. The set L_R is nonempty because $X \times X \subseteq X \times X$ is an equivalence relation containing R. Let S denote the set of pairs $(x_1, x_2) \in X \times X$ that appear in every element of L_R, that is, $S = \bigcap_{\ell \in L_R} \ell$. Note that $R \subseteq S$ by definition. We need only show that S is an equivalence relation.

Clearly, S is reflexive, because each $\ell \in L_R$ is. If $(x, y) \in S$, then $(x, y) \in \ell$ for all $\ell \in L_R$. But since each ℓ is an equivalence relation, $(y, x) \in \ell$ too, so $(y, x) \in S$. This shows that S is symmetric. The proof that it is transitive is similar: if $(x, y) \in S$ and $(y, z) \in S$, then they are both in each ℓ, which puts (x, z) in each ℓ, which puts it in S.

□

Exercise 3.3.1.8.

Consider the set $\mathbb{R}$ of real numbers. Draw the coordinate plane $\mathbb{R} \times \mathbb{R}$, and give it coordinates x and y. A binary relation on $\mathbb{R}$ is a subset $S \subseteq \mathbb{R} \times \mathbb{R}$, which can be graphed as a set of points in the (x, y) plane.

a. Draw the relation $\{(x, y) \mid y = x^2\}$.

b. Draw the relation $\{(x, y) \mid y \geqslant x^2\}$.

c. Let S_0 be the equivalence relation on $\mathbb{R}$ generated (in the sense of Proposition 3.3.1.7) by the empty set. Draw S_0 as a subset of the plane.

d. Consider the equivalence relation S_1 generated by $\{(1, 2), (1, 3)\}$. Draw S_1 in the plane. Highlight the equivalence class containing $(1, 2)$.

e. The reflexivity property and the symmetry property (from Definition 3.3.1.1) have pleasing visualizations in $\mathbb{R} \times \mathbb{R}$; what are they?

f. Can you think of a heuristic for visualizing the transitivity property?

◊

Solution 3.3.1.8.

a.

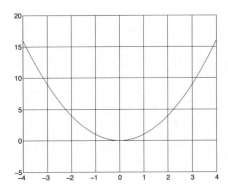

b.

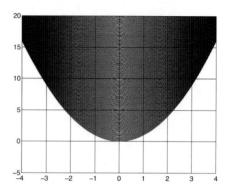

c.

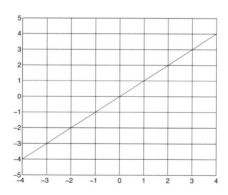

d.

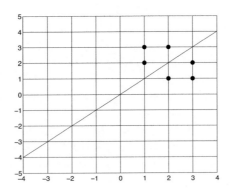

e. A relation R is reflexive if its graph in the (x, y) plane contains the line $y = x$. The relation is symmetric if its graph is symmetric about the line $y = x$.

f. I cannot think of a good one.

♦

Exercise 3.3.1.9.

Let X be a set, and consider the empty relation $R = \varnothing \subseteq X \times X$.

a. What is the equivalence relation $\sim$ generated by R (called the *trivial equivalence relation* on X)?

b. Is the quotient function $X \to X/\sim$ always an isomorphism?

◇

Solution 3.3.1.9.

a. It is the smallest reflexive relation $R := \{(x, x) \mid x \in X\}$.

b. Yes. We have $x \sim y$ if and only if $x = y$, so each equivalence class contains precisely one element.

♦

Exercise 3.3.1.10.

Consider the binary relation $R = \{(n, n + 1) \mid n \in \mathbb{Z}\} \subseteq \mathbb{Z} \times \mathbb{Z}$.

a. What is the equivalence relation $\sim$ generated by R?

b. How many equivalence classes are there?

◊

Solution 3.3.1.10.

a. For every two elements $m, n \in \mathbb{Z}$, we have $m \sim n$. Why? Think about what happens if m and n are spaced k apart, i.e., $|m - n| = k$. Then if $k = 0$, we have $m \sim n$ by reflexivity; if $k = 1$, we have $m \sim n$ by definition and symmetry. We can get the $k = 2$ case by transitivity. But then we get the $k = 3$ case by another transitivity, and continuing in this way, we find that $m \sim n$ regardless of the distance k.

b. There is one equivalence class.

◆

Exercise 3.3.1.11.

Suppose N is a network (system of nodes and edges). Let X be the nodes of the network, and let $R \subseteq X \times X$ denote the relation such that $(x, y) \in R$ iff there exists an edge connecting x to y.[4]

a. What is the equivalence relation $\sim$ generated by R?

b. What is the quotient $X/\sim$?

◊

Solution 3.3.1.11.

a. Node x is equivalent to node y if and only if one can get from x to y by moving along some finite number of edges (including no edges if $x = y$). In other words, if nodes are street addresses in a city, and each edge is like a street, then two addresses are equivalent if a pedestrian can get from one to the other.

b. It is called the set of "connected components" of the network. Think of a connected component as an island within the network. A pedestrian can get from everywhere on the island to everywhere else on the island but cannot get off the island.

◆

---

[4]The meaning of *iff* is "if and only if." In this case we are saying that the pair (x, y) is in R if and only if there exists an arrow connecting x and y.

Remark 3.3.1.12. Let X be a set and $R \subseteq X \times X$ a relation. The proof of Proposition 3.3.1.7 has the benefit of working even if $|X| \geqslant \infty$, but it has the cost that it is not very intuitive nor useful in practice when X is finite. The intuitive way to think about the idea of equivalence relation generated by R is as follows:

1. First add to R what is demanded by reflexivity, $R_1 := R \cup \{(x,x) \mid x \in X\}$.

2. To the result, add what is demanded by symmetry, $R_2 := R_1 \cup \{(x,y) \mid (y,x) \in R_1\}$.

3. Finally, to the result, add what is demanded by transitivity,

$$S = R_2 \cup \{(x,z) \mid (x,y) \in R_2, \text{ and } (y,z) \in R_2\}.$$

Then S is an equivalence relation, the smallest one containing R.

3.3.2 Pushouts

Equivalence relations are used to define pushouts.

Definition 3.3.2.1 (Pushout). Suppose given the following diagram of sets and functions:

$$\begin{array}{ccc} W & \xrightarrow{f} & X \\ {\scriptstyle g}\downarrow & & \\ Y & & \end{array} \qquad (3.15)$$

Its *fiber sum*, denoted $X \sqcup_W Y$, is defined as the quotient of $X \sqcup W \sqcup Y$ by the equivalence relation $\sim$ generated by $w \sim f(w)$ and $w \sim g(w)$ for all $w \in W$.

$$X \sqcup_W Y := (X \sqcup W \sqcup Y)/\sim, \quad \text{where } \forall w \in W, \ w \sim f(w), \text{ and } w \sim g(w).$$

There are obvious functions $i_1 : X \to X \sqcup_W Y$ and $i_2 : Y \to X \sqcup_W Y$, called *inclusions*.[5] The following diagram commutes:

$$\begin{array}{ccc} W & \xrightarrow{g} & Y \\ {\scriptstyle f}\downarrow & & \downarrow{\scriptstyle i_2} \\ X & \xrightarrow{i_1} & X \sqcup_W Y \end{array} \qquad (3.16)$$

---

[5]Note that the term *inclusion* is not too good because it seems to suggest that i_1 and i_2 are injective (see Definition 3.4.5.1) and this is not always the case. The reason we use *inclusion* terminology is to be consistent with the terminology of coproducts. The functions i_1 and i_2 are sometimes called *coprojections*.

Given the setup of diagram (3.15), we define a *pushout of X and Y over W* to be any set Z for which we have an isomorphism $X \sqcup_W Y \xrightarrow{\cong} Z$. The corner symbol $\ulcorner$ in diagram (3.16) indicates that $X \sqcup_W Y$ is a pushout.

Example 3.3.2.2. Let $X = \{x \in \mathbb{R} \mid 0 \leqslant x \leqslant 1\}$ be the set of numbers between 0 and 1, inclusive, and let $Y = \{y \in \mathbb{R} \mid 1 \leqslant y \leqslant 2\}$ be the set of numbers between 1 and 2, inclusive. We can form $X \sqcup Y$, but it has two copies of 1. This is weird, so we use pushouts; let $W = \{1\}$. Then the pushout $X \xleftarrow{f} W \xrightarrow{g} Y$, where f and g are the inclusions $(1 \mapsto 1)$, is $X \sqcup_W Y \cong \{z \in \mathbb{R} \mid 0 \leqslant z \leqslant 2\}$, as desired.

$$
\begin{array}{ccc}
\{1\} & \xrightarrow{\ g\ } & [1,2] \\
\scriptstyle f\downarrow & \ulcorner & \downarrow \\
[0,1] & \longrightarrow & [0,2]
\end{array}
$$

Example 3.3.2.3 (Pushout). In ologs (3.17) and (3.18) right-hand diagram is a pushout of the left-hand diagram. The new object, D, is the union of B and C, but instances of A are equated to their B and C aspects.

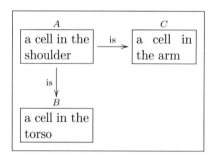 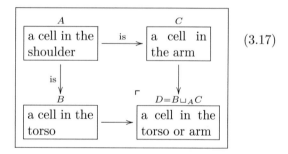   (3.17)

In diagram (3.17), the two arrows in the left-hand olog are inclusions: its author considers every cell in the shoulder to be both part of the arm and part of the torso. The pushout is then the union. In olog (3.17), the shoulder is seen as part of the arm and part of the torso. When taking the union of these two parts, we do not want to double-count cells in the shoulder (as would be done in the coproduct $B \sqcup C$; see Example 3.1.2.14). Thus we create a new type A for cells in the shoulder, which are considered the same whether viewed as cells in the arm or cells in the torso. In general, if one wishes to take two things and glue them together, with A as the glue and B and C as the two things to be glued, the result is the pushout $B \sqcup_A C$. (A nice image of this can be seen in the setting of topological spaces; see Example 6.1.3.39.)

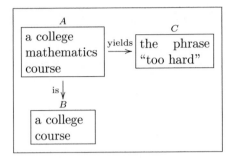 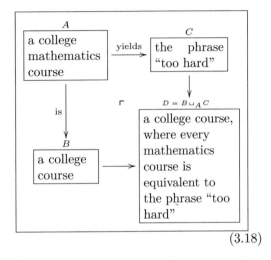

(3.18)

In olog (3.18), if every mathematics course is simply "too hard," then when reading off a list of courses, each math course may either be read aloud or simply be read as "too hard." To form D we begin by taking the union of B and C, and then we consider everything in A to be the same whether one looks at it as a course or as the phrase "too hard." The math courses are all blurred together as one thing. Thus we see that the power to equate different things can be exercised with pushouts.

Exercise 3.3.2.4.

Let W, X, Y be as drawn and $f: W \to X$ and $g: W \to Y$ the indicated functions.

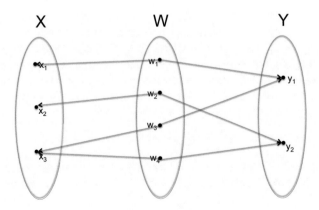

The pushout of the diagram $X \xleftarrow{f} W \xrightarrow{g} Y$ is a set P. Write the cardinality $|P|$ of P (see Definition 2.1.2.23).   ◊

Solution 3.3.2.4.

We have $x_1 \sim w_1 \sim y_1 \sim w_3 \sim x_3 \sim w_4 \sim y_2 \sim w_2 \sim x_2$. Since everything is equivalent to everything else, $|P| = 1$.   ◆

Exercise 3.3.2.5.

Suppose that $W = \varnothing$; what can you say about $X \sqcup_W Z$?   ◊

Solution 3.3.2.5.

If $W = \varnothing$, then $X \sqcup W \sqcup Y = X \sqcup Y$ and the equivalence relation $\sim$ is trivial, so $X \sqcup_W Z \cong X \sqcup Z$ is just the coproduct.   ◆

Exercise 3.3.2.6.

Let $W := \mathbb{N} = \{0, 1, 2, \ldots\}$ denote the set of natural numbers, let $X = \mathbb{Z}$ denote the set of integers, and let $Y = \{\odot\}$ denote a one-element set. Define $f \colon W \to X$ by $f(w) = -(w+1)$, and define $g \colon W \to Y$ to be the unique map. Describe the set $X \sqcup_W Y$.   ◊

Solution 3.3.2.6.

We start with $X \sqcup W \sqcup Y$ and write

$$\mathbb{Z}_{\text{``}X\text{''}} \sqcup \mathbb{N}_{\text{``}W\text{''}} \sqcup \{\odot\}_{\text{``}Y\text{''}},$$

where the subscripts indicate where things come from. Every element of $\mathbb{N}_{\text{``}W\text{''}}$ is made equivalent to $\odot$, so they are all equivalent to each other, forming a giant equivalence class. Each $n \in \mathbb{N}_{\text{``}W\text{''}}$ is also equivalent to $-(n+1) \in \mathbb{Z}_{\text{``}X\text{''}}$, so every negative number in $\mathbb{Z}_{\text{``}W\text{''}}$ is also in this giant equivalent class. But the rest of the elements in $\mathbb{Z}_{\text{``}X\text{''}}$, the non-negative numbers, are left alone—they are not forced to be equivalent to anything. So in the end, the pushout $X \sqcup_W Y$ is the set $\{\odot\} \sqcup \mathbb{N}$, where $\{\odot\}$ represents the negative integers and $\mathbb{N}$ represents the non-negative integers in $\mathbb{Z}_{\text{``}X\text{''}}$ ◆

Exercise 3.3.2.7.

Let $i \colon R \subseteq X \times X$ be an equivalence relation (see Example 2.1.2.4 for notation). Composing with the projections $\pi_1, \pi_2 \colon X \times X \to X$, we have two maps, $\pi_1 \circ i, \colon R \to X$ and $\pi_2 \circ i \colon R \to X$.

a. Consider the pushout $X \sqcup_R X$ of the diagram

$$X \xleftarrow{\pi_1 \circ i} R \xrightarrow{\pi_2 \circ i} X.$$

How should one think about $X \amalg_R X$? That is, before we defined pushouts, we went through some work to define something we can now call $X \sqcup_R X$—what was it?

b. If $i \colon R \subseteq X \times X$ is not assumed to be an equivalence relation, we can still define this pushout. Is there a relationship between the pushout $X \xleftarrow{\pi_1 \circ i} R \xrightarrow{\pi_2 \circ i} X$ and the equivalence relation generated by $R \subseteq X \times X$?

$\diamond$

Solution 3.3.2.7.

a. $X \sqcup_R X$ is isomorphic to the quotient X/R.

b. Yes, $X \sqcup_R X$ is isomorphic to the quotient $X/\sim$, where $\sim$ is the equivalence relation generated by R.

$\blacklozenge$

Proposition 3.3.2.8 (Universal property for pushout). *Suppose given the following diagram of sets and functions:*

$$
\begin{array}{ccc}
W & \xrightarrow{u} & Y \\
{\scriptstyle t}\Big\downarrow & & \\
X & &
\end{array}
$$

The pushout, $X \sqcup_W Y$ together with the inclusions i_1 and i_2, satisfies the following property. For any set A and commutative solid arrow diagram (i.e., functions $f \colon X \to A$ and

$g\colon Y \to A$ *such that* $f \circ t = g \circ u$*),*

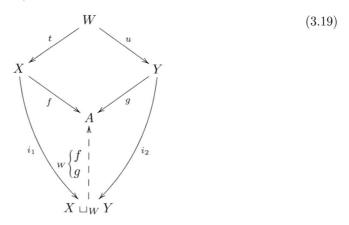

(3.19)

there exists a unique arrow $_W\begin{Bmatrix} f \\ g \end{Bmatrix} \colon X \sqcup_W Y \to A$ *making everything commute,*

$$f = {}_W\begin{Bmatrix} f \\ g \end{Bmatrix} \circ i_1 \qquad and \qquad g = {}_W\begin{Bmatrix} f \\ g \end{Bmatrix} \circ i_2.$$

3.3.3   Other finite colimits

Definition 3.3.3.1 (Coequalizer). Suppose given two parallel arrows

$$X \underset{g}{\overset{f}{\rightrightarrows}} Y. \tag{3.20}$$

The *coequalizer of* f *and* g is the commutative diagram

$$X \underset{g}{\overset{f}{\rightrightarrows}} Y \overset{q}{\longrightarrow} Coeq(f,g) \,,$$

where we define

$$Coeq(f,g) := Y/\!\sim, \qquad \text{where} \qquad f(x) \sim g(x) \text{ for all } x \in X,$$

and q is the quotient function $q\colon Y \to Y/\!\sim$.

Exercise 3.3.3.2.

   Let $X = \mathbb{R}$ be the set of real numbers. What is the coequalizer of the two maps $X \to X$ given by $x \mapsto x$ and $x \mapsto (x+1)$ respectively?   ◊

Solution 3.3.3.2.

Thus $-1 \sim 0 \sim 1 \sim 2 \sim \cdots$ and $3.14 \sim 4.14 \sim 5.14$, and so on. Everything is equivalent to itself plus 1. It is like coiling the real number line round and round into a helix of period 1, and making everything equivalent to whatever is directly above and below it. The resulting quotient is topologically a circle. Another, more concrete way to write this coequalizer (but which might be called topologically distasteful) is as the half-open interval $[0, 1)$,

$$Coeq \left(\mathbb{R} \underset{x \mapsto x}{\overset{x \mapsto x+1}{\rightrightarrows}} \mathbb{R} \right) \cong \{x \in \mathbb{R} \mid 0 \leqslant x < 1\}.$$

◆

Exercise 3.3.3.3.

Find a universal property enjoyed by the coequalizer of two arrows. ◊

Solution 3.3.3.3.

The coequalizer of f and g is a set $Coeq(f, g)$ together with a function $q\colon Y \to Eq(f, g)$, such that $q \circ f = q \circ g$:

$$X \underset{g}{\overset{f}{\rightrightarrows}} Y \overset{q}{\longrightarrow} Coeq(f, g).$$

The universal property is that for any other set C and function $p\colon Y \to C$ for which $p \circ f = p \circ g$, there exists a unique function $c\colon Coeq(f, g) \to C$ such that $c \circ q = p$. ◆

Exercise 3.3.3.4.

An initial set is a set S such that for every set A, there exists a unique function $S \to A$.

a. Find an initial set.

b. Do you think that the notion *initial set* belongs here in Section 3.3, i.e., in the same world as coproducts, pushouts, and coequalizers? Why? Another way to ask this is, If coproducts, pushouts, and coequalizers are all colimits, what do colimits have in common?

◊

Solution 3.3.3.4.

a. Let $S = \varnothing$. Then S is the initial set. This was the content of Exercise 2.1.2.13 part (b).

b. The notion of an initial set does fit well into Section 3.3 because it has a similar kind of universal property. Namely, for any other set S' that might fill the position of S, there is a unique map $S \to S'$. See Section 6.1.3.

♦

3.4   Other notions in Set

This section discusses some additional notions in the category **Set**.

3.4.1   Retractions

Definition 3.4.1.1. Suppose given a function $f \colon X \to Y$ and a function $g \colon Y \to X$ such that $g \circ f = \mathrm{id}_X$. In this case we call f a *retract section* and we call g a *retract projection*.

Exercise 3.4.1.2.

Create an olog that includes sets X and Y and functions $f \colon X \to Y$ and $g \colon Y \to X$ such that $g \circ f = \mathrm{id}_X$, but such that $f \circ g \neq \mathrm{id}_Y$; that is, such that f is a retract section but not an isomorphism. ◊

Solution 3.4.1.2.

$$\boxed{\begin{array}{c} X \\ \text{a mother} \end{array}} \xrightleftharpoons[\text{has}]{\text{has as first-born}} \boxed{\begin{array}{c} Y \\ \text{a child} \end{array}}$$

Every mother is the mother of her first-born child, but not every child is the first-born child of its mother.

♦

3.4.2   Currying

Currying is the idea that when a function takes many inputs, we can input them one at a time or all at once. For example, consider the function that takes a material M and an extension E and returns the force transmitted through material M when it is pulled to extension E. This is a function $e\colon \ulcorner\text{a material}\urcorner \times \ulcorner\text{an extension}\urcorner \to \ulcorner\text{a force}\urcorner$. This function takes two inputs at once, but it is convenient to curry the second input. Recall that $\text{Hom}_{\textbf{Set}}(\ulcorner\text{an extension}\urcorner, \ulcorner\text{a force}\urcorner)$ is the set of theoretical force-extension curves. Currying transforms e into a function

$$e'\colon \ulcorner\text{a material}\urcorner \to \text{Hom}_{\textbf{Set}}(\ulcorner\text{an extension}\urcorner, \ulcorner\text{a force}\urcorner).$$

This is a more convenient way to package the same information: each material M has a force-extension curve $e'(M)$. This will be made precise in Proposition 3.4.2.3.

Notation 3.4.2.1. Let A and B be sets. We sometimes denote by B^A the set of functions from A to B,

$$B^A := \text{Hom}_{\textbf{Set}}(A, B). \tag{3.21}$$

Exercise 3.4.2.2.

For a finite set A, let $|A| \in \mathbb{N}$ denote the cardinality of (number of elements in) A. If A and B are both finite (including the possibility that one or both are empty), is it always true that $|B^A| = |B|^{|A|}$? ◊

Solution 3.4.2.2.

If both A and B are empty, the answer to this question may be controversial; otherwise it is true that $|B^A| = |B|^{|A|}$. Back to the controversy, some people say that 0^0 is undefined, probably because of what happens with limits in calculus ($\lim_{x\to 0} x^0 = 1$, whereas $\lim_{y\to 0} 0^y = 0$). But if we think of the natural numbers as isomorphism classes of finite sets, then for $0 = |\varnothing|$ it is certainly best to think of $0^0 = 1$ because $\text{Hom}_{\textbf{Set}}(\varnothing, \varnothing) = 1$. Taking the convention that $0^0 = 1$ in $\mathbb{N}$, the answer to this question becomes an unqualified yes. ♦

Proposition 3.4.2.3 (Currying). *Let A denote a set. For any sets X, Y there is a bijection*

$$\phi\colon \text{Hom}_{\textbf{Set}}(X \times A, Y) \overset{\cong}{\to} \text{Hom}_{\textbf{Set}}(X, Y^A). \tag{3.22}$$

Proof. Suppose given $f\colon X \times A \to Y$. Define $\phi(f)\colon X \to Y^A$ as follows: for any $x \in X$, let $\phi(f)(x)\colon A \to Y$ be defined as follows: for any $a \in A$, let $\phi(f)(x)(a) := f(x, a)$.

We now construct the inverse, $\psi\colon \text{Hom}_{\textbf{Set}}(X, Y^A) \to \text{Hom}_{\textbf{Set}}(X \times A, Y)$. Suppose given $g\colon X \to Y^A$. Define $\psi(g)\colon X \times A \to Y$ as follows: for any pair $(x, a) \in X \times A$ let $\psi(g)(x, a) := g(x)(a)$.

Then for any $f \in \mathrm{Hom}_{\mathbf{Set}}(X \times A, Y)$, we have $\psi \circ \phi(f)(x, a) = \phi(f)(x)(a) = f(x, a)$, and for any $g \in \mathrm{Hom}_{\mathbf{Set}}(X, Y^A)$, we have $\phi \circ \psi(g)(x)(a) = \psi(g)(x, a) = g(x)(a)$. Thus we see that ϕ is an isomorphism as desired.

$\square$

Exercise 3.4.2.4.

Let $X = \{1, 2\}, A = \{a, b\}$, and $Y = \{x, y\}$.

a. Write three distinct elements of $L := \mathrm{Hom}_{\mathbf{Set}}(X \times A, Y)$.

b. Write all the elements of $M := \mathrm{Hom}_{\mathbf{Set}}(A, Y)$.

c. For each of the three elements $\ell \in L$ you chose in part (a), write the corresponding function $\phi(\ell) \colon X \to M$ guaranteed by Proposition 3.4.2.3.

$\Diamond$

Solution 3.4.2.4.

a. We write each of our choices $\ell_1, \ell_2, \ell_3 \in L$ in a tabular format:

ℓ_1	1	2
a	x	x
b	x	y

ℓ_2	1	2
a	y	x
b	x	y

ℓ_3	1	2
a	y	x
b	y	x

Here, for example, $\ell_1(1, a) = x$ and $\ell_1(2, b) = y$.

b. We write each of $m_1, m_2, m_3, m_4 \in M$ in tabular format:

m_1	
a	x
b	x

m_2	
a	x
b	y

m_3	
a	y
b	x

m_4	
a	y
b	y

c. We write each of $\phi(\ell_1), \phi(\ell_2), \phi(\ell_3)$ in a tabular format:

	1	2
$\phi(\ell_1)$	m_1	m_2

	1	2
$\phi(\ell_2)$	m_3	m_2

	1	2
$\phi(\ell_3)$	m_4	m_1

$\blacklozenge$

Exercise 3.4.2.5.

Let A and B be sets. We defined $B^A := \mathrm{Hom}_{\mathbf{Set}}(A, B)$, so we can write the identity function as $\mathrm{id}_{B^A}\colon \mathrm{Hom}_{\mathbf{Set}}(A, B) \to B^A$. Proposition 3.4.2.3, make the substitutions $X = \mathrm{Hom}_{\mathbf{Set}}(A, B)$, $Y = B$, and $A = A$. Consider the function

$$\phi^{-1}\colon \mathrm{Hom}_{\mathbf{Set}}(\mathrm{Hom}_{\mathbf{Set}}(A, B), B^A) \to \mathrm{Hom}_{\mathbf{Set}}(\mathrm{Hom}_{\mathbf{Set}}(A, B) \times A, B)$$

obtained as the inverse of (3.22). We have a canonical element id_{B^A} in the domain of ϕ^{-1}. We can apply the function ϕ^{-1} and obtain an element $ev = \phi^{-1}(\mathrm{id}_{B^A}) \in \mathrm{Hom}_{\mathbf{Set}}(\mathrm{Hom}_{\mathbf{Set}}(A, B) \times A, B)$, which is itself a function,

$$ev\colon \mathrm{Hom}_{\mathbf{Set}}(A, B) \times A \to B. \tag{3.23}$$

a. Describe the function ev in terms of how it operates on elements in its domain.

b. Why might one be tempted to denote this function ev?

$\Diamond$

Solution 3.4.2.5.

a. An element in $\mathrm{Hom}_{\mathbf{Set}}(A, B) \times A$ is a pair (f, a), where $f\colon A \to B$ is a function and $a \in A$ is an element. Applying ev to (f, a) returns $f(a)$, an element of B as desired.

b. One might be tempted because they are the first two letters of the word *evaluate*—we evaluate the function f on the input a.

$\blacklozenge$

If $n \in \mathbb{N}$ is a natural number, recall from (2.4) that there is a set $\underline{n} = \{1, 2, \ldots, n\}$. If A is a set, we often make the abbreviation

$$A^n := A^{\underline{n}}. \tag{3.24}$$

Exercise 3.4.2.6.

Example 3.1.1.7 said that $\mathbb{R}^2$ is an abbreviation for $\mathbb{R} \times \mathbb{R}$, but (3.24) says that $\mathbb{R}^2$ is an abbreviation for $\mathbb{R}^{\underline{2}} = \mathrm{Hom}_{\mathbf{Set}}(\underline{2}, \mathbb{R})$. Use Exercise 2.1.2.20, Exercise 3.1.2.12, and the fact that $1+1=2$, to prove that these are isomorphic, $\mathbb{R}^{\underline{2}} \cong \mathbb{R} \times \mathbb{R}$.

(The answer to Exercise 2.1.2.20 was $A = \{\odot\}$; i.e., $\mathrm{Hom}_{\mathbf{Set}}(\{\odot\}, X) \cong X$ for all X. The answer to Exercise 3.1.2.12 was $\mathrm{Hom}_{\mathbf{Set}}(X \sqcup Y, A) \overset{\cong}{\to} \mathrm{Hom}_{\mathbf{Set}}(X, A) \times \mathrm{Hom}_{\mathbf{Set}}(Y, A)$.)

$\Diamond$

Solution 3.4.2.6.

We have $\{\odot\} \cong \underline{1}$, which is a more convenient notation. We have

$$\mathbb{R}^{\underline{2}} \cong \mathbb{R}^{\underline{1}\sqcup\underline{1}} \cong \mathbb{R}^{\underline{1}} \times \mathbb{R}^{\underline{1}} \cong \mathbb{R} \times \mathbb{R}.$$

We obtain these three isomorphisms using $1+1 = 2$, then Exercise 3.1.2.12, then Exercise 2.1.2.20.   ♦

3.4.3   Arithmetic of sets

Proposition 3.4.3.1 summarizes some properties of products, coproducts, and exponentials, and shows them all in a familiar light, namely, that of elementary school arithmetic. In fact, one can think of the natural numbers as literally being the isomorphism classes of finite sets—that is what they are used for in counting.

Consider the standard procedure for counting the elements of a set S, say, cows in a field. One points to an element in S and simultaneously says "one", points to another element in S and simultaneously says "two", and so on until finished. By pointing at a cow as you speak a number, you are drawing an imaginary line between the number and the cow. In other words, this procedure amounts to nothing more than creating an isomorphism (one-to-one mapping) between S and some set $\{1, 2, 3, \ldots, n\}$.

Again, the natural numbers are the isomorphism classes of finite sets. Their behavior, i.e., the arithmetic of natural numbers, reflects the behavior of sets. For example, the fact that multiplication distributes over addition is a fact about grids of dots, as in Example 3.1.1.2. The following proposition lays out such arithmetic properties of sets.

This proposition denotes the coproduct of two sets A and B by the notation $A + B$ rather than $A \sqcup B$. It is a reasonable notation in general, and one that is often used.

Proposition 3.4.3.1. *The following isomorphisms exist for any sets $A, B,$ and C (except for one caveat; see Exercise 3.4.3.2).*

- $A + \underline{0} \cong A$

- $A + B \cong B + A$

- $(A + B) + C \cong A + (B + C)$

- $A \times \underline{0} \cong \underline{0}$

- $A \times \underline{1} \cong A$

- $A \times B \cong B \times A$

- $(A \times B) \times C \cong A \times (B \times C)$

- $A \times (B + C) \cong (A \times B) + (A \times C)$

- $A^{\underline{0}} \cong \underline{1}$

- $A^{\underline{1}} \cong A$

- $\underline{0}^A \cong \underline{0}$

- $\underline{1}^A \cong \underline{1}$

- $A^{B+C} \cong A^B \times A^C$

- $(A^B)^C \cong A^{B \times C}$

- $(A \times B)^C \cong A^C \times B^C$

Exercise 3.4.3.2.

Everything in Proposition 3.4.3.1 is true except in one case, namely, that of

$$\underline{0}^{\underline{0}}.$$

In this case we get conflicting answers, because for any set A, including $A = \varnothing = \underline{0}$, we have claimed both that $A^{\underline{0}} \cong \underline{1}$ and that $\underline{0}^A \cong \underline{0}$.

What is the correct answer for $\underline{0}^{\underline{0}}$, based on the definitions of $\underline{0}$ and $\underline{1}$, given in (2.4), and of A^B, given in (3.21)?   ◇

Solution 3.4.3.2.

$\mathrm{Hom}_{\mathbf{Set}}(\varnothing, \varnothing)$ has one element, so $\underline{0}^{\underline{0}} \cong \underline{1}$.   ◆

Exercise 3.4.3.3.

It is also true of natural numbers that if $a, b \in \mathbb{N}$ and $ab = 0$, then either $a = 0$ or $b = 0$. Is the analogous statement true of all sets?   ◇

Solution 3.4.3.3.

Yes; if A and B are sets and $A \times B \cong \varnothing$, then either $A = \varnothing$ or $B = \varnothing$.   ◆

Proposition 3.4.3.1 is in some sense about isomorphisms. It says that understanding isomorphisms of finite sets reduces to understanding natural numbers. But note that there is much more going on in **Set** than isomorphisms; in particular, there are functions that are not invertible.

In grade school you probably never saw anything that looked like this:

$$5^3 \times 3 \longrightarrow 5$$

And yet in Exercise 3.4.2.5 we found a function $ev\colon B^A \times A \to B$ that exists for any sets A, B. This function ev is not an isomorphism, so it somehow does not show up as an equation of natural numbers. But it still has important meaning.[6] In terms of mere number, it looks like we are being told of an important function $\underline{575} \to \underline{5}$, which is bizarre. The issue here is precisely the one confronted in Exercise 2.1.2.19.

Exercise 3.4.3.4.

Explain why there is a canonical function $\underline{5}^{\underline{3}} \times \underline{3} \longrightarrow \underline{5}$, but not a canonical function $\underline{575} \to \underline{5}$. ◊

Solution 3.4.3.4.

This is a more sophisticated version of Exercise 2.1.2.19: knowing that *there exists* an isomorphism between two sets is far inferior to *having* an isomorphism between them. So while it is true that $\underline{5}^{\underline{3}} \times \underline{3}$ is isomorphic to $\underline{575}$, we do not have an isomorphism between them; there is no canonical one. If one chooses an arbitrary isomorphism $f\colon \underline{575} \xrightarrow{\cong} \underline{5}^{\underline{3}} \times \underline{3}$, it would indeed compose with the evaluation function to give a function

$$\underline{575} \xrightarrow{f} \underline{5}^{\underline{3}} \times \underline{3} \xrightarrow{ev} 5,$$

but the function will be as arbitrary as the choice of isomorphism. ◆

Slogan 3.4.3.5.

> It is true that a set is isomorphic to any other set with the same number of elements, but do not be fooled into thinking that the study of sets reduces to the study of numbers. Functions that are not isomorphisms cannot be captured within the framework of numbers.

3.4.4 Subobjects and characteristic functions

Definition 3.4.4.1. For any set B, define the *power-set of B*, denoted $\mathbb{P}(B)$, to be the set of subsets of B.

---

[6]Roughly, the existence of $ev\colon \underline{5}^{\underline{3}} \times \underline{3} \longrightarrow \underline{5}$ says that given a dot in a $5 \times 5 \times 5$ grid of dots, and given one of the three axes, one can tell the coordinate of that dot along that axis.

Exercise 3.4.4.2.

a. How many elements does $\mathbb{P}(\varnothing)$ have?

b. How many elements does $\mathbb{P}(\{\odot\})$ have?

c. How many elements does $\mathbb{P}(\{1, 2, 3, 4, 5, 6\})$ have?

d. Why it be named "power-set"?

$\diamond$

Solution 3.4.4.2.

a. $|\mathbb{P}(\varnothing)| = 1$.

b. $|\mathbb{P}(\{\odot\})| = 2$.

c. $|\mathbb{P}(\{1, 2, 3, 4, 5, 6\})| = 64$.

d. For any finite set X, we find that $|\mathbb{P}(X)| = 2^{|X|}$, i.e., 2 to the *power* $|X|$.

$\blacklozenge$

3.4.4.3 Simplicial complexes

Definition 3.4.4.4. Let V be a set, let $\mathbb{P}(V)$ be its power-set. Since each element $x \in \mathbb{P}(V)$ is a subset $x \subseteq U$, we can make sense of the expression $x \subseteq x'$ for $x, x' \in \mathbb{P}(V)$. A subset $X \subseteq \mathbb{P}(V)$ is called *downward-closed* if for every $u \in X$ and every $u' \subseteq u$, we have $u' \in X$. We say that X *contains all atoms* if for every $v \in V$, the singleton set $\{v\}$ is an element of X.

A *simplicial complex* is a pair (V, X), where V is a set and $X \subseteq \mathbb{P}(V)$ is a downward-closed subset that contains all atoms. The elements of X are called *simplices* (singular: *simplex*). Any subset $u \subseteq V$ has a cardinality $|u|$, so we have a function $X \to \mathbb{N}$ sending each simplex to its cardinality. The set of simplices with cardinality $n + 1$ is denoted X_n, and each element $x \in X_n$ is called an *n-simplex*.[7] Since X contains all atoms (subsets of cardinality 1), we have an isomorphism $X_0 \cong V$, and we may also call the 0-simplices *vertices*. We sometimes call the 1-simplices *edges*.[8]

Since $X_0 \cong V$, a simplicial complex (V, X) may simply be denoted X.

---

[7]It seems anomalous that the set of subsets with cardinality 2 is denoted X_1, and so on. But this is standard convention because it fits with the standard notion of dimension: each element of X_1 corresponds to a two-dimensional shape, and more generally, each element of X_n is n-dimensional.

[8]The reason I write $X_0 \cong V$ rather than $X_0 = V$ is that X_0 is the set of one-element subsets of V. So if $V = \{a, b, c\}$, then $X_0 = \{\{a\}, \{b\}, \{c\}\}$. This is really just pedantry.

Example 3.4.4.5. Let $n \in \mathbb{N}$ be a natural number, and let $V = \underline{n+1}$. Define *the n-simplex*, denoted Δ^n, to be the simplicial complex $\mathbb{P}(V) \subseteq \mathbb{P}(V)$, i.e., the whole power-set, which indeed is downward-closed and contains all atoms.

We can draw a simplicial complex X by first putting all the vertices on the page as dots. Then for every $x \in X_1$, we see that $x = \{v, v'\}$ consists of two vertices, and we draw an edge connecting v and v'. For every $y \in X_2$ we see that $y = \{w, w', w''\}$ consists of three vertices, and we draw a (filled-in) triangle connecting them. All three edges will be drawn too, because X is assumed to be downward-closed.

The 0-simplex Δ^0, the 1-simplex Δ^1, the 2-simplex Δ^2, and the 3-simplex Δ^3 are drawn here:

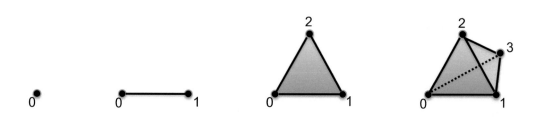

The n-simplices for various $n \in \mathbb{N}$ are not the only simplicial complexes. In general, a simplicial complex is a union, or gluing together of simplices in a prescribed manner. For example, consider the simplicial complex X with vertices $X_0 = \{1, 2, 3, 4\}$, edges $X_1 = \{\{1, 2\}, \{2, 3\}, \{2, 4\}\}$, and no higher simplices $X_2 = X_3 = \cdots = \emptyset$. We might draw X as follows:

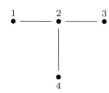

Exercise 3.4.4.6.

Let X be the following simplicial complex, so that $X_0 = \{A, B, \ldots, M\}$.

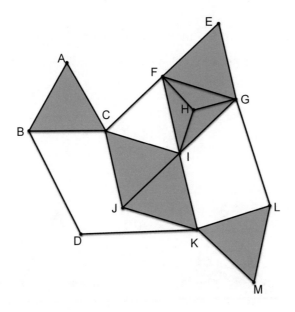

In this case X_1 consists of elements like $\{A, B\}$ and $\{D, K\}$, but not $\{D, J\}$.

Write X_2, X_3, and X_4. Hint: The drawing of X is supposed to indicate that X_3 should have one element. ◊

Solution 3.4.4.6.

X_2 is the set of triangles, X_3 is the set of tetrahedra, and X_4 is the set of 4-simplices:

$$X_2 = \{\{A, B, C\}, \{C, I, J\}, \{E, F, G\}, \{F, G, H\}, \{F, G, I\}, \{F, H, I\}, \{G, H, I\},$$
$$\{I, J, K\}, \{K, L, M\}\}.$$
$$X_3 = \{\{F, G, H, I\}\}.$$
$$X_4 = \varnothing.$$

Exercise 3.4.4.7.

The 2-simplex Δ^2 is drawn as a filled-in triangle with vertices $V = \{1, 2, 3\}$. There is a simplicial complex, often denoted $\partial\Delta^2$, that would be drawn as an empty triangle with the same set of vertices.

a. Draw Δ^2 and $\partial\Delta^2$ side by side and make clear the difference.

b. Write $X = \partial\Delta^2$ as a simplicial complex. In other words, what are the elements of the sets $X_0, X_1, X_2, X_3, \ldots$?

◇

Solution 3.4.4.7.

a.

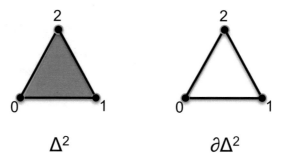

Δ^2 $\partial\Delta^2$

b. There are three 0-simplices, $X_0 = \{1, 2, 3\}$, and there are three 1-simplices, $X_1 = \{\{0, 1\}, \{1, 2\}, \{0, 2\}\}$. There are no simplices of dimension 2 or more, so for $n \geqslant 2$, we have $X_n = \varnothing$.

♦

3.4.4.8   Subobject classifier

Given a subset $A \subseteq X$, we can decide for every element of X whether it is in A or not. This is a true/false question for X.

Definition 3.4.4.9. We define the *subobject classifier* for **Set**, denoted Ω, to be the set $\Omega := \{True, False\}$, together with the function $\{\odot\} \to \Omega$ sending the unique element to *True*.

Proposition 3.4.4.10. *Let X be a set. There is an isomorphism*

$$\phi \colon \mathrm{Hom}_{\mathbf{Set}}(X, \Omega) \xrightarrow{\cong} \mathbb{P}(X).$$

Proof. Given a function $f \colon X \to \Omega$, let $\phi(f) = \{x \in X \mid f(x) = True\} \subseteq X$. We now construct a function $\psi \colon \mathbb{P}(X) \to \mathrm{Hom}_{\mathbf{Set}}(X, \Omega)$ to serve as the inverse of ϕ. Given a subset $A \subseteq X$, we define

$$\psi(A) \colon X \to \Omega \qquad \text{by} \qquad \psi(i)(x) = \begin{cases} True & \text{if } x \in A, \\ False & \text{if } x \notin A. \end{cases} \tag{3.25}$$

One checks easily that ϕ and ψ are mutually inverse. $\qquad\square$

Slogan 3.4.4.11.

> *A function X to $\Omega = \{True, False\}$ is like a roll call. We are interested in the subset that calls out True.*

Definition 3.4.4.12 (Characteristic function). Given a subset $A \subseteq X$, we define its *characteristic function of A in X* to be the function $\psi(A) \colon X \to \Omega$, from (3.25).

Let X be any set, and let $\mathbb{P}(X)$ be its power-set. By Proposition 3.4.4.10 there is a bijection between $\mathbb{P}(X)$ and Ω^X. Since Ω has cardinality 2, the cardinality of $\mathbb{P}(X)$ is $2^{|X|}$, which explains the correct answer to Exercise 3.4.4.2.

Exercise 3.4.4.13.

Let $f \colon X \to \Omega$ denote the characteristic function of some subset $A \subseteq X$, and define $A' = X - A$ to be its complement, i.e., $A' = \{x \in X \mid x \notin A\}$.

a. What is the characteristic function of $A' \subseteq X$?

b. Can you phrase it in terms of f and some function $\Omega \to \Omega$?

$\diamond$

Solution 3.4.4.13.

We solve both parts of this question at once. Let $c \colon \Omega \to \Omega$ be the function sending *True* to *False* and sending *False* to *True*. Then the characteristic function of $A' \subseteq X$ is the composite

$$X \xrightarrow{f} \Omega \xrightarrow{c} \Omega.$$

$\blacklozenge$

3.4.5 Surjections, injections

The classical definition of injections and surjections, given in Definition 3.4.5.1 involves elements. But a more robust notion involves functions; it is given in Proposition 3.4.5.8.

Definition 3.4.5.1. Let $f\colon X \to Y$ be a function.

- We say that f is *injective* if for all $x, x' \in X$ with $f(x) = f(x')$, we have $x = x'$.

- We say that f is *surjective* if for all $y \in Y$, there exists some $x \in X$ such that $f(x) = y$.

- We say that f is *bijective* if it is both injective and surjective.

We sometimes denote an injective function $X \hookrightarrow Y$, a surjective function $X \twoheadrightarrow Y$, and a bijective function $X \xrightarrow{\cong} Y$ (see Proposition 3.4.5.4).

Exercise 3.4.5.2.

a. Is the function $f\colon \mathbb{Z} \to \mathbb{N}$, given by $f(n) = n^2$, injective, surjective, or neither?

b. Is the function $g\colon \mathbb{N} \to \mathbb{N}$, given by $g(n) = n^2$, injective, surjective, or neither?

c. Is the function $h\colon \mathbb{Z} \to \mathbb{N}$, given by $h(n) = |n|$ (the absolute value), injective, surjective, or neither?

d. Is the function $i\colon \mathbb{Z} \to \mathbb{Z}$, given by $i(n) = -n$, injective, surjective, or neither?

$\Diamond$

Solution 3.4.5.2.

a. It is neither. Because $f(-1) = f(1)$, it cannot be injective. Because no integer $n \in \mathbb{Z}$ satisfies $f(n) \neq 2$, it cannot be surjective.

b. It is injective but not surjective.

c. It is surjective but not injective.

d. It is bijective (i.e., both injective and surjective).

$\blacklozenge$

Exercise 3.4.5.3.

Let $f\colon X \to Y$ and $g\colon Y \to Z$ be functions.

a. Show that if f and g are injections, then so is $g \circ f$.

b. Show that if f and g are both surjections, then so is $g \circ f$.

c. Show that if $g \circ f$ is an injection, then so is f.

d. Show that if $g \circ f$ is a surjection, then so is g.

◊

Solution 3.4.5.3.

a. Let $x, x' \in X$ and suppose that $g \circ f(x) = g \circ f(x')$. Then $g(f(x)) = g(f(x'))$, so the injectivity of g implies that $f(x) = f(x')$; the injectivity of f implies that $x = x'$.

b. Let $z \in Z$ be an element. The surjectivity of g implies that there is some $y \in Y$ with $g(y) = z$; the surjectivity of f implies that there is some $x \in X$ with $f(x) = y$.

c. Let $x, x' \in X$ and suppose that $f(x) = f(x')$. Because g is a function, $g \circ f(x) = g \circ f(x')$, and now the injectivity of $g \circ f$ implies that $x = x'$.

d. Let $z \in Z$ be an element. The surjectivity of $g \circ f$ implies that there is some $x \in X$ with $g \circ f(x) = z$. But then we have found $y := f(x) \in Y$ with $g(y) = z$.

♦

Proposition 3.4.5.4. *A function $f \colon X \to Y$ is bijective if and only if it is an isomorphism.*

Proof. Suppose that f is bijective; we define an inverse $g \colon Y \to X$. For each $y \in Y$, the preimage $f^{-1}(y) \subseteq X$ is a set with exactly one element. Indeed, it has at least one element because f is surjective, and it has at most one element because f is injective. Define $g(y)$ to be the unique element of $f^{-1}(y)$. It is easy to see that f and g are mutually inverse.

Note that for every set X, the identity function $\mathrm{id}_X \colon X \to X$ is bijective. Suppose now that f is an isomorphism, and let g be its inverse. The composition $g \circ f = \mathrm{id}_X$ is injective, and the composition $f \circ g = \mathrm{id}_Y$ is surjective, so f is injective and surjective by Exercise 3.4.5.3. □

Proposition 3.4.5.5. *Let $m, n \in \mathbb{N}$ be natural numbers. Then $m \leqslant n$ if and only if there exists an injection $m \hookrightarrow n$.*

Sketch of proof. If $m \leqslant n$, then there is an inclusion $\{1, 2, \ldots, m\} \to \{1, 2, \ldots, n\}$. Suppose now that we are given an injection $f \colon m \to n$; we assume that $m > n$ and derive a contradiction. If $m > n$, then $n + 1 \leqslant m$, and we have already shown that there exists an injection $g \colon \underline{n+1} \hookrightarrow \underline{m}$. Composing, we have an injection $h := g \circ f \colon \underline{n+1} \hookrightarrow \underline{n}$ by Exercise 3.4.5.3. One can show by induction on n that this is impossible.

□

Corollary 3.4.5.6. *Let $m, n \in \mathbb{N}$ be natural numbers. Then $m = n$ if and only if there exists an isomorphism $f \colon \underline{m} \xrightarrow{\cong} \underline{n}$.*

Proof. If $m = n$, then the identity $\mathrm{id}_m \colon \underline{m} \to \underline{n}$ is an isomorphism.

On the other hand, if we have an isomorphism $f \colon \underline{m} \xrightarrow{\cong} \underline{n}$, then both it and its inverse are injective by Proposition 3.4.5.4. Thus $m \leqslant n$ and $n \leqslant m$ by Proposition 3.4.5.5, which implies $m = n$.

□

Definition 3.4.5.7 (Monomorphisms, epimorphisms). Let $f \colon X \to Y$ be a function. We say that f is a *monomorphism* if for all sets A and pairs of functions $g, g' \colon A \to X$,

$$A \underset{g'}{\overset{g}{\rightrightarrows}} X \xrightarrow{f} Y,$$

if $f \circ g = f \circ g'$, then $g = g'$.

We say that f is an *epimorphism* if for all sets B and pairs of functions $h, h' \colon Y \to B$,

$$X \xrightarrow{f} Y \underset{h'}{\overset{h}{\rightrightarrows}} B,$$

if $h \circ f = h' \circ f$, then $h = h'$.

Proposition 3.4.5.8. *Let $f \colon X \to Y$ be a function. Then f is injective if and only if it is a monomorphism; f is surjective if and only if it is an epimorphism.*

Proof. We use notation as in Definition 3.4.5.7.

If f is a monomorphism, it is clearly injective by putting $A = \{☺\}$. Suppose that f is injective, and let $g, g' \colon A \to X$ be functions such that $f \circ g = f \circ g'$, but suppose for contradiction that $g \neq g'$. Then there is some element $a \in A$ such $g(a) \neq g'(a) \in X$. But by injectivity $f(g(a)) \neq f(g'(a))$, contradicting the fact that $f \circ g = f \circ g'$.

Suppose that $f \colon X \to Y$ is an epimorphism, and choose some $y_0 \in Y$ (noting that if Y is empty, then the claim is vacuously true). Let $B = \Omega$, and let $h \colon Y \to \Omega$ denote the characteristic function of the subset $\{y_0\} \subseteq Y$, and let $h' \colon Y \to \Omega$ denote the characteristic function of $\varnothing \subseteq Y$. Note that $h(y) = h'(y)$ for all $y \neq y_0$. Then since f is an epimorphism and $h \neq h'$, we must have $h \circ f \neq h' \circ f$, so there exists $x \in X$ with $h(f(x)) \neq h'(f(x))$, which implies that $f(x) = y_0$. This proves that f is surjective.

Finally, suppose that f is surjective, and let $h, h' \colon Y \to B$ be functions with $h \circ f = h' \circ f$. For any $y \in Y$, there exists some $x \in X$ with $f(x) = y$, so $h(y) = h(f(x)) = h'(f(x)) = h'(y)$. This proves that f is an epimorphism.

$\square$

Proposition 3.4.5.9. *Let $g \colon A \to Y$ be a monomorphism. Then for any function $f \colon X \to Y$, the left-hand map $g' \colon X \times_Y A \to X$ in the diagram*

$$
\begin{array}{ccc}
X \times_Y A & \xrightarrow{\ f'\ } & A \\
{\scriptstyle g'}\downarrow & \lrcorner & \downarrow{\scriptstyle g} \\
X & \xrightarrow[\ f\ ]{} & Y
\end{array}
$$

is a monomorphism.

Proof. To show that g' is a monomorphism, we take an arbitrary set B and two maps $m, n \colon B \to X \times_Y A$ such that $g' \circ m = g' \circ n$, denoting that function $p := g' \circ m \colon B \to X$. Now let $q = f' \circ m$ and $r = f' \circ n$. The diagram looks like this:

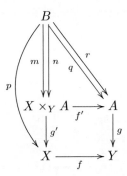

We have that

$$
g \circ q = g \circ f' \circ m = f \circ g' \circ m = f \circ p = f \circ g' \circ n = g \circ f' \circ n = g \circ r
$$

But we assumed that g is a monomorphism, so this implies that $q = r$. By the universal property for pullbacks, Proposition 3.2.1.15, we have $m = n = \langle q, p \rangle_Y \colon B \to X \times_Y A$.

$\square$

Example 3.4.5.10. Suppose an olog has a fiber product square

$$X \times_Y A \xrightarrow{f'} A$$

$$g' \downarrow \qquad \qquad \downarrow g$$

$$X \xrightarrow{f} Y$$

such that g is intended to be a monomorphism and f is any map.[9] In this case, there are labeling systems for f', g', and $X \times_Y A$. Namely,

- "is" is an appropriate label for g and g';

- the label for f is an appropriate label for f';

- $\langle\!\langle X \times_Y A \rangle\!\rangle := "\langle\!\langle X \rangle\!\rangle$, which $\langle\!\langle f \rangle\!\rangle \langle\!\langle A \rangle\!\rangle$" is an appropriate label for $X \times_Y A$.

To give an explicit example,

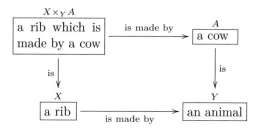

Corollary 3.4.5.11. *Let $i: A \to X$ be a monomorphism, and let True: $\{\odot\} \to \Omega$ be the subobject classifier (see Definition 3.4.4.9). Then there is a fiber product square of the form*

$$A \xrightarrow{f'} \{\odot\} \tag{3.26}$$

$$i \downarrow \qquad \qquad \downarrow True$$

$$X \xrightarrow{f} \Omega$$

Proof. Let $X' \subseteq X$ denote the image of i, and let $f: X \to \Omega$ denote the characteristic function of $X' \subseteq X$, given by Proposition 3.4.4.10. Then it is easy to check that diagram (3.26) is a pullback.

□

---

[9]Of course, this diagram is symmetrical, so the same ideas hold if f is a monomorphism and g is any map.

Exercise 3.4.5.12.

Consider the subobject classifier $\Omega = \{True, False\}$, the singleton $\{\odot\}$, and the map $\{\odot\} \xrightarrow{True} \Omega$ from Definition 3.4.4.9. In diagram (3.26), in the spirit of Example 3.4.5.10, devise a label for Ω, a label for $\{\odot\}$, and a label for $True$. Given a subobject $A \subseteq X$, both labeled, devise a label for f, a label for i, and a label for f' such that the English smoothly fits the mathematics. ◊

Solution 3.4.5.12.

Let's take the label for Ω to be the question mark, $\langle\!\langle \Omega \rangle\!\rangle = \ulcorner ? \urcorner$, the label for $\{\odot\}$ to be blank $\langle\!\langle \{\odot\} \rangle\!\rangle = \ulcorner \urcorner$, and the label for $True$ to be "is." Given any subobject $A \subseteq X$, we define $\langle\!\langle f \rangle\!\rangle = $ "is $\langle\!\langle A \rangle\!\rangle$". The rules from Example 3.4.5.10 imply that $\langle\!\langle i \rangle\!\rangle = $ "is" and $\langle\!\langle f' \rangle\!\rangle = \langle\!\langle f \rangle\!\rangle$. This all looks very abstract, so here is an example:

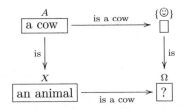

Note that the unique map to the terminal object from any other object, e.g., $\ulcorner$a cow$\urcorner$, now reads "a cow is a cow." The characteristic function for cows as a subset of any other type is now "is a cow?" The rules of Example 3.4.5.10 tell us that $\ulcorner$a cow$\urcorner$ should be semantically equivalent to $\ulcorner$an animal which is a cow$\urcorner$, which it is. ◆

Exercise 3.4.5.13.

Show, in analogy to Proposition 3.4.5.9, that pushouts preserve epimorphisms. ◊

Solution 3.4.5.13.

We want to prove the following:

Let $g \colon Y \to A$ be an epimorphism. Then for any function $f \colon Y \to X$, the right-hand map $g' \colon X \to A \sqcup_Y X$ in the diagram

$$
\begin{array}{ccc}
Y & \xrightarrow{\ f\ } & X \\
{\scriptstyle g}\downarrow & \ulcorner & \downarrow{\scriptstyle g'} \\
A & \xrightarrow[\ f'\ ]{} & A \sqcup_Y X
\end{array}
$$

is an epimorphism.

Proof. To show that g' is an epimorphism, we take an arbitrary set B and two maps $m, n\colon A \sqcup_Y X \to B$ such that $m \circ g' = n \circ g'$, denoting that function $p := m \circ g'\colon A \to B$. Now let $q = m \circ g'$ and $r = n \circ g'$. The diagram looks like this:

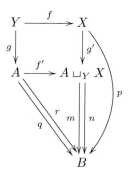

We have that

$$
q \circ g = m \circ f' \circ g = m \circ g' \circ f = p \circ f = n \circ g' \circ f = n \circ f' \circ g = r \circ g.
$$

But we assumed that g is an epimorphism, so this implies that $q = r$. By the universal property for pushouts, Proposition 3.3.2.8, we have $m = n =\ _Y \begin{cases} q \\ p \end{cases}$.

$\square$

$\blacklozenge$

3.4.6   Multisets, relative sets, and set-indexed sets

In this section we prepare to consider categories other than **Set** by looking at some categories related to **Set**.

3.4.6.1 Multisets

Consider the set X of words in a given document. If $WC(X)$ is the word count of the document, we do not generally have $WC(X) = |X|$. The reason is that a set cannot contain the same element more than once, so words like *the* might be undercounted in $|X|$. A *multiset* X consists of a set of names, N_X, and each name is assigned a multiplicity, i.e., a positive finite number of times it is to be counted. For example, the multiset $X =$(The, man, just, ate, and, ate, and, ate) has names $N_X = \{$The, man, just, ate, and$\}$, and these names have multiplicity $1, 1, 1, 3, 2$ respectively.

But if X and Y are multisets, what is the appropriate type of mapping from X to Y? Since every set can be cast as a multiset (in which each element has multiplicity 1), let's restrict ourselves to notions of mapping that agree with the usual one on sets. That is, if multisets X and Y happen to be ordinary sets, then our mappings $X \to Y$ should just be functions.

In order to define what I believe is the appropriate notion of mapping of multisets, it is useful to take a step back from this definition. The role of the natural numbers in multisets is to count the number of occurrences of each element. The point perhaps is not the number, but the set of occurrences it counts. Each occurrence has a name, so we have a function from occurrences to names. The fact that every name has multiplicity at least 1 means that this function is surjective. So I suggest the following definition of multisets and mappings.

Definition 3.4.6.2. A *multiset* is a sequence $X := (Oc, N, \pi)$, where Oc and N are sets and $\pi\colon Oc \to N$ is a surjective function. We refer to Oc as the set of *occurrences in X*, to N as the set of *names in X*, and to π as the *naming function for X*. Given a name $x \in N$, let $\pi^{-1}(x) \subseteq Oc$ be the preimage; the number of elements in $\pi^{-1}(x)$ is called the *multiplicity of x*.

Suppose that $X = (Oc, N, \pi)$ and $X' = (Oc', N', \pi')$ are multisets. A *mapping from X to Y*, denoted $f\colon X \to Y$, consists of a pair (f_1, f_0) such that $f_1\colon Oc \to Oc'$ and $f_0\colon N \to N'$ are functions and such that the following diagram commutes:

$$\begin{array}{ccc} Oc & \xrightarrow{\ f_1\ } & Oc' \\ {\scriptstyle \pi}\downarrow & & \downarrow{\scriptstyle \pi'} \\ N & \xrightarrow[\ f_0\ ]{} & N' \end{array} \qquad (3.27)$$

Exercise 3.4.6.3.

Suppose that a pseudo-multiset is defined to be almost the same as a multiset, except that π is not required to be surjective.

a. Write a pseudo-multiset that is not a multiset.

b. Describe the difference between the two notions (multiset vs. pseudo-multiset) in terms of multiplicities.

◊

Solution 3.4.6.3.

a. $X = (\varnothing, \{\odot\}, !)$, where $! \colon \varnothing \to \{\odot\}$ is the unique function.

b. In a multiset the multiplicity of each element must be an integer $n \geqslant 1$, whereas in a pseudo-multiset the multiplicity of an element may be 0.

♦

Exercise 3.4.6.4.

Consider the multisets $X = (a, a, b, c)$ and $Y = (d, d, e, e, e)$.

a. Write each of them in the form (Oc, N, π), as in Definition 3.4.6.2.

b. In terms of the same definition, how many mappings $X \to Y$ are there?

c. If we were to remove the restriction that diagram (3.27) must commute, how many mappings $X \to Y$ would there be?

◊

Solution 3.4.6.4.

a. We use the sequence notation from (2.5) for functions out of $\underline{n}$:

$$Oc_X = \underline{4}, \quad N_X = (a, b, c), \quad \pi_X = (a, a, b, c)$$
$$Oc_Y = \underline{5}, \quad N_Y = (d, e), \quad \pi_Y = (d, d, e, e, e)$$

b. Computations are difficult to follow and explain. So the following will likely only make sense to those who have worked on the problem.

A map $X \to Y$ is obtained by choosing a function $f : N_X \to N_Y$ from X names to Y names and then, for each X name $x \in N_X$, choosing a function from the fiber $\pi_X^{-1}(x)$ over x to the fiber $\pi_Y^{-1}(f(x))$ over $f(x)$. There are eight maps $N_X \to N_Y$:

	1	2	3	4	5	6	7	8
a	d	d	d	d	e	e	e	e
b	d	d	e	e	d	d	e	e
c	d	e	d	e	d	e	d	e

For each such map f and for each $x \in \{a, b, c\}$, we count how many maps from the fiber over x to the fiber over $f(x)$. The sizes of the various fibers are 1, 2, and 3, because these are the multiplicities of a, b, c, d, e. So we have the following array:

	1	2	3	4	5	6	7	8
a	2^2	2^2	2^2	2^2	3^2	3^2	3^2	3^2
b	2^1	2^1	3^1	3^1	2^1	2^1	3^1	3^1
c	2^1	3^1	2^1	3^1	2^1	3^1	2^1	3^1

The result now comes by multiplying the entries in each column and adding:

$$16 + 24 + 24 + 36 + 36 + 54 + 54 + 81 = 325.$$

c. There would be $5^4 * 2^3 = 5000$.

♦

3.4.6.5   Relative sets

Continuing with ideas from multisets, let's suppose that we have a fixed set N of names that we want to keep once and for all. Whenever someone discusses a set, each of its elements must have a name in N. And whenever someone discusses a mapping, it must preserve the naming. For example, if N is the set of English words, then every document consists of a set $\{1, 2, 3, \ldots, n\}$ mapping to N (e.g., $1 \mapsto$ Continuing, $2 \mapsto$ with, $3 \mapsto$ ideas, $\ldots$). A mapping from document A to document B would send each word found somewhere in A to the same word found somewhere in B. This notion is defined in the following definition.

Definition 3.4.6.6 (Relative set). Let N be a set. A *relative set over N*, or simply a *set over N*, is a pair (E, π) such that E is a set and $\pi \colon E \to N$ is a function. A *mapping of relative sets over N*, denoted $f \colon (E, \pi) \to (E', \pi')$, is a function $f \colon E \to E'$ such that the following triangle commutes, i.e., $\pi = \pi' \circ f$:

Exercise 3.4.6.7.

     Given sets X, Y, Z and functions $f \colon X \to Y$ and $g \colon Y \to Z$, we can compose them to get a function $X \to Z$. If N is a set, if $(X, p), (Y, q)$, and (Z, r) are relative sets over N, and if $f \colon (X, p) \to (Y, q)$ and $g \colon (Y, q) \to (Z, r)$ are mappings of relative sets, is there a reasonable notion of composition such that we get a mapping of relative sets $(X, p) \to (Z, r)$? Hint: Draw diagrams.   ◊

Solution 3.4.6.7.

Yes. We are given two commutative triangles, $p = q \circ f$ and $q = r \circ g$:

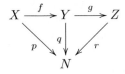

The composite function $g \circ f \colon X \to Z$ works as a mapping of relative sets over N because $p = q \circ f = r \circ (g \circ f)$, as required. ♦

Exercise 3.4.6.8.

a. Let $\{\odot\}$ denote a set with one element. What is the difference between sets relative to $N := \{\odot\}$ and simply sets?

b. Describe the sets relative to $\varnothing$. How many are there?

◊

Solution 3.4.6.8.

a. There is no real difference, because a relative set over $\{\odot\}$ is given by specifying a set X and a function $X \to \{\odot\}$, which is no more information than simply specifying the set X.

b. A set relative to $\varnothing$ is a set X and a function $X \to \varnothing$. But the only set for which there exists such a function is $X = \varnothing$. So there is only one set relative to $\varnothing$.

♦

3.4.6.9 Indexed sets

Let A be a set. Suppose we want to assign to each element $a \in A$ a set S_a. This is called an A-indexed set. In category theory we are always interested in the legal mappings between two different objects of the same sort of structure, so we need a notion of A-indexed mappings.

Example 3.4.6.10. Let C be a set of classrooms. For each $c \in C$, let P_c denote the set of people in classroom c, and let S_c denote the set of seats (chairs) in classroom c. Then P

and S are C-indexed sets. The appropriate kind of mapping between them respects the indices. That is, a mapping of C-indexed sets $P \to S$ should, for each classroom $c \in C$, be a function $P_c \to S_c$.[10]

Definition 3.4.6.11. Let A be a set. An *A-indexed set* is a collection of sets S_a, one for each element $a \in A$; for now we denote this $(S_a)_{a \in A}$. Each element $a \in A$ is called an *index*. If $(S'_a)_{a \in A}$ is another A-indexed set, an *A-indexed function from* $(S_a)_{a \in A}$ *to* $(S'_a)_{a \in A}$, denoted

$$(f_a)_{a \in A} \colon (S_a)_{a \in A} \to (S'_a)_{a \in A},$$

is a collection of functions $f_a \colon S_a \to S'_a$, one for each element $a \in A$.

Exercise 3.4.6.12.

Let $\{\odot\}$ denote a one-element set. What are $\{\odot\}$-indexed sets and $\{\odot\}$-indexed functions? ◊

Solution 3.4.6.12.

A $\{\odot\}$-indexed set is just a collection of exactly one set. So there is no real difference between $\{\odot\}$-indexed sets and simply sets, nor is there a difference between $\{\odot\}$-indexed functions and simply functions. ♦

Exercise 3.4.6.13.

There is a strong relationship between A-indexed sets and relative sets over A. What is it? ◊

Solution 3.4.6.13.

These two notions are equivalent. Given an A-indexed set $(S_a)_{a \in A}$, we can turn it into a relative set as follows. Consider $S := \coprod_{a \in A} S_a$, the disjoint union of all sets in the collection. The obvious function $S \to A$, given by sending each element to the index from which it came, is a relative set. Going the other way, given a relative set $E \xrightarrow{\pi} N$, we create an N-indexed set $(P_n)_{n \in N}$, putting $P_n := \pi^{-1}(n)$. ♦

---

[10]If we wanted to allow people from any classroom to choose a chair from just any classroom, category theory would tell us to reconsider P and S as sets, forgetting their C-indices. See Section 7.1.4.6.

Chapter 4

Categories and Functors, Without Admitting It

In this chapter we begin to use our understanding of sets to examine more interesting mathematical worlds, each of which organizes understanding of a certain kind of domain. For example, monoids organize thoughts about agents acting on objects. Groups are monoids except restricted to only allow agents to act in reversible ways. We then study graphs, which are systems of nodes and arrows that can capture ideas like information flow through a network or model connections between building blocks in a material. We discuss orders, which can be used to study taxonomies or hierarchies. Finally we take a mathematical look at databases, which actually subsume everything else in the chapter. Databases are connection patterns for structuring information.

Everything studied in this chapter is an example of a category (see Chapter 5). So is **Set**, the category of sets studied in Chapters 2 and 3. One way to think of a category is as a bunch of objects and a connection pattern between them. The category **Set** has individual sets as objects, with functions serving as the connections between them. But there is a certain self-similarity here—each set, thought of as a bag of dots, can itself be viewed as a category: the objects inside it are just disconnected. Each set *is* a category, but there is also a category *of* sets. In this way, sets have an interior view and an exterior view, as do all the categories in this chapter. Each monoid *is* a category, but there is also a category *of* monoids.

However, the word *category* is not used much in this chapter. It seems preferable to let the ideas arise as interesting structures in their own right before explaining how everything fits into a single framework.

4.1   Monoids

A common way to interpret phenomena around us is to say that agents are acting on objects. For example, the user of a computer drawing program *acts on* the canvas in certain prescribed ways. Choices of actions from an available list can be performed in sequence to transform one image into another. As another example, one might investigate the notion that time *acts on* the position of hands on a clock in a prescribed way. A first rule for actions is captured in the following slogan.

Slogan 4.1.0.14.

> *The performance of a sequence of several actions is itself the performance of an action—a more complex action, but an action nonetheless.*

Mathematical objects called *monoids* and *groups* are tasked with encoding the agent's perspective, i.e., what the agent can do, and what happens when she does a sequence of actions in succession. A monoid can be construed as a set of actions together with a formula that encodes how a sequence of actions is itself considered an action. A group is the same as a monoid except that every action is required to be reversible.

4.1.1   Definition and examples

Definition 4.1.1.1 (Monoid). A *monoid* is a sequence $(M, e, \star)$, where M is a set, $e \in M$ is an element, and $\star \colon M \times M \to M$ is a function, such that the following *monoid laws* hold for all $m, n, p \in M$:

- $m \star e = m$.

- $e \star m = m$.

- $(m \star n) \star p = m \star (n \star p)$.

We refer to e as the *unit element* and to $\star$ as the *multiplication formula* for the monoid.[1] We call the first two rules *unit laws* and the third rule the *associativity law* for monoids.

Remark 4.1.1.2. To be pedantic, the conditions from Definition 4.1.1.1 should be stated

- $\star(m, e) = m$.

- $\star(e, m) = m$.

---

[1]Although the function $\star \colon M \times M \to M$ is called the multiplication formula, it may have nothing to do with multiplication. It is just a formula for taking two inputs and returning an output.

- $\star(\star(m,n),p) = \star(m,(\star(n,p)))$.

The way they are written in Definition 4.1.1.1 is called *infix notation,*. Given a function $\star\colon A \times B \to C$, we may write $a \star b$ rather than $\star(a,b)$.

Example 4.1.1.3 (Additive monoid of natural numbers). Let $M = \mathbb{N}$ be the set of natural numbers. Let $e = 0$, and let $\star\colon M \times M \to M$ denote addition, so that $\star(4,18) = 4 \star 18 = 22$. Then the equations $m \star 0 = m$ and $0 \star m = m$ hold, and $(m \star n) \star p = m \star (n \star p)$ because, as we learned in grade school, addition is associative. By assigning e and $\star$ in this way, we have given $\mathbb{N}$ the structure of a monoid. We usually denote it $(\mathbb{N}, 0, +)$.

Remark 4.1.1.4. Sometimes we are working with a monoid $(M, e, \star)$, and the unit e and multiplication $\star$ are somehow clear from context. In this case we might refer to the set M as though it were the whole monoid. For example, if we were discussing the monoid from Example 4.1.1.3, we might refer to it as $\mathbb{N}$. The danger comes because sets may have multiple monoid structures (see Exercise 4.1.1.6).

Example 4.1.1.5 (Nonmonoid). If M is a set, we might call a function $f\colon M \times M \to M$ an *operation on* M. For example, if $M = \mathbb{N}$ is the set of natural numbers, we can consider the operation $f\colon \mathbb{N} \times \mathbb{N} \to \mathbb{N}$ called exponentiation e.g., $f(2,5) = 2 * 2 * 2 * 2 * 2 = 32$ and $f(7,2) = 49$. This is indeed an operation, but it is not the multiplication formula for any monoid. First, there is no possible unit. Trying the obvious choice of $e = 1$, we see that $a^1 = a$ (good), but that $1^a = 1$ (bad: we need it to be a). Second, this operation is not associative because in general $a^{(b^c)} \neq (a^b)^c$. For example, $2^{(1^2)} = 2$, but $(2^1)^2 = 4$.

One might also attempt to consider an operation $f\colon M \times M \to M$ that upon closer inspection is not even an operation. For example, if $M = \mathbb{Z}$, then exponentiation is not even an operation. Indeed, $f(2,-1) = 2^{-1} = \frac{1}{2}$, and this is not an integer. To have a function $f\colon M \times M \to M$, it is required that every element of the domain—in this case every pair of integers—have an output under f. So there is no exponentiation function on $\mathbb{Z}$.

Exercise 4.1.1.6.

Let $M = \mathbb{N}$ be the set of natural numbers. Taking $e = 1$ as the unit, devise a formula for $\star$ that gives $\mathbb{N}$ the structure of a monoid. ◊

Solution 4.1.1.6.

Let $\star$ denote the usual multiplication of natural numbers, e.g., $5 \star 7 = 35$. Then for any $m, n, p \in \mathbb{N}$, we have $1 \star m = m \star 1 = m$ and $(m \star n) \star p = m \star (n \star p)$, as required. ♦

Exercise 4.1.1.7.

Find an operation on the set $M = \{1, 2, 3, 4\}$, i.e., a legitimate function $f\colon M \times M \to M$, such that f cannot be the multiplication formula for a monoid on M. That is, either it is not associative or no element of M can serve as a unit.   ◊

Solution 4.1.1.7.

Here is an example: $f(m, n) = 4$ for all $m, n \in M$. This fails to be a multiplication formula because no element of M can serve as a unit.

Here are multiplication formulas that have a unit (namely, 4) but that have been prevented from being associative.

⋆	1	2	3	4
1	?	2	2	1
2	?	?	3	2
3	?	?	?	3
4	1	2	3	4

Here $1 \star 3 = 2$, and so on. Each question mark (?) can be filled with any element of M, and the result will fail to be a multiplication formula because the following argument shows that it cannot be associative:

$$(1 \star 2) \star 3 = 2 \star 3 = 3 \neq 2 = 1 \star 3 = 1 \star (2 \star 3).$$

◆

Exercise 4.1.1.8.

In both Example 4.1.1.3 and Exercise 4.1.1.6, the monoids $(M, e, \star)$ satisfied an additional rule called *commutativity*, namely, $m \star n = n \star m$ for every $m, n \in M$. There is a monoid $(M, e, \star)$ in linear algebra that is not commutative; if you have background in linear algebra, what monoid $(M, e, \star)$ might I be referring to?   ◊

Solution 4.1.1.8.

Matrix multiplication is not commutative. Let M be the set of 2×2 matrices, let $e \in M$ be the identity matrix, and let $\star$ be matrix multiplication. It is not commutative:

$$\begin{pmatrix} 1 & 1 \\ 0 & 0 \end{pmatrix} \begin{pmatrix} 1 & 0 \\ 1 & 0 \end{pmatrix} \neq \begin{pmatrix} 1 & 0 \\ 1 & 0 \end{pmatrix} \begin{pmatrix} 1 & 1 \\ 0 & 0 \end{pmatrix}.$$

◆

Exercise 4.1.1.9.

Recall the notion of commutativity for monoids from Exercise 4.1.1.8.

a. What is the smallest set M that you can give the structure of a noncommutative monoid?

b. What is the smallest set M that you can give the structure of a monoid?

◇

Solution 4.1.1.9.

a. Take $M = \{1, 2, 3\}$ with unit 1 and multiplication given as follows:

$\star$	1	2	3
1	1	2	3
2	2	2	3
3	3	2	3

Then $2 \star 3 = 3 \neq 2 = 3 \star 2$, so it is not commutative. One can check that it is associative. There are two monoid structures on any set with two elements, and they are both commutative.

b. The set $M = \{\odot\}$ has a unique possibility for unit and for multiplication formula, and these give it the structure of a monoid.

♦

Example 4.1.1.10 (Trivial monoid). There is a monoid with only one element, $M = (\{e\}, e, \star)$, where $\star\colon \{e\} \times \{e\} \to \{e\}$ is the unique function. We call this monoid *the trivial monoid* and sometimes denote it $\underline{1}$.

Example 4.1.1.11. Suppose that $(M, e, \star)$ is a monoid. Given elements m_1, m_2, m_3, m_4, there are five different ways to parenthesize the product $m_1 \star m_2 \star m_3 \star m_4$, and the associativity law for monoids will show them all to be the same. We have

$$
\begin{aligned}
((m_1 \star m_2) \star m_3) \star m_4 &= (m_1 \star m_2) \star (m_3 \star m_4) \\
&= (m_1 \star (m_2 \star m_3)) \star m_4 \\
&= m_1 \star (m_2 \star (m_3 \star m_4)) \\
&= m_1 \star ((m_2 \star m_3) \star m_4).
\end{aligned}
$$

In fact, the product of any list of monoid elements is the same, regardless of parenthesization. Therefore, we can unambiguously write $m_1 \star m_2 \star m_3 \star m_4 \star m_5$ rather than any given parenthesization of it. A substantial generalization of this is known as the *coherence theorem* and can be found in Mac Lane [29].

4.1.1.12   Free monoids and finitely presented monoids

Definition 4.1.1.13. Let X be a set. A *list in X* is a pair (n, f), where $n \in \mathbb{N}$ is a natural number (called the *length of the list*) and $f \colon \underline{n} \to X$ is a function, where $\underline{n} = \{1, 2, \ldots, n\}$. We may denote such a list

$$(n, f) = [f(1), f(2), \ldots, f(n)].$$

The set of lists in X is denoted $\mathrm{List}(X)$.

The *empty list* is the unique list in which $n = 0$; we may denote it $[\ ]$. Given an element $x \in X$, the *singleton list on x* is the list $[x]$. Given a list $L = (n, f)$ and a number $i \in \mathbb{N}$ with $i \leqslant n$, the *ith entry of L* is the element $f(i) \in X$.

Given two lists $L = (n, f)$ and $L' = (n', f')$, define the *concatenation of L and L'*, denoted $L +\!+ L'$, to be the list $(n + n', f +\!+ f')$, where $f +\!+ f' \colon \underline{n + n'} \to X$ is given on $1 \leqslant i \leqslant n + n'$ by

$$(f +\!+ f')(i) := \begin{cases} f(i) & \text{if } 1 \leqslant i \leqslant n, \\ f'(i - n) & \text{if } n + 1 \leqslant i \leqslant n + n'. \end{cases}$$

Example 4.1.1.14. Let $X = \{a, b, c, \ldots, z\}$. The following are elements of $\mathrm{List}(X)$:

$$[a, b, c], \quad [p], \quad [p, a, a, a, p], \quad [\ ], \quad \ldots.$$

The concatenation of $[a, b, c]$ and $[p, a, a, a, p]$ is $[a, b, c, p, a, a, a, p]$. The concatenation of any list ℓ with $[\ ]$ is just ℓ.

Definition 4.1.1.15. Let X be a set. The *free monoid generated by X* is the sequence $F_X := (\mathrm{List}(X), [\ ], +\!+)$, where $\mathrm{List}(X)$ is the set of lists of elements in X, $[\ ] \in \mathrm{List}(X)$ is the empty list, and $+\!+$ is the operation of list concatenation. We refer to X as the set of *generators* for the monoid F_X.

Exercise 4.1.1.16.

Let $\{\odot\}$ denote a one-element set.

a. What is the free monoid generated by the set $\{\odot\}$?

b. What is the free monoid generated by $\varnothing$?

$\Diamond$

Solution 4.1.1.16.

a. The set List($\{\odot\}$) of lists in which every entry is $\odot$ can be identified with the set $\mathbb{N}$ of natural numbers, because such a list has a length but no additional information. The empty list corresponds to $0 \in \mathbb{N}$, and concatenation of lists corresponds to addition of natural numbers. So the free monoid $F_\odot$ on one generator is $(\mathbb{N}, 0, +)$, as in Example 4.1.1.3.

b. An element of List($\emptyset$) is a pair (n, f), where $n \in \mathbb{N}$ and $f \colon \underline{n} \to \emptyset$. But the only time there is a function $X \to \emptyset$ is when $X = \emptyset$, so we must have $n = 0$. That is, List($\emptyset$) consists of one element, the empty list [ ], which serves as the identity. Thus the free monoid $F_\emptyset$ on an empty set of generators is the trivial monoid $\underline{1}$ (see Example 4.1.1.10).

$\blacklozenge$

An equivalence relation that interacts well with the multiplication formula of a monoid is called a congruence on that monoid.

Definition 4.1.1.17. Let $\mathcal{M} := (M, e, \star)$ be a monoid. A *congruence* on $\mathcal{M}$ is an equivalence relation $\sim$ on M, such that for any $m, m' \in M$ and any $n, n' \in M$, if $m \sim m'$ and $n \sim n'$, then $m \star n \sim m' \sim n'$.

Proposition 4.1.1.18. *Suppose that* $\mathcal{M} := (M, e, \star)$ *is a monoid. Then the following facts hold:*

1. *Given any relation* $R \subseteq M \times M$, *there is a smallest congruence* S *containing* R. *We call* S *the* congruence generated by R.

2. *If* $R = \emptyset$ *and* $\sim$ *is the congruence it generates, then there is an isomorphism* $M \overset{\cong}{\to} (M/\sim)$.

3. *Suppose that* $\sim$ *is a congruence on* $\mathcal{M}$. *Then there is a monoid structure* $\mathcal{M}/\sim$ *on the quotient set* $M/\sim$, *compatible with* $\mathcal{M}$.

Proof.   1. Let L_R be the set of all congruences on $\mathcal{M}$ that contain R. Using reasoning similar to that used in the proof of Proposition 3.3.1.7, one sees that L_R is nonempty and that its intersection, $S = \bigcap_{\ell \in L_R} \ell$, serves.

2. If $R = \emptyset$, then the minimal reflexive relation $\{(m, m) \mid m \in M\} \subseteq M \times M$ is the congruence generated by M. We have an isomorphism $M \overset{\cong}{\to} M/\sim$ by Exercise 3.3.1.9.

3. Let $Q\colon M \to M/\sim$ be the quotient function (as in Definition 3.3.1.1); note that it is surjective. We first want to give a monoid structure on $M/\sim$, i.e., we need a unit element e' and a multiplication formula $\star'$. Let $e' = Q(e)$. Suppose given $p, q \in M/\sim$ and respectively let $m, n \in M$ be a pair of representatives, so $Q(m) = p$ and $Q(n) = q$. Define $p\star'q := Q(m\star n)$. If we chose a different pair of representatives $Q(m') = p$ and $Q(n') = q$, then we would have $m \sim m'$ and $n \sim n'$ so $(m \star n) \sim (m' \star n')$, which implies $Q(m \star n) = Q(m' \star n')$; hence the composition formula is well defined. It is easy to check that $\mathcal{M}/\sim := (M/\sim, e', \star')$ is a monoid. It follows that $Q\colon M \to M/\sim$ extends to a monoid homomorphism $Q\colon \mathcal{M} \to \mathcal{M}/\sim$, as in Definition (4.1.4.1), which makes precise the *compatibility* claim.

□

Definition 4.1.1.19 (Presented monoid). Let G be a finite set, and let $R \subseteq \mathrm{List}(G) \times \mathrm{List}(G)$ be a relation. The *monoid presented by generators G and relations R* is the monoid $\mathcal{M} = (M, e, \star)$, defined as follows. Begin with the free monoid $F_G = (\mathrm{List}(G), [\ ], +\!+)$ generated by G. Let $\sim$ denote the congruence on F_G generated by R, as in Proposition 4.1.1.18, and define $\mathcal{M} := F_G/\sim$.

Each element $r \in R$ is of the form $r = (\ell, \ell')$ for lists $\ell, \ell' \in \mathrm{List}(G)$. For historical reasons we call the each of the resulting expressions $\ell \sim \ell'$ a *relation* in R.

Slogan 4.1.1.20.

> *A presented monoid is a set of buttons you can press and some facts about when different button sequences have the same results.*

Remark 4.1.1.21. Every free monoid is a presented monoid, because we can just take the set of relations to be empty.

Example 4.1.1.22. Let $G = \{a, b, c, d\}$. Think of these as buttons that can be pressed. The free monoid $F_G = (\mathrm{List}(G), [\ ], +\!+)$ is the set of all ways of pressing buttons, e.g., pressing a, then a, then c, then c, then d corresponds to the list $[a, a, c, c, d]$. The idea of presented monoids is that we can assert that pressing $[a, a, c]$ always gives the same result as pressing $[d, d]$ and that pressing $[c, a, c, a]$ is the same thing as doing nothing.

In this case, the relation $R \subseteq \mathrm{List}(G) \times \mathrm{List}(G)$ would be

R	
$[a, a, c]$	$[d, d]$
$[a, c, a, c]$	$[\ ]$

As in Proposition 4.1.1.18, the relation R generates a congruence $\sim$ on $\mathrm{List}(G)$, and this can be complex. For example, would you guess that $[b, c, b, d, d, a, c, a, a, c, d] \sim$

$[b, c, b, a, d, d, d]$? Here is the calculation in $M = \text{List}(G)/\sim$:

$$
\begin{aligned}
[b, c, b, d, d, a, c, a, a, c, d] &= [b, c, b] \star [d, d] \star [a, c, a, a, c, d] \\
&= [b, c, b, a] \star [a, c, a, c] \star [a, a, c, d] \\
&= [b, c, b, a, a, a, c, d] \\
&= [b, c, b, a] \star [a, a, c] \star [d] \\
&= [b, c, b, a, d, d, d].
\end{aligned}
$$

Exercise 4.1.1.23.

Let $K := \{BS, a, b, c, \ldots, z\}$, a set having 27 elements. Suppose one thinks of $BS \in K$ as the backspace key and the elements $a, b, \ldots z \in K$ as the letter keys on a keyboard. Then the free monoid $\text{List}(K)$ is not quite appropriate for modeling the keyboard because we want, e.g., $[a, b, d, BS] = [a, b]$.

a. Choose a set of relations for which the monoid presented by generators K and the chosen relations is appropriate to this application.

b. Under your relations, how does the singleton list $[BS]$ compare with the empty list $[\,]$? Is that suitable?

$\diamond$

Solution 4.1.1.23.

a. We need a relation $R \subseteq \text{List}(K) \times \text{List}(K)$. Let

$$
R = \{([x, BS], [\,]) \mid x \in \text{List}(K), x \neq BS\}.
$$

The idea is that, for every non-backspace key $x \neq BS$, if we press x, then BS, we get the same result as doing nothing. That is, $[x, BS] \sim [\,]$.

b. Note that we have $[BS] \neq [\,]$, which might seem strange because one normally thinks of pressing the backspace key on an empty string as yielding the empty string. But this is required because if we were to have $[BS] =^? [\,]$, then we would have $[\,] = [x_1, BS] = [x_1] \star [BS] =^? [x_1]$, which would kill everything, i.e., make the presented monoid trivial.

$\blacklozenge$

4.1.1.24   Cyclic monoids

Definition 4.1.1.25. A monoid is called *cyclic* if it has a presentation involving only one generator.

Example 4.1.1.26. Let Q be a symbol; we look at some cyclic monoids generated by $\{Q\}$. With no relations the monoid would be the free monoid on one generator and would have underlying set $\{[\ ], [Q], [Q, Q], [Q, Q, Q], \ldots\}$, with unit element $[\ ]$ and multiplication given by concatenation (e.g., $[Q, Q, Q] ++ [Q, Q] = [Q, Q, Q, Q, Q]$). This is just $\mathbb{N}$, the additive monoid of natural numbers.

With the really strong relation $[Q] \sim [\ ]$ we would get the trivial monoid, as in Example 4.1.1.10.

Another possibility is given in the first part of Example 4.1.2.3, where the relation $Q^{12} \sim [\ ]$ is used, where Q^{12} is shorthand for $[Q, Q, Q, Q, Q, Q, Q, Q, Q, Q, Q, Q]$. This monoid has 12 elements.

Example 4.1.1.27. Consider the cyclic monoid with generator Q and relation $Q^7 = Q^4$. This monoid has seven elements,

$$\{Q^0, Q^1, Q^2, Q^3, Q^4, Q^5, Q^6\},$$

where $Q^0 = e$ and $Q^1 = Q$. As an example of the multiplication formula, we have:

$$Q^6 \star Q^5 = Q^7 * Q^4 = Q^4 * Q^4 = Q^7 * Q = Q^5.$$

One might depict the cyclic monoid with relation $Q^7 = Q^4$ as follows:

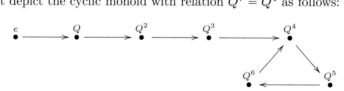

To see the mathematical source of this intuitive depiction, see Example 7.2.1.19.

Exercise 4.1.1.28.

Classify all the cyclic monoids up to isomorphism. That is, construct a naming system such that every cyclic monoid can be given a name in your system, no two nonisomorphic cyclic monoids have the same name, and no name exists in the system unless it refers to a cyclic monoid.

Hint: One might see a pattern in which the three monoids in Example 4.1.1.26 correspond respectively to ∞, 1, and 12, and think that Cyclic monoids can be classified by (i.e., systematically named by elements of) the set $\mathbb{N} \sqcup \{\infty\}$. That idea is on the right track, but it is not complete.   ◊

Solution 4.1.1.28.

Cyclic monoids are either finite or infinite. The free monoid on one generator, $(\mathbb{N}, 0, +)$ is the only infinite cyclic monoid, because once one makes a relation $Q^m \sim Q^n$ on $\mathrm{List}(Q)$ for some $n > m$, it is ensured that there are only finitely many elements (in fact, n-many). Finite cyclic monoids can be drawn as backward σ's (i.e., as $\mathfrak{v}$'s), with varying loop lengths and total lengths. The finite cyclic monoids can be classified by the set

$$FCM := \{(n, k) \in \mathbb{N} \times \mathbb{N} \mid 1 \leqslant k \leqslant n\}.$$

For each $(n, k) \in FCM$, there is a cyclic monoid with n elements and a loop of length k. For example, we can draw $(8, 6)$ and $(5, 1)$ respectively as

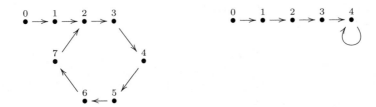

How do these pictures correspond to monoids? The nodes represent elements, so $(8, 6)$ has eight elements. The unit element is the leftmost node (the only one with no arrow pointing to it). Each node is labeled by the length of the shortest path from the unit (so 0 is the unit). To multiply $m \star n$, we see where the path of length $m + n$, starting at 0, ends up. So in the cyclic monoid of type $(8, 6)$, we have $4 + 4 = 2$, whereas in $(5, 1)$, we have $4 + 4 = 4$. ♦

4.1.2 Monoid actions

Definition 4.1.2.1 (Monoid action). Let $(M, e, \star)$ be a monoid, and let S be a set. An *action of* $(M, e, \star)$ *on* S, or simply an *action of M on S*, or an *M action on S*, is a function

$$\circlearrowright : M \times S \to S$$

such that the following *monoid action laws* hold for all $m, n \in M$ and all $s \in S$:

- $e \circlearrowright s = s$

- $m \circlearrowright (n \circlearrowright s) = (m \star n) \circlearrowright s.$[2]

---

[2] Definition 4.1.2.1 actually defines a *left action* of $(M, e, \star)$ on S. A *right action* is like a left action

Remark 4.1.2.2. To be pedantic (and because it is sometimes useful), we may decide not to use infix notation. That is, we may rewrite $\circlearrowright$ as $\alpha\colon M \times S \to S$ and restate the conditions from Definition 4.1.2.1 as

- $\alpha(e, s) = s$;

- $\alpha(m, \alpha(n, s)) = \alpha(m \star n, s)$.

Example 4.1.2.3. Let $S = \{0, 1, 2, \ldots, 11\}$, and let $N = (\mathbb{N}, 0, +)$ be the additive monoid of natural numbers (see Example 4.1.1.3). We define a function $\circlearrowright\colon \mathbb{N} \times S \to S$ by taking a pair (n, s) to the remainder that appears when $n + s$ is divided by 12. For example, $4 \circlearrowright 2 = 6$ and $8 \circlearrowright 9 = 5$. This function has the structure of a monoid action because the monoid laws from Definition 4.1.2.1 hold.

Similarly, let T denote the set of points on a circle, elements of which are denoted by a real number in the interval $[0, 12)$, i.e.,

$$T = \{x \in \mathbb{R} \mid 0 \leqslant x < 12\},$$

and let $R = (\mathbb{R}, 0, +)$ denote the additive monoid of real numbers. Then there is an action $R \times T \to T$, similar to the preceding one (see Exercise 4.1.2.4).

One can think of this as an action of the monoid of time on the clock. Here T is the set of positions at which the hour hand may be pointing. Given any number $r \in R$, we can go around the clock by r many hours and get a new hour-hand position. For example, $7.25 \circlearrowright 8.5 = 3.75$, meaning that 7.25 hours after 8:30 is 3:45.

Exercise 4.1.2.4.

Warning: This exercise is abstract.

a. Realize the set $T := [0, 12) \subseteq \mathbb{R}$ as a coequalizer of some pair of arrows $\mathbb{R} \rightrightarrows \mathbb{R}$.

b. For any $x \in \mathbb{R}$, realize the mapping $x+\colon T \to T$, implied by Example 4.1.2.3, using the universal property for coequalizers.

c. Prove that it is an action.

$\diamond$

---

except the order of operations is somehow reversed. We focus on left actions is in this text, but right actions are briefly defined here for completeness. The only difference is in the second condition. Using the same notation, we replace it by the condition that for all $m, n \in M$ and all $s \in S$, we have

$$m \circlearrowright (n \circlearrowright s) = (n \star m) \circlearrowright s.$$

Solution 4.1.2.4.

a. Let $f\colon \mathbb{R} \to \mathbb{R}$ be given by $f(x) = x + 12$. Then $\mathrm{id}_{\mathbb{R}}$ and f are a pair of arrows $\mathbb{R} \to \mathbb{R}$, and their coequalizer is T.

b. Let $x \in \mathbb{R}$ be a real number. We want a function $x+\colon T \to T$, but we begin with a function (by the same name) $x+\colon \mathbb{R} \to \mathbb{R}$, given by adding x to any real number. The following solid-arrow diagram commutes because $12 + x = x + 12$ for any $x \in \mathbb{R}$:

$$
\begin{array}{ccc}
\mathbb{R} \underset{f}{\overset{\mathrm{id}_{\mathbb{R}}}{\rightrightarrows}} \mathbb{R} \longrightarrow T \\
x+\downarrow \qquad x+\downarrow \qquad \vdots \\
\mathbb{R} \underset{f}{\overset{\mathrm{id}_{\mathbb{R}}}{\rightrightarrows}} \mathbb{R} \longrightarrow T
\end{array}
$$

By the universal property for coequalizers, there is a unique dotted arrow $T \to T$ making the diagram commute, and this is $x+\colon T \to T$. It represents the action "add $x \in \mathbb{R}$ hours to clock position $t \in T$."

c. Clearly, if $x = 0$, then the $x+$ function is $\mathrm{id}_{\mathbb{R}}$, and it follows from the universal property that $0+ = \mathrm{id}_T$. We see that $x + (y + t) = (x + y) + t$ using the commutative diagram

$$
\begin{array}{ccc}
\mathbb{R} \underset{f}{\overset{\mathrm{id}_{\mathbb{R}}}{\rightrightarrows}} \mathbb{R} \longrightarrow T \\
x+\downarrow \qquad x+\downarrow \qquad x+\downarrow \\
\mathbb{R} \underset{f}{\overset{\mathrm{id}_{\mathbb{R}}}{\rightrightarrows}} \mathbb{R} \longrightarrow T \\
y+\downarrow \qquad y+\downarrow \qquad y+\downarrow \\
\mathbb{R} \underset{f}{\overset{\mathrm{id}_{\mathbb{R}}}{\rightrightarrows}} \mathbb{R} \longrightarrow T
\end{array}
$$

The universal property for coequalizers implies the result.

◆

Exercise 4.1.2.5.

Let B denote the set of buttons (or positions) of a video game controller (other than, say, "start" and "select"), and consider the free monoid $\mathrm{List}(B)$ on B.

a. What would it mean for $\mathrm{List}(B)$ to act on the set of states of some (single-player) video game? Imagine a video game G' that uses the controller, but for which $\mathrm{List}(B)$

would not be said to act on the states of G'. Now imagine a simple game G for which List(B) would be said to act. Describe the games G and G'.

b. Can you think of a state s of G, and two distinct elements $\ell, \ell' \in \mathrm{List}(B)$ such that $\ell \curvearrowright s = \ell' \curvearrowright s$?

c. In video game parlance, what would you call a monoid element $b \in B$ such that for every state $s \in G$, one has $b \curvearrowright s = s$?

d. In video game parlance, what would you call a state $s \in S$ such that for every sequence of buttons $\ell \in \mathrm{List}(B)$, one has $\ell \curvearrowright s = s$?

e. Define $\mathbb{R}_{>0}$ to be the set of positive real numbers, and consider the free monoid $M := \mathrm{List}(\mathbb{R}_{>0} \times B)$. An element of this monoid can be interpreted as a list in which each entry is a button $b \in B$ being pressed after a wait time $t \in \mathbb{R}_{>0}$. Can you find a game that uses the controller but for which M does not act?

◊

Solution 4.1.2.5.

a. Suppose $B = \{up, right\}$. Then B acts on the set of states of a game if pressing either *up* or *right* will act the same way every time on a given state, sending you to a new state, and if when you do nothing with the controller, the state stays exactly the same. Pressing buttons very quickly would end up with the same result as pressing them slowly.

Most games one can think of are not going to be modeled by such an action. But a simple game for which the controller would be said to act is just a game where a character can walk around an arena, as though with time stopped. If the arena in G' had the feel of time progressing, the state of the game would change even when the controller was not pushed (and hence G' would not be modeled by an action of this monoid).

But for example, there was an old "Streetfighter" game, in which one was sometimes tasked with destroying a car by kicking it and punching it. If the speed at which one pressed the buttons had no effect, this would constitute an action. Let's call this G, with $B = \{punch, kick\}$.

b. Yes, when the car is completely destroyed, then whether you punch it or kick it, the result is the same.

c. I would call it "a useless button."

d. I would call it "game over."

e. This is an excellent model. A state is now an unfolding situation in which the agent is not acting. The whole future-history of his last action is identified as a single state. He acts on that state by waiting a certain amount of time and then pressing a button, hence "changing the future."

To my thinking, for any single-player video game in existence with controller B, the monoid $\text{List}(\mathbb{R}_{>0} \times B)$ can be made to act in accordance with the actual game play. (Even the Wii or Kinect should only be taking as data a finite number of samples within one continuous movement.)

♦

Application 4.1.2.6. Let $f \colon \mathbb{R} \to \mathbb{R}$ be a differentiable function of which we want to find roots (points $x \in \mathbb{R}$ such that $f(x) = 0$). Let $x_0 \in \mathbb{R}$ be a starting point. For any $n \in \mathbb{N}$, we can apply Newton's method to x_n to get

$$x_{n+1} = x_n - \frac{f(x_n)}{f'(x_n)}.$$

This is a monoid (namely, $\mathbb{N}$, the free monoid on one generator) acting on a set (namely, $\mathbb{R}$).

However, Newton's method can get into trouble. For example, at a critical point it causes division by zero, and sometimes it can oscillate or overshoot. In these cases we want to perturb a bit to the left or right. To have these actions available to us, we would add "perturb" elements to our monoid. Now we have more available actions at any point, but at the cost of using a more complicated monoid.

When publishing an experimental finding, there may be some deep methodological questions that are not considered suitably important to mention. For example, one may not publish the kind of solution-finding method (e.g., Newton's method or Runge-Kutta) that was used, or the set of available actions, e.g., what kinds of perturbation were used by the researcher. However, these may actually influence the reproducibility of results. By using a language such as that of monoid actions, we can align our data model with our unspoken assumptions about how functions are analyzed.

◊◊

Remark 4.1.2.7. A monoid is useful for understanding how an agent acts on the set of states of an object, but there is only one *context* for action—at any point, all actions are available. In reality, it is often the case that contexts can change and different actions are available at different times. For example, on a computer the commands available in one application have no meaning in another. This points us to categories, which are generalizations of monoids (see Chapter 5).

4.1.2.8   Monoid actions as ologs

If monoids are understood in terms of how they act on sets, then it is reasonable to think of them in terms of ologs. In fact, the ologs associated to monoids are precisely those ologs that have exactly one type (and possibly many arrows and commutative diagrams).

Example 4.1.2.9. This example shows how to associate an olog to a monoid action. Consider the monoid M generated by the set $\{u, d, r\}$, standing for "up, down, right," and subject to the relations

$$[u, d] \sim [\ ], \qquad [d, u] \sim [\ ], \qquad [u, r] = [r, u], \qquad \text{and} \qquad [d, r] = [r, d].$$

We might imagine that M acts on the set of positions for a character in an old video game. In that case the olog corresponding to this action should look something like Figure 4.1.

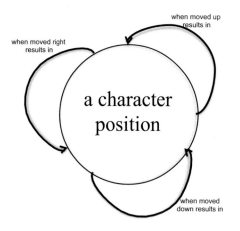

Given x, a character position, consider the following. We know that x is a character position, which when moved up results in a character position, which when moved down results in a character position that we'll call P(x). We also know that x is a character position that we'll call Q(x). Fact: whenever x is a character position we will have P(x)=Q(x). **Summary: [up, down] = [ ]**

Given x, a character position, consider the following. We know that x is a character position, which when moved down results in a character position, which when moved up results in a character position that we'll call P(x). We also know that x is a character position that we'll call Q(x). Fact: whenever x is a character position we will have P(x)=Q(x). **Summary: [down, up] = [ ]**

Given x, a character position, consider the following. We know that x is a character position, which when moved up results in a character position, which when moved right results in a character position that we'll call P(x). We also know that x is a character position, which when moved right results in a character position, which when moved up results in a character position that we'll call Q(x). Fact: whenever x is a character position we will have P(x)=Q(x). **Summary: [up, right] = [right, up]**

Given x, a character position, consider the following. We know that x is a character position, which when moved down results in a character position, which when moved right results in a character position that we'll call P(x). We also know that x is a character position, which when moved right results in a character position, which when moved down results in a character position that we'll call Q(x). Fact: whenever x is a character position we will have P(x)=Q(x). **Summary: [down, right] = [right, down]**

Figure 4.1

4.1.2.10   Finite state machines

According to Wikipedia, a *deterministic finite state machine* is a quintuple $(\Sigma, S, s_0, \delta, F)$, where

1. Σ is a finite nonempty set of symbols, called the *input alphabet*;

2. S is a finite, nonempty set, called *the state set*;

3. $\delta\colon \Sigma \times S \to S$ is a function, called the *state-transition function*;

4. $s_0 \in S$ is an element, called *the initial state*;

5. $F \subseteq S$ is a subset, called the *set of final states*.

Here we focus on the state transition function δ, by which the alphabet Σ acts on the set S of states (see Figure 4.2).

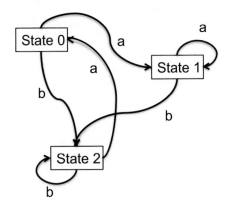

Figure 4.2 A finite state machine with alphabet $\Sigma = \{a, b\}$ and state set $S = \{\text{State 0, State 1, State 2}\}$.

The following proposition expresses the notion of finite state automata in terms of free monoids and their actions on finite sets.

Proposition 4.1.2.11. *Let Σ, S be finite nonempty sets. Giving a function $\delta\colon \Sigma \times S \to S$ is equivalent to giving an action of the free monoid $\mathrm{List}(\Sigma)$ on S.*

Proof. The proof is sketched here, leaving two details for Exercise 4.1.2.13. By Definition 4.1.2.1, we know that function $\epsilon\colon \mathrm{List}(\Sigma) \times S \to S$ constitutes an action of the monoid $\mathrm{List}(\Sigma)$ on the set S if and only if, for all $s \in S$, we have $\epsilon([\,], s) = s$, and for any two elements $m, m' \in \mathrm{List}(\Sigma)$, we have $\epsilon(m, \epsilon(m', s)) = \epsilon(m +\!\!+ m', s)$, where $m +\!\!+ m'$ is the concatenation of lists. Let

$$A := \{\epsilon\colon \mathrm{List}(\Sigma) \times S \to S \mid \epsilon \text{ constitutes an action}\}.$$

We need to prove that there is an isomorphism of sets

$$\phi\colon A \xrightarrow{\cong} \mathrm{Hom}_{\mathbf{Set}}(\Sigma \times S, S).$$

Given an element $\epsilon\colon \mathrm{List}(\Sigma) \times S \to S$ in A, define $\phi(\epsilon)$ on an element $(\sigma, s) \in \Sigma \times S$ by $\phi(\epsilon)(\sigma, s) := \epsilon([\sigma], s)$, where $[\sigma]$ is the one-element list. We now define

$$\psi\colon \mathrm{Hom}_{\mathbf{Set}}(\Sigma \times S, S) \to A.$$

Given an element $f \in \mathrm{Hom}_{\mathbf{Set}}(\Sigma \times S, S)$, define $\psi(f)\colon \mathrm{List}(\Sigma) \times S \to S$ on a pair $(L, s) \in \mathrm{List}(\Sigma) \times S$, where $L = [\ell_1, \ldots, \ell_n]$ as follows. By induction, if $n = 0$, put $\psi(f)(L, s) = s$; if $n \geqslant 1$, let $\partial L = [\ell_1, \ldots, \ell_{n-1}]$ and put $\psi(f)(L, s) = \psi(f)(\partial L, f(\ell_n, s))$.

One checks easily that $\psi(f)$ satisfies these two rules, making it an action of $\mathrm{List}(\Sigma)$ on S. It is also easy to check that ϕ and ψ are mutually inverse, completing the proof. (See Exercise 4.1.2.13).

$\square$

The idea of this section is summed up as follows:

Slogan 4.1.2.12.

    A finite state machine is an action of a free monoid on a finite set.

Exercise 4.1.2.13.

Consider the functions ϕ and ψ as defined in the proof of Proposition 4.1.2.11.

a. Show that for any $f\colon \Sigma \times S \to S$, the map $\psi(f)\colon \mathrm{List}(\Sigma) \times S \to S$ constitutes an action.

b. Show that ϕ and ψ are mutually inverse functions (i.e., $\phi \circ \psi = \mathrm{id}_{\mathrm{Hom}(\Sigma \times S, S)}$ and $\psi \circ \phi = \mathrm{id}_A$).

$\diamond$

Solution 4.1.2.13.

a. Let $s \in S$ be an arbitrary element. By the base of the induction, $\psi(f)([\ ], s) = s$, so $\psi(f)$ satisfies the unit law. Now let $L_1, L_2 \in \mathrm{List}(\Sigma)$ be two lists with $L = L_1 \mathbin{++} L_2$ their concatenation. We need to show that $\psi(f)(L_1, \psi(f)(L_2, s)) = \psi(f)(L, s)$. We do this by induction on the length of L_2. If $|L_2| = 0$, then $L = L_1$ and we have that $\psi(f)(L_1, \psi(f)(L_2, s)) = \psi(f)(L_1, s) = \psi(f)(L, s)$.

Now suppose the result is true for all lists of length $|L_2| - 1 \geqslant 0$. We have $\partial L = L_1 \mathbin{++} \partial L_2$, where ∂ removes the last entry of a nonempty list. If ℓ is the last entry of L and L_2, then we have

$$\psi(f)(L_1, \psi(f)(L_2, s)) = \psi(f)(L_1, \psi(f)(\partial L_2, f(\ell, s))) = \psi(f)(\partial L, f(\ell, s))$$
$$= \psi(f)(L, s).$$

b. We first show that for $f \in \text{Hom}(\Sigma \times S, S)$, we have $\phi \circ \psi(f) = f$. To do so, we choose $(\sigma, s) \in \Sigma \times S$, and the formulas for ϕ and ψ from the proof of Proposition 4.1.2.11 give

$$\phi(\psi(f))(\sigma, s) = \psi(f)([\sigma], s) = f(\sigma, s).$$

We next show that for $\epsilon \in A$, we have $\psi \circ \phi(\epsilon) = \epsilon$. To do so, we choose $(L, s) \in$ List$(\Sigma) \times S$ and show that $\psi(\phi(\epsilon))(L, s) = \epsilon(L, s)$. We do this by induction on the length $n = |L|$ of L. If $n = 0$, then $\psi(\phi(\epsilon))([\,], s) = s = \epsilon([\,], s)$. We may now assume that $n \geqslant 1$ and that the result holds for ∂L. Let ℓ be the last entry of L. We use the formulas for ϕ and ψ, and the fact that ϵ is an action, to get the following derivation:

$$\begin{aligned}
\psi(\phi(\epsilon))(L, s) = \psi(\phi(\epsilon))(\partial L, \phi(\epsilon)(\ell, s)) &= \psi(\phi(\epsilon))(\partial L, \epsilon([\ell], s)) \\
&= \epsilon(\partial L, \epsilon([\ell], s)) \\
&= \epsilon(\partial L +\!+ [\ell], s) = \epsilon(L, s).
\end{aligned}$$

◆

4.1.3 Monoid action tables

Let M be a monoid generated by the set $G = \{g_1, \ldots, g_m\}$, and with some relations, and suppose that $\alpha \colon M \times S \to S$ is an action of M on a set $S = \{s_1, \ldots, s_n\}$. We can represent the action α using an *action table* whose columns are the generators $g \in G$ and whose rows are the elements of S. In each cell (row, col), where $row \in S$ and $col \in G$, we put the element $\alpha(col, row) \in S$.

Example 4.1.3.1 (Action table). If Σ and S are the sets from Figure 4.2, the displayed action of List(Σ) on S would be given by action table (4.1)

Action from Fig. 4.2		
ID	**a**	**b**
State 0	State 1	State 2
State 1	State 2	State 1
State 2	State 0	State 0

(4.1)

Example 4.1.3.2 (Multiplication action table). Every monoid $(M, e, \star)$ acts on itself by its multiplication formula, $\star \colon M \times M \to M$. If G is a generating set for M, we can write the elements of G as the columns and the elements of M as rows, and call this a multiplication table. For example, let $(\mathbb{N}, 1, *)$ denote the multiplicative monoid of natural numbers. The multiplication table is the usual multiplication table from grade school:

Multiplication of natural numbers							
$\mathbb{N}$	**0**	**1**	**2**	**3**	**4**	**5**	$\cdots$
0	0	0	0	0	0	0	$\cdots$
1	0	1	2	3	4	5	$\cdots$
2	0	2	4	6	8	10	$\cdots$
3	0	3	6	9	12	15	$\cdots$
4	0	4	8	12	16	20	$\cdots$
$\vdots$	$\vdots$	$\vdots$	$\vdots$	$\vdots$	$\vdots$	$\vdots$	$\ddots$
21	0	21	42	63	84	105	$\cdots$
$\vdots$	$\vdots$	$\vdots$	$\vdots$	$\vdots$	$\vdots$	$\vdots$	$\ddots$

$$(4.2)$$

Try to understand what is meant by this: "Applying column 2 and then column 2 returns the same thing as applying column 4."

Table (4.2) implicitly takes every element of $\mathbb{N}$ as a generator (since there is a column for every natural number). In fact, there is a smallest generating set for the monoid $(\mathbb{N}, 1, *)$, so that every element of the monoid is a product of some combination of these generators, namely, the primes and 0.

Multiplication of natural numbers							
$\mathbb{N}$	**0**	**2**	**3**	**5**	**7**	**11**	$\cdots$
0	0	0	0	0	0	0	$\cdots$
1	0	2	3	5	7	11	$\cdots$
2	0	4	6	10	14	22	$\cdots$
3	0	6	9	15	21	33	$\cdots$
4	0	8	12	20	28	44	$\cdots$
$\vdots$	$\vdots$	$\vdots$	$\vdots$	$\vdots$	$\vdots$	$\vdots$	$\ddots$
21	0	42	63	105	147	231	$\cdots$
$\vdots$	$\vdots$	$\vdots$	$\vdots$	$\vdots$	$\vdots$	$\vdots$	$\ddots$

Exercise 4.1.3.3.

Let $\mathbb{N}$ be the additive monoid of natural numbers, let $S = \{0, 1, 2, \ldots, 11\}$, and let Clock: $\mathbb{N} \times S \to S$ be the clock action given in Example 4.1.2.3. Using a small generating set for the monoid, write the corresponding action table. ◇

Solution 4.1.3.3.

Since $(\mathbb{N}, 0, +)$ is the free monoid on one generator, we only need one column for the table. We denote it $+1$ because this is a descriptive name for the action in Example 4.1.2.3. The action table is

Clock	
S	**+1**
0	1
1	2
2	3
3	4
4	5
5	6
6	7
7	8
8	9
9	10
10	11
11	0

♦

4.1.4 Monoid homomorphisms

A monoid $(M, e, \star)$ involves a set, a unit element, and a multiplication formula. For two monoids to be comparable, their sets, unit elements, and multiplication formulas should be appropriately comparable. For example, the additive monoids $\mathbb{N}$ and $\mathbb{Z}$ should be comparable because $\mathbb{N} \subseteq \mathbb{Z}$ is a subset, the unit elements in both cases are the same $e = 0$, and the multiplication formulas are both integer addition.

Definition 4.1.4.1. Let $\mathcal{M} := (M, e, \star)$ and $\mathcal{M}' := (M', e', \star')$ be monoids. A *monoid homomorphism* f *from* $\mathcal{M}$ *to* $\mathcal{M}'$, denoted $f \colon \mathcal{M} \to \mathcal{M}'$, is a function $f \colon M \to M'$ satisfying two conditions:

- $f(e) = e'$.

- $f(m_1 \star m_2) = f(m_1) \star' f(m_2)$, for all $m_1, m_2 \in M$.

The set of monoid homomorphisms from $\mathcal{M}$ to $\mathcal{M}'$ is denoted $\mathrm{Hom}_{\mathbf{Mon}}(\mathcal{M}, \mathcal{M}')$.

Example 4.1.4.2 (From $\mathbb{N}$ to $\mathbb{Z}$). As stated, the inclusion map $i \colon \mathbb{N} \to \mathbb{Z}$ induces a monoid homomorphism $(\mathbb{N}, 0, +) \to (\mathbb{Z}, 0, +)$ because $i(0) = 0$ and $i(n_1 + n_2) = i(n_1) + i(n_2)$.

Let $i_5\colon \mathbb{N} \to \mathbb{Z}$ denote the function $i_5(n) = 5*n$, so $i_5(4) = 20$. This is also a monoid homomorphism because $i_5(0) = 5*0 = 0$ and $i_5(n_1+n_2) = 5*(n_1+n_2) = 5*n_1+5*n_2 = i_5(n_1) + i_5(n_2)$.

Application 4.1.4.3. Let $R = \{a,c,g,u\}$, and let $T = R^3$, the set of triplets in R. Let $\mathcal{R} = \mathrm{List}(R)$ be the free monoid on R, and let $\mathcal{T} = \mathrm{List}(T)$ denote the free monoid on T. There is a monoid homomorphism $F\colon \mathcal{T} \to \mathcal{R}$ given by sending $t = (r_1,r_2,r_3)$ to the list $[r_1,r_2,r_3]$.[3]

If A is the set of amino acids and $\mathcal{A} = \mathrm{List}(A)$ is the free monoid on A, the process of translation gives a monoid homomorphism $G\colon \mathcal{T} \to \mathcal{A}$, turning a list of RNA triplets into a polypeptide. But how do we go from a list of RNA nucleotides to a polypeptide, i.e., from $\mathcal{R}$ to $\mathcal{A}$? It seems that there is no good way to do this mathematically. So what is going wrong?

The answer is that there should not be a monoid homomorphism $\mathcal{R} \to \mathcal{A}$ because not all sequences of nucleotides produce a polypeptide; for example, if the sequence has only two elements, it does not code for a polypeptide. There are several possible remedies to this problem. One is to take the image of $F\colon \mathcal{T} \to \mathcal{R}$, which is a submonoid $\mathcal{R}' \subseteq \mathcal{R}$. It is not hard to see that there is a monoid homomorphism $F'\colon \mathcal{R}' \to \mathcal{T}$, and we can compose it with G to get the desired monoid homomorphism $G \circ F'\colon \mathcal{R}' \to \mathcal{A}$. [4]

$\lozenge\lozenge$

Example 4.1.4.4. Given any monoid $\mathcal{M} = (M,e,\star)$, there is a unique monoid homomorphism from $\mathcal{M}$ to the trivial monoid $\underline{1}$ (see Example 4.1.1.10). There is also a unique homomorphism $\underline{1} \to \mathcal{M}$ because a monoid homomorphism must send the unit to the unit. These facts together means that between any two monoids $\mathcal{M}$ and $\mathcal{M}'$ we can always construct a homomorphism

$$\mathcal{M} \xrightarrow{\ !\ } \underline{1} \xrightarrow{\ !\ } \mathcal{M}',$$

called the *trivial homomorphism* $\mathcal{M} \to \mathcal{M}'$. It sends everything in M to $e \in M'$. A homomorphism $\mathcal{M} \to \mathcal{M}'$ that is not trivial is called a *nontrivial homomorphism*.

Proposition 4.1.4.5. *Let* $\mathcal{M} = (\mathbb{Z},0,+)$ *and* $\mathcal{M}' = (\mathbb{N},0,+)$. *The only monoid homomorphism* $f\colon \mathcal{M} \to \mathcal{M}'$ *is trivial, i.e., it sends every element* $m \in \mathbb{Z}$ *to* $0 \in \mathbb{N}$.

Proof. Let $f\colon \mathcal{M} \to \mathcal{M}'$ be a monoid homomorphism, and let $n = f(1)$ and $n' = f(-1)$ in $\mathbb{N}$. Then we know that since $0 = 1+(-1)$ in $\mathbb{Z}$, we must have $0 = f(0) = f(1+(-1)) = f(1)+f(-1) = n+n' \in \mathbb{N}$. But if $n \geq 1$, then this is impossible, so $n = 0$. Similarly, $n' = 0$. Any element $m \in \mathbb{Z}$ can be written as $m = 1 + 1 + \cdots + 1$ or as $m = -1 + -1 + \cdots + -1$,

---

[3]More precisely, the monoid homomorphism F sends a list $[t_1,t_2,\ldots,t_n]$ to the list $[r_{1,1},r_{1,2},r_{1,3},r_{2,1},r_{2,2},r_{2,3},\ldots,r_{n,1},r_{n,2},r_{n,3}]$, where for each $0 \leq i \leq n$, we have $t_i = (r_{i,1},r_{i,2},r_{i,3})$.

[4]Adding stop-codons to the mix, we can handle more of $\mathcal{R}$, e.g., sequences that do not have a multiple-of-three many nucleotides.

and it is easy to see that $f(1) + f(1) + \cdots + f(1) = 0 = f(-1) + f(-1) + \cdots + f(-1)$. Therefore, $f(m) = 0$ for all $m \in \mathbb{Z}$.

$\square$

Exercise 4.1.4.6.

For any $m \in \mathbb{Z}$, let $i_m \colon \mathbb{N} \to \mathbb{Z}$ be the function $i_m(n) = m * n$, so $i_6(7) = -42$. All such functions are monoid homomorphisms $(\mathbb{N}, 0, +) \to (\mathbb{Z}, 0, +)$. Do any monoid homomorphisms $(\mathbb{N}, 0, +) \to (\mathbb{Z}, 0, +)$ not come in this way? For example, what about using $n \mapsto (5n - 1)$ or $n \mapsto n^2$ or some other function? $\diamond$

Solution 4.1.4.6.

All monoid homomorphisms $(\mathbb{N}, 0, +) \to (\mathbb{Z}, 0, +)$ come in this way. To see this, let $f \colon (\mathbb{N}, 0, +) \to (\mathbb{Z}, 0, +)$ be a monoid homomorphism. Then $f(1) = m$ for some $m \in \mathbb{Z}$. But then the multiplication law for monoid homomorphisms says we must have $f(1 + 1) = m + m$ and $f(1 + 1 + 1) = m + m + m$, and so on.

The function $n \mapsto n^2$ is not a monoid homomorphism because it does not respect multiplication: $(m + n)^2 \neq m^2 + n^2$. The function $n \mapsto (5n - 1)$ is not a monoid homomorphism because it respects neither the unit nor the multiplication. $\blacklozenge$

Exercise 4.1.4.7.

Let $\mathcal{M} := (\mathbb{N}, 0, +)$ be the additive monoid of natural numbers, let $\mathcal{N} = (\mathbb{R}_{\geq 0}, 0, +)$ be the additive monoid of nonnegative real numbers, and let $\mathcal{P} := (\mathbb{R}_{> 0}, 1, *)$ be the multiplicitive monoid of positive real numbers. Can you think of any nontrivial monoid homomorphisms (Example 4.1.4.4) of the following sorts:

a. $f \colon \mathcal{M} \to \mathcal{N}$?

b. $g \colon \mathcal{M} \to \mathcal{P}$?

c. $h \colon \mathcal{N} \to \mathcal{P}$?

d. $i \colon \mathcal{N} \to \mathcal{M}$?

e. $j \colon \mathcal{P} \to \mathcal{N}$?

$\diamond$

Solution 4.1.4.7.

a. The scalar multiplication function $f(n) = 17.5 * n$ works.

b. The exponentiation function $g(n) = 3.5^n$ works.

c. The exponentiation function $h(x) = 3.65^x$ works.

d. No, there are none. Suppose $i \colon \mathcal{N} \to \mathcal{M}$ is a candidate. For i to be nontrivial, there must be some $x \in \mathbb{R}_{>0}$ with $i(x) = n \neq 0$. Then $i(\frac{x}{2})$ would have to be $\frac{n}{2}$, which is forced to be a natural number, and then we have $\frac{n}{4} \in \mathbb{N}$ and $\frac{n}{8} \in \mathbb{N}$, and so on. There is no such $n \in \mathbb{N}$.

e. The base 10 logarithm function $j(x) = log_{10}(x)$ works.

$\blacklozenge$

4.1.4.8 Homomorphisms from free monoids

Recall that $(\mathbb{N}, 0, +)$ is the free monoid on one generator. It turns out that for any other monoid $\mathcal{M} = (M, e, \star)$, the set of monoid homomorphisms $\mathbb{N} \to \mathcal{M}$ is in bijection with the set M. This is a special case (in which G is a set with one element) of the following proposition.

Proposition 4.1.4.9. *Let G be a set, let $F(G) := (\mathrm{List}(G), [\ ], +\!\!+\ )$ be the free monoid on G, and let $\mathcal{M} := (M, e, \star)$ be any monoid. There is a natural bijection*

$$\mathrm{Hom}_{\mathbf{Mon}}(F(G), \mathcal{M}) \xrightarrow{\cong} \mathrm{Hom}_{\mathbf{Set}}(G, M).$$

Proof. We provide a function $\phi \colon \mathrm{Hom}_{\mathbf{Mon}}(F(G), \mathcal{M}) \to \mathrm{Hom}_{\mathbf{Set}}(G, M)$ and a function $\psi \colon \mathrm{Hom}_{\mathbf{Set}}(G, M) \to \mathrm{Hom}_{\mathbf{Mon}}(F(G), \mathcal{M})$ and show that they are mutually inverse. Let us first construct ϕ. Given a monoid homomorphism $f \colon F(G) \to \mathcal{M}$, we need to provide $\phi(f) \colon G \to M$. Given any $g \in G$, we define $\phi(f)(g) := f([g])$.

Now let us construct ψ. Given $p \colon G \to M$, we need to provide $\psi(p) \colon \mathrm{List}(G) \to \mathcal{M}$ such that $\psi(p)$ is a monoid homomorphism. For a list $L = [g_1, \ldots, g_n] \in \mathrm{List}(G)$, define $\psi(p)(L) := p(g_1) \star \cdots \star p(g_n) \in M$. In particular, $\psi(p)([\ ]) = e$. It is not hard to see that this is a monoid homomorphism. Also, $\phi \circ \psi(p) = p$ for all $p \in \mathrm{Hom}_{\mathbf{Set}}(G, M)$. We show that $\psi \circ \phi(f) = f$ for all $f \in \mathrm{Hom}_{\mathbf{Mon}}(F(G), \mathcal{M})$. Choose $L = [g_1, \ldots, g_n] \in \mathrm{List}(G)$. Then

$$\psi(\phi f)(L) = (\phi f)(g_1) \star \cdots \star (\phi f)(g_n) = f[g_1] \star \cdots \star f[g_n] = f([g_1, \ldots, g_n]) = f(L).$$

$\square$

Exercise 4.1.4.10.

Let $G = \{a, b\}$, let $\mathcal{M} := (M, e, \star)$ be any monoid, and let $f \colon G \to M$ be given by $f(a) = m$ and $f(b) = n$, where $m, n \in M$. If $\psi \colon \mathrm{Hom}_{\mathbf{Set}}(G, M) \to \mathrm{Hom}_{\mathbf{Mon}}(F(G), \mathcal{M})$ is the function constructed in the proof of Proposition 4.1.4.9 and $L = [a, a, b, a, b]$, what is $\psi(f)(L)$?

$\diamond$

Solution 4.1.4.10.

We have $\psi(f)([a,a,b,a,b]) = [m,m,n,m,n]$. ♦

4.1.4.11 Restriction of scalars

A monoid homomorphism $f\colon M \to M'$ (see Definition 4.1.4.1) ensures that the elements of M have a reasonable interpretation in M'; they act the same way over in M' as they did in M. If we have such a homomorphism f and we have an action $\alpha\colon M' \times S \to S$ of M' on a set S, then we have a method for allowing M to act on S as well. Namely, we take an element of M, send it to M', and use that to act on S. In terms of functions, we define $\Delta_f(\alpha)$ to be the composite:

$$M \times S \xrightarrow{\;f\times \mathrm{id}_S\;} M' \times S \xrightarrow{\;\alpha\;} S$$
$$\underbrace{\hspace{5cm}}_{\Delta_f(\alpha)}$$

After Proposition 4.1.4.12 we will know that $\Delta_f(\alpha)\colon M \times S \to S$ is indeed a monoid action, and we say that it is given by *restriction of scalars along f*.

Proposition 4.1.4.12. *Let $\mathcal{M} := (M, e, \star)$ and $\mathcal{M}' := (M', e', \star')$ be monoids, $f\colon \mathcal{M} \to \mathcal{M}'$ a monoid homomorphism, S a set, and suppose that $\alpha\colon M' \times S \to S$ is an action of $\mathcal{M}'$ on S. Then $\Delta_f(\alpha)\colon M \times S \to S$, as defined, is a monoid action as well.*

Proof. Refer to Remark 4.1.2.2, We assume α is a monoid action and want to show that $\Delta_f(\alpha)$ is too. We have $\Delta_f(\alpha)(e, s) = \alpha(f(e), s) = \alpha(e', s) = s$. We also have

$$\begin{aligned}
\Delta_f(\alpha)(m, \Delta_f(\alpha)(n, s)) &= \alpha(f(m), \alpha(f(n), s)) = \alpha(f(m) \star' f(n), s) \\
&= \alpha(f(m \star n), s) \\
&= \Delta_f(\alpha)(m \star n, s).
\end{aligned}$$

Then the unit law and the multiplication law hold. □

Example 4.1.4.13. Let $\mathbb{N}$ and $\mathbb{Z}$ denote the additive monoids of natural numbers and integers respectively, and let $i\colon \mathbb{N} \to \mathbb{Z}$ be the inclusion, which Example 4.1.4.2 showed is a monoid homomorphism. There is an action $\alpha\colon \mathbb{Z} \times \mathbb{R} \to \mathbb{R}$ of the monoid $\mathbb{Z}$ on the set $\mathbb{R}$ of real numbers, given by $\alpha(n, x) = n + x$. Clearly, this action works just as well if we restrict the scalars to $\mathbb{N} \subseteq \mathbb{Z}$, and allow only adding natural numbers to real numbers. This is the action $\Delta_i\alpha\colon \mathbb{N} \times \mathbb{R} \to \mathbb{R}$, because for $(n, x) \in \mathbb{N} \times \mathbb{R}$, we have $\Delta_i\alpha(n, x) = \alpha(i(n), x) = \alpha(n, x) = n + x$, just as expected.

Example 4.1.4.14. Suppose that V is a complex vector space. In particular, this means that the monoid $\mathbb{C}$ of complex numbers (under multiplication) acts on the elements of V. The elements of $\mathbb{C}$ are called *scalars* in this context. If $i\colon \mathbb{R} \to \mathbb{C}$ is the inclusion of the real line inside $\mathbb{C}$, then i is a monoid homomorphism. Restriction of scalars in the preceding sense turns V into a real vector space, so the name "restriction of scalars" is apt.

Exercise 4.1.4.15.

Let $\mathbb{N}$ be the free monoid on one generator, and let $\Sigma = \{a, b\}$. Consider the map of monoids $f\colon \mathbb{N} \to \mathrm{List}(\Sigma)$ given by sending $1 \mapsto [a, b, b, b]$. Consider the state set $S = \{\text{State } 0, \text{State } 1, \text{State } 2\}$. The monoid action $\alpha\colon \mathrm{List}(\Sigma) \times S \to S$ given in Example 4.1.3.1 can be transformed by restriction of scalars along f to an action $\Delta_f(\alpha)$ of $\mathbb{N}$ on S. Write its action table.   ◊

Solution 4.1.4.15.

Recall the action α of Σ on S given in Example 4.1.3.1 (or see left-hand side of (4.3)). The action $\Delta_f(\alpha)$ allows every natural number $n \in \mathbb{N}$ to act on S by "doing $[a, b, b, b]$ again and again n times." Since $\mathbb{N}$ is generated by 1, it suffices to record what happens when we do it once, i.e., follow a, then b, then b, then b (see right-hand side of (4.3)).

Action α from Ex. 4.1.3.1		
ID	**a**	**b**
State 0	State 1	State 2
State 1	State 2	State 1
State 2	State 0	State 0

Action $\Delta_f(\alpha)$	
ID	**1**
State 0	State 1
State 1	State 0
State 2	State 2

(4.3)

♦

4.2   Groups

Groups are monoids with the property that every element has an inverse. If we think of these structures in terms of how they act on sets, the difference between groups and monoids is that the action of every group element can be undone. One way of thinking about groups is in terms of symmetries. For example, the rotations and reflections of a square form a group because they can be undone.

Another way to think of the difference between monoids and groups is in terms of time. Monoids are likely useful in thinking about diffusion, in which time plays a role and things cannot be undone. Groups are more likely useful in thinking about mechanics, where actions are time-reversible.

4.2.1 Definition and examples

Definition 4.2.1.1. Let $(M, e, \star)$ be a monoid. An element $m \in M$ is said to *have an inverse* if there exists an $m' \in M$ such that $mm' = e$ and $m'm = e$. A *group* is a monoid $(M, e, \star)$ in which every element $m \in M$ has an inverse.

Proposition 4.2.1.2. *Suppose that* $\mathcal{M} := (M, e, \star)$ *is a monoid, and let* $m \in M$ *be an element. Then* m *has at most one inverse.*[5]

Proof. Suppose that both m' and m'' are inverses of m; we want to show that $m' = m''$. This follows by the associative law for monoids:

$$m' = m'(mm'') = (m'm)m'' = m''.$$

$\square$

Example 4.2.1.3. The additive monoid $(\mathbb{N}, 0, +)$ is not a group because none of its elements are invertible, except for 0. However, the monoid of integers $(\mathbb{Z}, 0, +)$ is a group. The monoid of clock positions from Example 4.1.1.26 is also a group. For example, the inverse of Q^5 is Q^7 because $Q^5 \star Q^7 = e = Q^7 \star Q^5$.

Example 4.2.1.4. Consider a square centered at the origin in $\mathbb{R}^2$. It has rotational and mirror symmetries. There are eight of these, denoted

$$\{e, \rho, \rho^2, \rho^3, \phi, \phi\rho, \phi\rho^2, \phi\rho^3\},$$

where ρ stands for $90°$ counterclockwise rotation and ϕ stands for horizontal flip (across the vertical axis). So relations include $\rho^4 = e$, $\phi^2 = e$, and $\rho^3\phi = \phi\rho$. This group is called the *dihedral group of order eight*.

Example 4.2.1.5. The set of 3×3 matrices can be given the structure of a monoid, where the unit element is the 3×3 identity matrix, the multiplication formula is given by matrix multiplication. It is a monoid but not a group because not all matrices are invertible.

The subset of invertible matrices does form a group, called *the general linear group of degree 3* and denoted GL_3. Inside of GL_3 is the *orthogonal group*, denoted O_3, of matrices M such that $M^{-1} = M^\top$. These matrices correspond to symmetries of the two-dimensional sphere centered at the origin in $\mathbb{R}^2$.

Another interesting group is the Euclidean group $E(3)$, which consists of all *isometries* of $\mathbb{R}^3$, i.e., all functions $\mathbb{R}^3 \to \mathbb{R}^3$ that preserve distances.

Application 4.2.1.6. In crystallography one is often concerned with the symmetries that arise in the arrangement A of atoms in a molecule. To think about symmetries in terms of

---

[5]If $\mathcal{M}$ is a group, then every element m has one and only one inverse.

groups, we first define an *atom arrangement* to be a finite subset $i \colon A \subseteq \mathbb{R}^3$. A symmetry in this case is an isometry of $\mathbb{R}^3$ (see Example 4.2.1.5), say, $f \colon \mathbb{R}^3 \to \mathbb{R}^3$, such that there exists a dotted arrow making the following diagram commute:

$$
\begin{array}{ccc}
A & \dashrightarrow & A \\
\downarrow{\scriptstyle i} & & \downarrow{\scriptstyle i} \\
\mathbb{R}^3 & \xrightarrow{\ f\ } & \mathbb{R}^3
\end{array}
$$

That is, it is an isometry of $\mathbb{R}^3$ such that each atom of A is sent to a position currently occupied by an atom of A. It is not hard to show that the set of such isometries forms a group, called the *space group* of the crystal.

◊◊

Exercise 4.2.1.7.

Let X be a finite set. A *permutation of X* is an isomorphism $f \colon X \xrightarrow{\cong} X$. Let $\mathrm{Iso}(X) := \{f \colon X \to X \mid f \text{ is an isomorphism}\}$ be the set of permutations of X. Here is a picture of an element in $\mathrm{Iso}(S)$, where $S = \{s_1, s_2, s_3, s_4\}$:

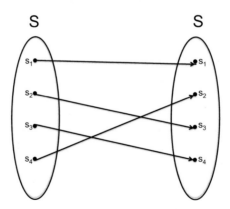

a. Devise a unit and a multiplication formula, such that the set $\mathrm{Iso}(X)$ of permutations of X forms a monoid.

b. Is the monoid $\mathrm{Iso}(X)$ always in fact a group?

◊

Solution 4.2.1.7.

a. We can take the unit to be the identity function $\mathrm{id}_S \colon S \xrightarrow{\cong} S$ and the multiplication formula to be a composition of isomorphisms $f \star g = f \circ g$. Clearly, $\mathrm{id}_S \circ f = f \circ \mathrm{id}_S = f$ and $(f \circ g) \circ h = f \circ (g \circ h)$, so this formula satisfies the unit and multiplication laws. In other words, we have put a monoid structure on the set $Iso(S)$.

b. Yes, $\mathrm{Iso}(X)$ is a group because every element of $f \in Iso(S)$ is invertible. Namely, the fact that f is an isomorphism means that there is some $f^{-1} \in Iso(S)$ with $f \circ f^{-1} = f^{-1} \circ f = \mathrm{id}_S$.

$\blacklozenge$

Exercise 4.2.1.8.

In Exercise 4.1.1.28 you classified the cyclic monoids. Which of them are groups? $\Diamond$

Solution 4.2.1.8.

The infinite cyclic monoid $(\mathbb{N}, 0, +)$ is not a group. The finite cyclic monoids are classified by the set $\{(n, k) \in \mathbb{N} \times \mathbb{N} \mid 1 \leqslant k \leqslant n\}$. Among these, the groups are precisely those with $n = k$, i.e., the o's among the σ's $\blacklozenge$

Definition 4.2.1.9 (Group action). Let $(G, e, \star)$ be a group and S a set. An *action* of G on S is a function $\mathtt{G} \colon G \times S \to S$ such that for all $s \in S$ and $g, g' \in G$, we have

- $e \mathbin{\mathtt{G}} s = s$;

- $g \mathbin{\mathtt{G}} (g' \mathbin{\mathtt{G}} s) = (g \star g') \mathbin{\mathtt{G}} s$.

In other words, considering G as a monoid, it is an action in the sense of Definition 4.1.2.1.

Example 4.2.1.10. When a group acts on a set, it has the character of symmetry. For example, consider the group whose elements are angles θ. This group may be denoted $U(1)$ and is often formalized as the unit circle in $\mathbb{C}$, i.e., the set of complex numbers $z = a + bi$ such that $|z| = a^2 + b^2 = 1$. The set of such points is given the structure of a group $(U(1), 1 + 0i, \star)$ by defining the unit element to be $1 + 0i$ and the group law to be complex multiplication. But for those unfamiliar with complex numbers, this is simply angle addition, where we understand that $360° = 0°$. If $\theta_1 = 190°$ and $\theta_2 = 278°$, then $\theta_1 \star \theta_2 = 468° = 108°$. In the language of complex numbers, $z = e^{i\theta}$.

The group $U(1)$ acts on any set that we can picture as having rotational symmetry about a fixed axis, such as the earth around the north-south axis. We will define $S =$

$\{(x, y, z) \in \mathbb{R}^3 \mid x^2 + y^2 + z^2 = 1\}$ to be the unit sphere in $\mathbb{R}^3$, and seek to understand the rotational action of $U(1)$ on S.

We first show that $U(1)$ acts on $\mathbb{R}^3$ by $\theta \circlearrowright (x, y, z) = (x \cos \theta + y \sin \theta, -x \sin \theta + y \cos \theta, z)$, or with matrix notation as

$$\theta \circlearrowright (x, y, z) := (x, y, z) \begin{pmatrix} \cos(\theta) & -\sin(\theta) & 0 \\ \sin(\theta) & \cos(\theta) & 0 \\ 0 & 0 & 1 \end{pmatrix}.$$

Trigonometric identities ensure that this is indeed an action.

In terms of action tables, we would need infinitely many rows and columns to express this action. Here is a sample:

Action of $U(1)$ on $\mathbb{R}^3$				
$\mathbb{R}^3$	$\theta = 45°$	$\theta = 90°$	$\theta = 100°$	$\cdots$
$(0, 0, 0)$	$(0, 0, 0)$	$(0, 0, 0)$	$(0, 0, 0)$	$\cdots$
$(1, 0, 0)$	$(0.71, 0.71, 0)$	$(0, 1, 0)$	$(-0.17, 0.98, 0)$	$\cdots$
$(0, 1, -4.2)$	$(-0.71, 0.71, -4.2)$	$(-1, 0, -4.2)$	$(-0.98, -0.17, -4.2)$	$\cdots$
$(3, 4, 2)$	$(4.95, 0.71, 2)$	$(-4, 3, 2)$	$(3.42, -3.65, 2)$	$\cdots$
$\vdots$	$\vdots$	$\vdots$	$\vdots$	$\ddots$

Since $S \subseteq \mathbb{R}^3$ consists of all vectors of length 1, we need to check that the action preserves length, i.e., that if $(x, y, z) \in S$, then $\theta \circlearrowright (x, y, z) \in S$. In this way we will have confirmed that $U(1)$ indeed acts on S. The calculation begins by assuming $x^2 + y^2 + z^2 = 1$, and one uses trigonometric identities to see that

$$(x \cos \theta + y \sin \theta)^2 + (-x \sin \theta + y \cos \theta)^2 + z^2 = x^2 + y^2 + z^2 = 1.$$

Exercise 4.2.1.11.

Let X be a set and consider the group $\mathrm{Iso}(X)$ of permutations of X (see Exercise 4.2.1.7). Find a canonical action of Iso_X on X.   ◊

Solution 4.2.1.11.

The elements of $\mathrm{Iso}(X)$ are isomorphisms $f \colon X \xrightarrow{\cong} X$. To get an action $\circlearrowright \colon \mathrm{Iso}(X) \times X \to X$, we need, for every pair (f, x), an element of X. The obvious choice is $f(x) \in X$.[6] Let's check that this really gives an action. For any $f, g \in \mathrm{Iso}(X)$ and any $x \in X$ we indeed have $\mathrm{id}_X(x) = x$ and we indeed have $f(g(x)) = (f \circ g)(x)$, so our choice works.   ♦

---

[6]It is worth noting the connection with $ev \colon \mathrm{Hom}_{\mathbf{Set}}(X, X) \times X \to X$ from (3.23).

Definition 4.2.1.12. Let G be a group acting on a set X. For any point $x \in X$, the *orbit of x*, denoted Gx, is the set

$$Gx := \{x' \in X \mid \exists g \in G \text{ such that } gx = x'\}.$$

Application 4.2.1.13. Let S be the surface of the earth, understood as a sphere, and let $G = U(1)$ be the group of angles acting on S by rotation as in Example 4.2.1.10. The orbit of any point $p = (x, y, z) \in S$ is the set of points on the same latitude line as p.

One may also consider a small band around the earth, i.e., the set $A = \{(x, y, z) \mid 1.0 \leqslant x^2 + y^2 + z^2 \leqslant 1.05\}$. The action of $U(1) \mathrel{\reflectbox{$\in$}} S$ extends to an action $U(1) \mathrel{\reflectbox{$\in$}} A$. The orbits are latitude-lines-at-altitude. A simplifying assumption in climatology may be given by assuming that $U(1)$ acts on all currents in the atmosphere in an appropriate sense. Thus, instead of considering movement within the whole space A, we only allow movement that behaves the same way throughout each orbit of the group action.

◇◇

Exercise 4.2.1.14.

a. Consider the $U(1)$ action on the sphere S given in Example 4.2.1.10. Describe the set of orbits of this action.

b. What are the orbits of the canonical action of the permutation group $\text{Iso}_{\{1,2,3\}}$ on the set $\{1, 2, 3\}$? (See Exercise 4.2.1.11.)

◇

Solution 4.2.1.14.

a. The orbits are the lines of latitude.

b. There is only one orbit: the whole set $\{1, 2, 3\}$.

◆

Exercise 4.2.1.15.

Let $(G, e, \star)$ be a group and X a set on which G acts. Is "being in the same orbit" an equivalence relation on X?   ◇

Solution 4.2.1.15.

Yes. Everything is in the same orbit as itself (because $e \cdot x = x$); if x is in the same orbit as y, then y is in the same orbit as x (because if $g \cdot x = y$, then $g^{-1} \cdot y = x$); and if x is in the same orbit as y, and y is in the same orbit as z, then x is in the same orbit as z (because if $g \cdot x = y$ and $h \cdot y = z$, then $(h \star g) \cdot x = z$).   ◆

Definition 4.2.1.16. Let G and G' be groups. A *group homomorphism* $f\colon G \to G'$ is defined to be a monoid homomorphism $G \to G'$, where G and G' are being regarded as monoids in accordance with Definition 4.2.1.1.

4.3   Graphs

Unless otherwise specified, whenever I speak of graphs in this book, I do not mean curves in the plane, such as parabolas, or pictures of functions generally, but rather systems of vertices and arrows.

Graphs are taken to be *directed*, meaning that every arrow points *from* a vertex *to* a vertex; rather than merely connecting vertices, arrows have direction. If a and b are vertices, there can be many arrows from a to b, or none at all. There can be arrows from a to itself. Here is the formal definition in terms of sets and functions.

4.3.1   Definition and examples

Definition 4.3.1.1. A *graph* G consists of a sequence $G := (V, A, src, tgt)$, where

- V is a set, called *the set of vertices of G* (singular: *vertex*);

- A is a set, called *the set of arrows of G*;

- $src\colon A \to V$ is a function, called *the source function for G*;

- $tgt\colon A \to V$ is a function, called *the target function for G*.

Given an arrow $a \in A$ we refer to $src(a)$ as the *source vertex* of a and to $tgt(a)$ as the *target vertex* of a.

To draw a graph, first draw a dot for every element of V. Then for every element $a \in A$, draw an arrow connecting dot $src(a)$ to dot $tgt(a)$.

Example 4.3.1.2 (Graph). Here is a picture of a graph $G = (V, A, src, tgt)$:

$$\tag{4.4}$$

We have $V = \{v, w, x, y, z\}$ and $A = \{f, g, h, i, j, k\}$. The source and target functions $src, tgt \colon A \to V$ are expressed in the following table (left-hand side):

A	src	tgt
f	v	w
g	w	x
h	w	x
i	y	y
j	y	z
k	z	y

V
v
w
x
y
z

In fact, all the data of the graph G is captured in these two tables—together they tell us the sets A and V and the functions src and tgt.

Example 4.3.1.3. Every olog has an underlying graph, in the sense of Definition 4.3.1.1. An olog has additional information, namely, information about which pairs of paths are declared equivalent as well as text that has certain English-readability rules.

Exercise 4.3.1.4.

a. Draw the graph corresponding to the following tables:

A	src	tgt
f	v	w
g	v	w
h	v	w
i	x	w
j	z	w
k	z	z

V
u
v
w
x
y
z

b. Write two tables like the ones in part (a) corresponding to the following graph:

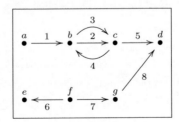

◊

Solution 4.3.1.4.

a.

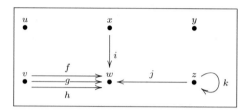

b.

A	**src**	**tgt**
1	a	b
2	b	c
3	b	c
4	c	b
5	c	d
6	f	e
7	f	g
8	g	d

V
a
b
c
d
e
f
g

♦

Exercise 4.3.1.5.

a. Let $A = \{1, 2, 3, 4, 5\}$ and $B = \{a, b, c\}$. Draw them, and choose an arbitrary function $f \colon A \to B$ and draw it.

b. Let $A \sqcup B$ be the coproduct of A and B (Definition 3.1.2.1), and let $A \xrightarrow{i_1} A \sqcup B \xleftarrow{i_2} B$ be the two inclusions. Consider the two functions $src, tgt \colon A \to A \sqcup B$, where $src = i_1$ and tgt is the composition $A \xrightarrow{f} B \xrightarrow{i_2} A \sqcup B$. Draw the associated graph $G := (A \sqcup B, A, src, tgt)$.

◊

Solution 4.3.1.5.

a. Here is a picture of $f: A \to B$:

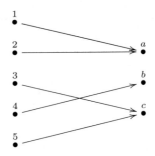

b. The graph G is drawn exactly as the one in part (a). The typical way we draw functions is by drawing nodes and arrows, i.e., a graph. This exercise has shown a formal way to obtain that graph given the function f.

◆

Exercise 4.3.1.6.

a. Let V be a set. Suppose we just draw the elements of V as vertices and have no arrows between them. Is this a graph?

b. Given V, is there any other canonical or somehow automatic nonrandom procedure for generating a graph with those vertices?

◇

Solution 4.3.1.6.

a. Yes. With arrows $A = \varnothing$, there is a unique function $!: A \to V$, so we have $(V, \varnothing, !, !)$. This is called the *discrete graph* on vertices V.

b. Yes. Choose as arrows $A = V \times V$, and let $src: A \to V$ and $tgt: A \to V$ be the projections. This gives the *indiscrete graph* $Ind(V) := (V, V \times V, \pi_1, \pi_2)$ on vertices V. An indiscrete graph is one in which each vertex is connected (backward and forward) to every other vertex and also points to itself.

Another would be $(V, V, \mathrm{id}_V, \mathrm{id}_V)$, which puts a loop at every vertex and has no other arrows.

◆

Example 4.3.1.7. Recall from Construction 3.2.2.6 the notion of a bipartite graph, defined to be a span (i.e., pair of functions; see Definition 3.2.2.1) $A \xleftarrow{f} R \xrightarrow{g} B$. Now that we have a formal definition of a graph, we might hope that the notion of bipartite graphs fits in as a particular sort of graph, and it does. Let $V = A \sqcup B$, and let $i \colon A \to V$ and $j \colon B \to V$ be the inclusions. Let $src = i \circ f \colon R \to V$, and let $tgt = j \circ g \colon R \to V$ be the composites:

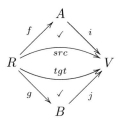

Then (V, R, src, tgt) is a graph that would be drawn exactly as specified the drawing of spans in Construction 3.2.2.6.

Example 4.3.1.8. Let $n \in \mathbb{N}$ be a natural number. The *chain graph of length n*, denoted $[n]$, is the following graph:

In general, $[n]$ has n arrows and $n + 1$ vertices. In particular, when $n = 0$, we have that $[0]$ is the graph consisting of a single vertex and no arrows.

Example 4.3.1.9. Let $G = (V, A, src, tgt)$ be a graph, Suppose that we want to spread it out over discrete time, so that each arrow does not occur within a given time slice but instead over a quantum unit of time.

Let $[\mathbb{N}] = (\mathbb{N}, \mathbb{N}, n \mapsto n, n \mapsto n + 1)$ be the graph depicted:

$$\overset{0}{\bullet} \overset{0}{\longrightarrow} \overset{1}{\bullet} \overset{1}{\longrightarrow} \overset{2}{\bullet} \overset{2}{\longrightarrow} \cdots$$

The discussion of limits in a category (see Chapter 6) clarifies that products can be taken in the category of graphs (see Example 6.1.1.5), so $[\mathbb{N}] \times G$ will make sense. For now, we construct it by hand.

Let $T(G) = (V \times \mathbb{N}, A \times \mathbb{N}, src', tgt')$ be a new graph, where for $a \in A$ and $n \in \mathbb{N}$, we have $src'(a, n) := (src(a), n)$ and $tgt'(a, n) = (tgt(a), n + 1)$.

Let G be the following graph:

Then $T(G)$ will be the graph

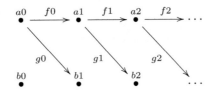

The f arrows still take a's to a's, and the g arrows still take a's to b's, but they always march forward in time.

Exercise 4.3.1.10.

Let G be the following graph:

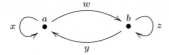

Draw the graph $T(G)$ defined in Example 4.3.1.9, using ellipses $(\cdots)$ if necessary. ◊

Solution 4.3.1.10.

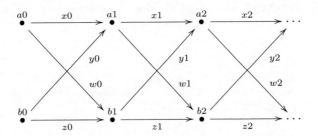

♦

Exercise 4.3.1.11.

Consider the following infinite graph $G = (V, A, src, tgt)$:

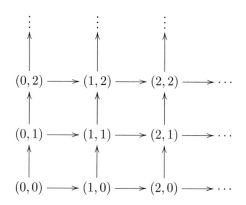

a. Write the sets A and V.

b. What are the source and target functions $A \to V$?

$\Diamond$

Solution 4.3.1.11.

a. Let $V = \mathbb{N} \times \mathbb{N}$ and $A = \mathbb{N} \times \mathbb{N} \times \{\mathtt{right}, \mathtt{up}\}$.

b. For all $m, n \in \mathbb{N}$, the source function $src \colon A \to V$ is given by

$$src(m, n, \mathtt{right}) = src(m, n, \mathtt{up}) = (m, n).$$

The target function $tgt \colon A \to V$ is given by

$$tgt(m, n, \mathtt{right}) = (m + 1, n) \qquad \text{and} \qquad tgt(m, n, \mathtt{up}) = (m, n + 1).$$

$\blacklozenge$

Exercise 4.3.1.12.

A graph is a pair of functions $A \rightrightarrows V$. This sets up the notion of equalizer and coequalizer (see Definitions 3.2.3.1 and 3.3.3.1).

a. What feature of a graph G is captured by the equalizer of its source and target functions?

b. What feature of a graph G is captured by the coequalizer of its source and target functions?

◊

Solution 4.3.1.12.

a. The equalizer of src, tgt is the set of loops in G, i.e., arrows pointing from a vertex to itself.

b. The coequalizer of srs, tgt is the set of connected components in G. See Exercise 3.3.1.11.

♦

4.3.2 Paths in a graph

One usually has some idea of what a path in a graph is, especially if one is is told that a path must always follow the direction of arrows. The following definition makes this idea precise. In particular, one can have paths of any finite length $n \in \mathbb{N}$, even length 0 or 1. Also, we want to be able to talk about the source vertex and target vertex of a path as well as about concatenation of paths.

Definition 4.3.2.1. Let $G = (V, A, src, tgt)$ be a graph. A *path of length n* in G, denoted $p \in \mathrm{Path}_G^{(n)}$, is a head-to-tail sequence

$$p = (v_0 \xrightarrow{a_1} v_1 \xrightarrow{a_2} v_2 \xrightarrow{a_3} \cdots \xrightarrow{a_n} v_n) \tag{4.5}$$

of arrows in G, denoted $_{v_0}[a_1, a_2, \ldots, a_n]$. A path is a list of arrows, so we use a variant of list notation, but the extra subscript at the beginning, which indicates the source vertex, reminds us that this list is actually a path. We have canonical isomorphisms $\mathrm{Path}_G^{(1)} \cong A$ and $\mathrm{Path}_G^{(0)} \cong V$: a path of length 1 is an arrow, and a path of length 0 is a vertex. We refer to the length 0 path $_v[\,]$ on vertex v as the *trivial path on v*.

We denote by Path_G the set of paths (of any length) in G, i.e.,

$$\mathrm{Path}_G := \bigsqcup_{n \in \mathbb{N}} \mathrm{Path}_G^{(n)}.$$

Every path $p \in \mathrm{Path}_G$ has a source vertex and a target vertex, and we may denote these $\overline{src}, \overline{tgt} : \mathrm{Path}_G \to V$. If p is a path with $\overline{src}(p) = v$ and $\overline{tgt}(p) = w$, we may denote it $p : v \to w$. Given two vertices $v, w \in V$, we write $\mathrm{Path}_G(v, w)$ to denote the set of all paths $p : v \to w$.

There is a concatenation operation on paths. Given a path $p\colon v \to w$ and $q\colon w \to x$, we define the concatenation, denoted $p\mathbin{+\!\!+} q\colon v \to x$, using concatenation of lists (see Definition 4.1.1.13). That is, if $p = {}_v[a_1, a_2, \ldots, a_m]$ and $q = {}_w[b_1, b_2, \ldots, b_n]$, then $p \mathbin{+\!\!+} q = {}_v[a_1, \ldots, a_m, b_1, \ldots, b_n]$. In particular, if $p = {}_v[\,]$ is the trivial path on vertex v (resp. if $r = {}_w[\,]$ is the trivial path on vertex w), then for any path $q\colon v \to w$, we have $p \mathbin{+\!\!+} q = q$ (resp. $q \mathbin{+\!\!+} r = q$).

Example 4.3.2.2. Let $G = (V, A, src, tgt)$ be a graph, and suppose $v \in V$ is a vertex. If $p\colon v \to v$ is a path of length $|p| \in \mathbb{N}$ with $\overline{src}(p) = \overline{tgt}(p) = v$, we call it a *loop of length* $|p|$. For $n \in \mathbb{N}$, we write $p^n\colon v \to v$ to denote the n-fold concatenation $p^n := p \mathbin{+\!\!+} p \mathbin{+\!\!+} \cdots \mathbin{+\!\!+} p$ (where p is written n times).

Example 4.3.2.3. In diagram (4.4), page 146, we see a graph G. In it, there are no paths from v to y, one path (namely, ${}_v[f]$) from v to w, two paths (namely, ${}_v[f, g]$ and ${}_v[f, h]$) from v to x, and infinitely many paths

$$\{{}_y[i]^{q_1} \mathbin{+\!\!+} {}_y[j, k]^{r_1} \mathbin{+\!\!+} \cdots \mathbin{+\!\!+} {}_y[i]^{q_n} \mathbin{+\!\!+} {}_y[j, k]^{r_n} \mid n, q_1, r_1, \ldots, q_n, r_n \in \mathbb{N}\}$$

from y to y. There are other paths as well in G, including the five trivial paths.

Exercise 4.3.2.4.

How many paths are there in the following graph?

$$\underset{\bullet}{\overset{1}{}} \overset{f}{\longrightarrow} \underset{\bullet}{\overset{2}{}} \overset{g}{\longrightarrow} \underset{\bullet}{\overset{3}{}}$$

$\diamondsuit$

Solution 4.3.2.4.

There are six: the length 0 paths ${}_1[\,]$, ${}_2[\,]$, and ${}_3[\,]$; the length 1 paths ${}_1[f]$ and ${}_2[g]$; and the length 2 path ${}_1[f, g]$. $\blacklozenge$

Exercise 4.3.2.5.

Let G be a graph, and consider the set Path_G of paths in G. Suppose someone claimed that there is a monoid structure on the set Path_G, where the multiplication formula is given by concatenation of paths. Are they correct? Why, or why not? $\diamondsuit$

Solution 4.3.2.5.

No, they are not correct, unless G has only one vertex. If G has exactly one vertex, then every path starts and ends there, so we can multiply paths by concatenating them, and we can take the trivial path as the unit of the monoid. But if G has no vertices, then Path$_G$ has no elements, so it is not a monoid (it is missing a unit). And if G has at least two vertices $a \neq b$, then the trivial paths at a and b are elements of Path$_G$, but they cannot be concatenated, so the purported multiplication formula is not defined. ♦

4.3.3 Graph homomorphisms

A graph (V, A, src, tgt) involves two sets and two functions. For two graphs to be comparable, their two sets and their two functions should be appropriately comparable.

Definition 4.3.3.1. Let $G = (V, A, src, tgt)$ and $G' = (V', A', src', tgt')$ be graphs. A *graph homomorphism f from G to G'*, denoted $f: G \to G'$, consists of two functions $f_0: V \to V'$ and $f_1: A \to A'$ such that the diagrams in (4.6) commute:

$$
\begin{array}{ccc}
A & \xrightarrow{f_1} & A' \\
{\scriptstyle src}\downarrow & & \downarrow{\scriptstyle src'} \\
V & \xrightarrow{f_0} & V'
\end{array}
\qquad
\begin{array}{ccc}
A & \xrightarrow{f_1} & A' \\
{\scriptstyle tgt}\downarrow & & \downarrow{\scriptstyle tgt'} \\
V & \xrightarrow{f_0} & V'
\end{array}
\qquad (4.6)
$$

Remark 4.3.3.2. The conditions (4.6) may look abstruse at first, but they encode a very important idea, roughly stated "arrows are bound to their endpoints." Under a map of graphs $G \to G'$, one cannot flippantly send an arrow of G any old arrow of G': it must still connect the vertices it connected before. Following is an example of a mapping that does not respect this condition: a connects 1 and 2 before but not after:

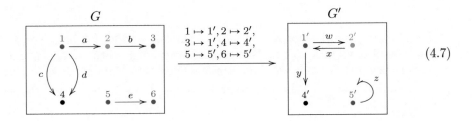

The commutativity of the diagrams in (4.6) is exactly what is needed to ensure that arrows are handled in the expected way by a proposed graph homomorphism.

Example 4.3.3.3 (Graph homomorphism). Let $G = (V, A, src, tgt)$ and $G' = (V', A', src', tgt')$ be the graphs drawn in (4.7):

$$(4.7)$$

The colors indicate the choice of function $f_0 \colon V \to V'$. Given that choice, condition (4.6) imposes in this case that there is a unique choice of graph homomorphism $f \colon G \to G'$. In other words, where arrows are sent is completely determined by where vertices are sent, in this particular case.

Exercise 4.3.3.4.

a. Where are a, b, c, d, e sent under $f_1 \colon A \to A'$ in diagram (4.7)?

b. Choose an element $x \in A$, and check that it behaves as specified by diagram (4.6).

◊

Solution 4.3.3.4.

a. We have:

$$f_1(a) = w, \qquad f_1(b) = x, \qquad f_1(c) = y, \qquad f_1(d) = y, \qquad f_1(e) = z.$$

b. In equation form, diagram (4.6) says that we need to check for any $x \in A = \{a, b, c, d, e\}$ that

$$src' \circ f_1(x) = f_0 \circ src(x) \qquad \text{and} \qquad tgt' \circ f_1(x) = f_0 \circ tgt(x).$$

We check these two criteria for $x = a$:

$$src'(f_1(a)) = src'(w) = 1' = f_0(1) = f_0(src(a)).$$
$$tgt'(f_1(a)) = tgt'(w) = 2' = f_0(2) = f_0(tgt(a)).$$

♦

Exercise 4.3.3.5.

Let G be a graph, let $n \in \mathbb{N}$ be a natural number, and let $[n]$ be the chain graph of length n, as in Example 4.3.1.8. Is a path of length n in G the same thing as a graph homomorphism $[n] \to G$, or are there subtle differences? More precisely, is there always an isomorphism between the set of graph homomorphisms $[n] \to G$ and the set $\mathrm{Path}_G^{(n)}$ of length n paths in G?

◊

Solution 4.3.3.5.

Yes, a path of length n in G is the same thing as a graph homomorphism $[n] \to G$. The discussion of categories in Chapter 5 makes clear how to write this fact formally as an isomorphism:

$$\mathrm{Hom}_{\mathbf{Grph}}([n], G) \cong \mathrm{Path}_G^{(n)}.$$

♦

Exercise 4.3.3.6.

Given a homomorphism of graphs $f \colon G \to G'$, there is an induced function between their sets of paths, $\mathrm{Path}(f) \colon \mathrm{Path}(G) \to \mathrm{Path}(G')$.

a. Explain how this works.

b. Is it the case that for every $n \in \mathbb{N}$, the function $\mathrm{Path}(f)$ carries $\mathrm{Path}^{(n)}(G)$ to $\mathrm{Path}^{(n)}(G')$, or can path lengths change in this process?

c. Suppose that f_0 and f_1 are injective (meaning no two distinct vertices in G are sent to the same vertex (resp. for arrows) under f). Does this imply that $\mathrm{Path}(f)$ is also injective (meaning no two distinct paths are sent to the same path under f)?

d. Suppose that f_0 and f_1 are surjective (meaning every vertex in G' and every arrow in G' is in the image of f). Does this imply that $\mathrm{Path}(f)$ is also surjective? Hint: At least one of the answers to parts (b)–(d) is no.

◊

Solution 4.3.3.6.

a. Given a path p in G, we apply f to it, node by node and arrow by arrow, to get a path in G'. A high-level way to think about this, given Exercise 4.3.3.5, is as follows. A path in G is a graph homomorphism $p \colon [n] \to G$, for some chain graph $[n]$ of length $n \in \mathbb{N}$. Composing, $f \circ p \colon [n] \to G'$ is a path of length n in G'.

b. Yes, it is the case that $\mathrm{Path}(f)$ carries $\mathrm{Path}^{(n)}(G)$ to $\mathrm{Path}^{(n)}(G')$.

c. Yes, it would be injective.

d. No; it is possible for f_0 and f_1 to be surjective while $\mathrm{Path}(f)$ is not surjective. Here is an example. Consider the following graph homomorphism $(a \mapsto a''; b, b' \mapsto b''; c \mapsto c')$:

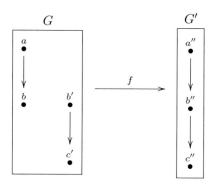

This homomorphism is surjective on vertices and arrows, but a new path of length 2 emerges.

◆

Exercise 4.3.3.7.

Given a graph (V, A, src, tgt), let $\langle src, tgt \rangle \colon A \to V \times V$ be the function guaranteed by the universal property for products. One might hope to summarize condition (4.6) for graph homomorphisms by the commutativity of the single square

$$
\begin{array}{ccc}
A & \xrightarrow{\;f_1\;} & A' \\
{\scriptstyle\langle src, tgt\rangle}\big\downarrow & & \big\downarrow{\scriptstyle\langle src', tgt'\rangle} \\
V \times V & \xrightarrow[f_0 \times f_0]{} & V' \times V'
\end{array}
\tag{4.8}
$$

Is the commutativity of the diagram in (4.8) indeed equivalent to the commutativity of the diagrams in (4.6)?   ◇

Solution 4.3.3.7.

Yes. This follows from the universal property for products, Proposition 3.1.1.10.   ◆

4.3.3.8   Binary relations and graphs

Definition 4.3.3.9. Let X be a set. A *binary relation on X* is a subset $R \subseteq X \times X$.

If $X = \mathbb{N}$ is the set of integers, then the usual $\leqslant$ defines a binary relation on X: given $(m, n) \in \mathbb{N} \times \mathbb{N}$, we put $(m, n) \in R$ iff $m \leqslant n$. As a table it might be written as in the left-hand table in (4.9):

(4.9)

$m \leqslant n$	
m	n
0	0
0	1
1	1
0	2
1	2
2	2
0	3
$\vdots$	$\vdots$

$n = 5m$	
m	n
0	0
1	5
2	10
3	15
4	20
5	25
6	30
$\vdots$	$\vdots$

$\lvert n - m \rvert \leqslant 1$	
m	n
0	0
0	1
1	0
1	1
1	2
2	1
2	2
$\vdots$	$\vdots$

The middle table is the relation $\{(m, n) \in \mathbb{N} \times \mathbb{N} \mid n = 5m\} \subseteq \mathbb{N} \times \mathbb{N}$, and the right-hand table is the relation $\{(m, n) \in \mathbb{N} \times \mathbb{N} \mid \lvert n - m \rvert \leqslant 1\} \subseteq \mathbb{N} \times \mathbb{N}$.

Exercise 4.3.3.10.

A relation on $\mathbb{R}$ is a subset of $\mathbb{R} \times \mathbb{R}$, and one can indicate such a subset of the plane by shading. Choose an error bound $\epsilon > 0$, and draw the relation one might refer to as ϵ-approximation. To say it another way, draw the relation "x is within ϵ of y." ◊

Solution 4.3.3.10.

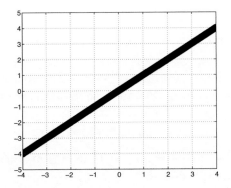

♦

Exercise 4.3.3.11.

Recall that (4.9) uses tables to express relations; it may help to use the terminology of tables in answering some of the following questions.

a. If $R \subseteq S \times S$ is a binary relation, find a natural way to make a graph G_R from it, having vertices S.

b. What is the set A of arrows in G_R?

c. What are the source and target functions $src, tgt \colon A \to S$ in G_R?

d. Consider the seven number rows in the left-hand table in (4.9), ignoring the elipses. Draw the corresponding graph.

e. Do the same for the right-hand table in (4.9).

◇

Solution 4.3.3.11.

a. We have two projections $\pi_1, \pi_2 \colon S \times S \to S$, and we have an inclusion $i \colon R \subseteq S \times S$. Thus we have a graph

$$R \underset{\pi_2 \circ i}{\overset{\pi_1 \circ i}{\rightrightarrows}} S$$

The idea is that for each row in the table, we draw an arrow from the first column's value to the second column's value.

b. It is R, which one could call "the number of rows in the table."

c. These are $\pi_1 \circ i$ and $\pi_2 \circ i$, which one could call "the first and second columns in the table." In other words, $G_R := (S, R, \pi_1 \circ i, \pi_2 \circ i)$.

d. The seven solid arrows in the following graph correspond to the seven displayed rows in the left-hand table, and we include 3 more dashed arrows to complete the picture (they still satisfy the $\leqslant$ relation).

Sample of $G_{m \leqslant n}$

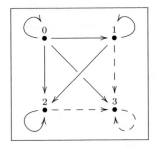

e. Seven rows, seven arrows:

Sample of $G_{|n-m| \leqslant 1}$

♦

Exercise 4.3.3.12.

a. If (V, A, src, tgt) is a graph, find a natural way to make a binary relation $R \subseteq V \times V$ from it.

b. For the left-hand graph G in (4.7), and write out the corresponding binary relation in table form.

◊

Solution 4.3.3.12.

a. Since we have functions $src, tgt \colon A \to V$, the universal property for products gives us a function $\langle src, tgt \rangle \colon A \to V \times V$. This is not a binary relation yet because it may not be injective. So let $R \subseteq V \times V$ be the image of $\langle src, tgt \rangle$; now we have the desired binary relation.

b.

1	2
1	4
2	3
5	6

◆

Exercise 4.3.3.13.

a. Given a binary relation $R \subseteq S \times S$, you know from Exercise 4.3.3.11 how to construct a graph out of it, and from Exercise 4.3.3.12 how to make a new binary relation out of that, making a roundtrip. How does the resulting relation compare with the original?

b. Given a graph $G = (V, A, src, tgt)$, you know from Exercise 4.3.3.12 how to make a new binary relation out of it, and from Exercise 4.3.3.11 how to construct a new graph out of that, making the other roundtrip. How does the resulting graph compare with the original?

◇

Solution 4.3.3.13.

a. It is the same.

b. It is different. The new graph G' never has two arrows in the same direction. That is, in the original graph G, we might have two different arrows $v_1 \to v_2$, but this cannot happen in the new graph. So the old graph and the new graph have the same number of vertices, and there exists an arrow from $v_1 \to v_2$ in G if and only if there exists an arrow $v_1 \to v_2$ in G'; but multiple arrows from one vertex to another in G are merged into a single arrow in G'.

◆

4.4 Orders

People usually think of certain sets as though they come with a canonical order. For example, one might think the natural numbers come with the ordering by which $3 < 5$, or that the letters in the alphabet come with the order by which $b < e$. But in fact we *put* orders on sets, and some orders are simply more commonly used. For instance, one could order the letters in the alphabet by frequency of use, in which case e would come before b. Given different purposes, we can put different orders on the same set. For example, in Example 4.4.3.2 we give a different ordering on the natural numbers that is useful in elementary number theory.

In science, we might order the set of materials in two different ways. In the first, we could consider material A to be less than material B if A is an ingredient or part of B, so water would be less than concrete. But we could also order materials based on how electrically conductive they are, whereby concrete would be less than water. This section is about different kinds of orders.

4.4.1 Definitions of preorder, partial order, linear order

Definition 4.4.1.1. Let S be a set and $R \subseteq S \times S$ a binary relation on S; if $(s, s') \in R$, we write $s \leqslant s'$. Then we say that R is a *preorder* if, for all $s, s', s'' \in S$, we have

Reflexivity: $s \leqslant s$, and

Transitivity: if $s \leqslant s'$ and $s' \leqslant s''$, then $s \leqslant s''$.

We say that R is a *partial order* if it is a preorder and, in addition, for all $s, s' \in S$, we have

Antisymmetry: If $s \leqslant s'$ and $s' \leqslant s$, then $s = s'$.

We say that R is a *linear order* if it is a partial order and, in addition, for all $s, s' \in S$, we have

Comparability: Either $s \leqslant s'$ or $s' \leqslant s$.

We denote such a preorder (or partial order or linear order) by $(S, \leqslant)$.

Exercise 4.4.1.2.

a. The relation in the left-hand table in (4.9) is a preorder. Is it a linear order?

b. Show that neither the middle table nor the right-hand table in (4.9) is even a preorder.

◊

Solution 4.4.1.2.

a. Yes.

b. If $\leqslant$ denotes the middle relation $n = 5m$, we have neither reflexivity ($1 \nleqslant 1$) nor transitivity ($25 = 5 * 1$ and $5 = 5 * 1$, but $25 \neq 5 * 1$). If $\leqslant$ denotes the right-hand relation $|n - m| \leqslant 1$, we do have reflexivity but not transitivity.

♦

Example 4.4.1.3 (Partial order not linear order). The following is an olog for playing cards:

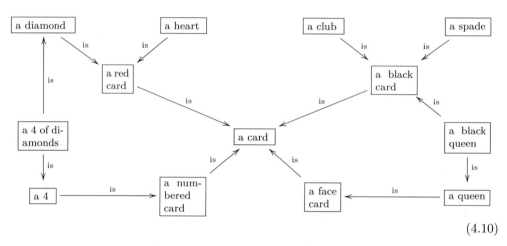

$$(4.10)$$

We can put a binary relation on the set of boxes here by saying $A \leqslant B$ if there is a path $A \to B$. One can see immediately that this is a preorder because length 0 paths give reflexivity, and concatenation of paths gives transitivity. To see that it is a partial order we only note that there are no loops of any length. But this partial order is not a linear order because there is no path (in either direction) between, e.g., ⌜a 4 of diamonds⌝ and ⌜a black queen⌝, so it violates the comparability condition.

Remark 4.4.1.4. Note that olog (4.10) in Example 4.4.1.3 is a good olog in the sense that given any collection of cards (e.g., choose 45 cards at random from each of seven decks and throw them in a pile), they can be classified according to it. In other words, each box in the olog will refer to some subset of the pile, and every arrow will refer to a function between these sets. For example, the arrow ⌜a heart⌝ $\xrightarrow{\text{is}}$ ⌜a red card⌝ is a function from the set of hearts in the pile to the set of red cards in the pile.

Example 4.4.1.5 (Preorder, not partial order). Every equivalence relation is a preorder, but rarely are they partial orders. For example, if $S = \{1, 2\}$ and we put $R = S \times S$, then this is an equivalence relation. It is a preorder but not a partial order (because $1 \leqslant 2$ and $2 \leqslant 1$, but $1 \neq 2$, so antisymmetry fails).

Application 4.4.1.6. Classically, we think of time as linearly ordered. A model is $(\mathbb{R}, \leqslant)$, the usual linear order on the set of real numbers. But according to the theory of relativity, there is not actually a single order to the events in the universe. Different observers correctly observe different orders on the set of events.

◊◊

Example 4.4.1.7 (Finite linear orders). Let $n \in \mathbb{N}$ be a natural number. Define a linear order $[n] = (\{0, 1, 2, \ldots, n\}, \leqslant)$ in the standard way. Pictorially,

$$[n] := \overset{0}{\bullet} \overset{\leqslant}{\longrightarrow} \overset{1}{\bullet} \overset{\leqslant}{\longrightarrow} \overset{2}{\bullet} \overset{\leqslant}{\longrightarrow} \cdots \overset{\leqslant}{\longrightarrow} \overset{n}{\bullet}$$

Every finite linear order, i.e., linear order on a finite set, is of the preceding form. That is, though the labels might change, the picture would be the same. This can be made precise when morphisms of orders are defined (see Definition 4.4.4.1)

Exercise 4.4.1.8.

Let $S = \{1, 2, 3\}$.

a. Find a preorder $R \subseteq S \times S$ such that the set R is as small as possible. Is it a partial order? Is it a linear order?

b. Find a preorder $R' \subseteq S \times S$ such that the set R' is as large as possible. Is it a partial order? Is it a linear order?

◊

Solution 4.4.1.8.

Write R and R' as tables:

R	
1	1
2	2
3	3

R'	
1	1
1	2
1	3
2	1
2	2
2	3
3	1
3	2
3	3

a. R is a partial order but not a linear order because $1 \nleqslant 2$ and $2 \nleqslant 1$.

b. R' is not a partial order because $1 \leqslant 2$ and $2 \leqslant 1$, so it cannot be a linear order.

♦

Exercise 4.4.1.9.

a. List all the preorder relations possible on the set $\{1, 2\}$.

b. For any $n \in \mathbb{N}$, how many linear orders exist on the set $\{1, 2, 3, \ldots, n\}$?

c. Does your formula work when $n = 0$?

◊

Solution 4.4.1.9.

a.

R_1	
1	1
2	2

R_2	
1	1
1	2
2	2

R_3	
1	1
2	1
2	2

R_4	
1	1
1	2
2	1
2	2

b. The factorial $n! = 1 * 2 * \cdots * n$.

c. Yes, there is one way to order the empty set, namely, $\varnothing \subseteq \varnothing \times \varnothing$, and $0! = 1$.

♦

Remark 4.4.1.10. We can draw any preorder $(S, \leqslant)$ as a graph with vertices S and with an arrow $a \to b$ if $a \leqslant b$. These are precisely the graphs with the following two properties for any vertices $a, b \in S$:

1. There is at most one arrow $a \to b$.

2. If there is a path from a to b, then there is an arrow $a \to b$.

If $(S, \leqslant)$ is a partial order, then the associated graph has an additional no-loops property:

3. If $n \in \mathbb{N}$ is an integer with $n \geqslant 2$, then there are no paths of length n that start at a and end at a.

If $(S, \leqslant)$ is a linear order then there is an additional comparability property:

4. For any two vertices a, b, there is an arrow $a \to b$ or an arrow $b \to a$.

Given a graph G, we can create a binary relation $\leqslant$ on its set S of vertices as follows. Put $a \leqslant b$ if there is a path in G from a to b. This relation will be reflexive and transitive, so it is a preorder. If the graph satisfies property 3, then the preorder will be a partial order, and if the graph also satisfies property 4, then the partial order will be a linear order. Thus graphs give us a nice way to visualize orders.

Slogan 4.4.1.11.

A graph generates a preorder: $v \leqslant w$ if there is a path $v \to w$.

Exercise 4.4.1.12.

Let $G = (V, A, src, tgt)$ be the following graph:

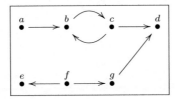

In the corresponding preorder, which of the following are true?

a. $a \leqslant b$.

b. $a \leqslant d$.

c. $c \leqslant b$.

d. $b = c$.

e. $e \leqslant f$.

f. $f \leqslant d$.

◊

Solution 4.4.1.12.

a. True.

b. True.

c. True.

d. False (though this would be true in the corresponding partial order, which has not been discussed).

e. False.

f. True.

♦

Exercise 4.4.1.13.

a. Let $S = \{1, 2\}$. The set $\mathbb{P}(S)$ of subsets of S form a partial order. Draw the associated graph.

b. Repeat this for $Q = \varnothing$, $R = \{1\}$, and $T = \{1, 2, 3\}$. That is, draw the partial orders on $\mathbb{P}(Q), \mathbb{P}(R)$, and $\mathbb{P}(T)$.

c. Do you see n-dimensional cubes?

◊

Solution 4.4.1.13.

a.

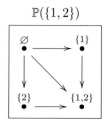

b.

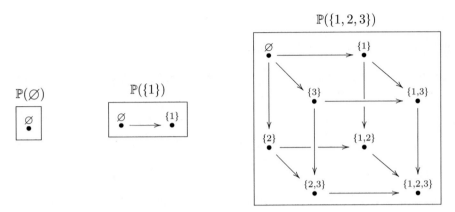

c. Yes. The graph associated to $\mathbb{P}(\underline{n})$ looks like an n-dimensional cube.

$\blacklozenge$

Definition 4.4.1.14. Let $(S, \leqslant)$ be a preorder. A *clique* is a subset $S' \subseteq S$ such that for each $a, b \in S'$, one has $a \leqslant b$.

Exercise 4.4.1.15.

True or false: A partial order is a preorder that has no cliques? (If false, is there a nearby true statement?) $\Diamond$

Solution 4.4.1.15.

False. Every element is always in its own clique, so if X is a partial order with at least one element, then it has a clique. But a nearby statement is true. Let's define a *nontrivial clique* to be a clique consisting of two or more elements.

Slogan.

A partial order is a preorder that has no nontrivial cliques.

$\blacklozenge$

Just as every relation generates an equivalence relation (see Proposition 3.3.1.7), every relation also generates a preorder.

Example 4.4.1.16. Let X be a set and $R \subseteq X \times X$ a relation. For elements $x, y \in X$, we say there is an *R-path* from x to y if there exists a natural number $n \in \mathbb{N}$ and elements $x_0, x_1, \ldots, x_n \in X$ such that

1. $x = x_0$;

2. $x_n = y$;

3. for all $i \in \mathbb{N}$, if $0 \leqslant i \leqslant n - 1$, then $(x_i, x_{i+1}) \in R$.

Let $\overline{R}$ denote the relation where $(x, y) \in \overline{R}$ if there exists an R-path from x to y. We call $\overline{R}$ the *preorder generated by* R. and note some facts about $\overline{R}$:

Containment. If $(x, y) \in R$, then $(x, y) \in \overline{R}$. That is, $R \subseteq \overline{R}$.

Reflexivity. For all $x \in X$, we have $(x, x) \in \overline{R}$.

Transitivity. For all $x, y, z \in X$, if $(x, y) \in \overline{R}$ and $(y, z) \in \overline{R}$, then $(x, z) \in \overline{R}$.

Let's write $x \leqslant y$ if $(x, y) \in \overline{R}$. To check the containment claim, use $n = 1$ so $x_0 = x$ and $x_n = y$. To check the reflexivity claim, use $n = 0$ so $x = x_0 = y$ and condition 3 is vacuously satisfied. To check transitivitiy, suppose given R-paths $x = x_0 \leqslant x_1 \leqslant \ldots \leqslant x_n = y$ and $y = y_0 \leqslant y_1 \leqslant \ldots \leqslant y_p = z$; then $x = x_0 \leqslant x_1 \leqslant \ldots \leqslant x_n \leqslant y_1 \leqslant \ldots \leqslant y_p = z$ will be an R-path from x to z.

We can turn any relation into a preorder in a canonical way. Here is a concrete case of this idea.

Let $X = \{a, b, c, d\}$ and suppose given the relation $\{(a, b), (b, c), (b, d), (d, c), (c, c)\}$. This is neither reflexive nor transitive, so it is not a preorder. To make it a preorder we follow the preceding prescription. Starting with R-paths of length $n = 0$, we put $\{(a, a), (b, b), (c, c), (d, d)\}$ into $\overline{R}$. The R-paths of length 1 add the original elements, $\{(a, b), (b, c), (b, d), (d, c), (c, c)\}$. Redundancy (e.g., (c, c)) is permissible, but from now on in this example we write only the new elements. The R-paths of length 2 add $\{(a, c), (a, d)\}$ to $\overline{R}$. One can check that R-paths of length 3 and above do not add anything new to $\overline{R}$, so we are done. The relation

$$\overline{R} = \{(a, a), (b, b), (c, c), (d, d), (a, b), (b, c), (b, d), (d, c), (a, c), (a, d)\}$$

is reflexive and transitive, hence a preorder.

Exercise 4.4.1.17.

Let $X = \{a, b, c, d, e, f\}$, and let $R = \{(a, b), (b, c), (b, d), (d, e), (f, a)\}$.

a. What is the preorder $\overline{R}$ generated by R?

b. Is it a partial order?

$\diamond$

Solution 4.4.1.17.

Start by drawing the associated graph, which helps with visualization.

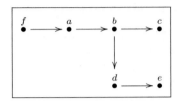

a. This is not a preorder right out of the box, because we need to include all the paths, of which there are 19. You can see the original R as the length 1 paths, a bit down the table:

\overline{R}	
a	a
b	b
c	c
d	d
e	e
f	f
a	b
b	c
b	d
d	e
f	a
a	c
a	d
b	e
f	b
a	e
f	c
f	d
f	e

b. Yes, it is. There are no nontrivial cliques.

◆

Exercise 4.4.1.18.

Let X be the set of people, and let $R \subseteq X \times X$ be the relation with $(x, y) \in R$ if x is the child of y. Describe the preorder generated by R in layperson's terms. ◇

Solution 4.4.1.18.

It is the descendant preorder: $x \leqslant y$ if x is a descendant of y. But be careful; everybody is considered to be a descendant of themselves in this preorder. ◆

4.4.2 Meets and joins

Let X be any set. Recall from Definition 3.4.4.9 that the power-set of X, denoted $\mathbb{P}(X)$, is the set of subsets of X. There is a natural order on $\mathbb{P}(X)$ given by the subset relationship, as exemplified in Exercise 4.4.1.13. Given two elements $a, b \in \mathbb{P}(X)$, we can consider them as subsets of X and take their intersection as an element of $\mathbb{P}(X)$, denoted $a \cap b$. We can also consider them as subsets of X and take their union as an element of $\mathbb{P}(X)$, denoted $a \cup b$. The intersection and union operations are generalized in the following definition.

Definition 4.4.2.1. Let $(S, \leqslant)$ be a preorder, and let $s, t \in S$ be elements. A *meet of s and t* is an element $w \in S$ satisfying the following universal property:

- $w \leqslant s$ and $w \leqslant t$,

- for any $x \in S$, if $x \leqslant s$ and $x \leqslant t$, then $x \leqslant w$.

If w is a meet of s and t, we write $w \cong s \wedge t$.
    A *join of s and t* is an element $w \in S$ satisfying the following universal property:

- $s \leqslant w$ and $t \leqslant w$,

- for any $x \in S$, if $s \leqslant x$ and $t \leqslant x$, then $w \leqslant x$.

If w is a join of s and t, we write $w \cong s \vee t$.

That is, the meet of s and t is the biggest thing that is smaller than both, i.e., a *greatest lower bound*, and the join of s and t is the smallest thing that is bigger than both, i.e., a *least upper bound*. Note that the meet of s and t might be s or t itself.
    It may happen that s and t have more than one meet (or more than one join). However, any two meets of s and t must be in the same clique, by the universal property (and the same for joins).

Exercise 4.4.2.2.

Consider the partial order from Example 4.4.1.3.

a. What is the join of ⌜a diamond⌝ and ⌜a heart⌝?

b. What is the meet of ⌜a black card⌝ and ⌜a queen⌝?

c. What is the meet of ⌜a diamond⌝ and ⌜a card⌝?

◊

Solution 4.4.2.2.

a. ⌜a diamond⌝ ∨ ⌜a heart⌝ = ⌜a red card⌝.

b. ⌜a black card⌝ ∧ ⌜a queen⌝ = ⌜a black queen⌝.

c. ⌜a diamond⌝ ∧ ⌜a card⌝ = ⌜a diamond⌝.

♦

Not every two elements in a preorder need have a meet, nor need they have a join.

Exercise 4.4.2.3.

a. If possible, find two elements in the partial order from Example 4.4.1.3 that do not have a meet.[7]

b. If possible, find two elements that do not have a join (in that preorder).

◊

Solution 4.4.2.3.

a. There is no meet for ⌜a heart⌝ and ⌜a club⌝; no card is both.

b. Every two elements have a join here. But note that some of these joins are "wrong" because the olog is not complete. For example, we have ⌜a 4⌝ ∨ ⌜a queen⌝ = ⌜a card⌝, whereas the correct answer would be ⌜a card that is either a 4 or a queen⌝.

♦

Exercise 4.4.2.4.

As mentioned, the power-set $S := \mathbb{P}(X)$ of any set X naturally has the structure of a partial order. Its elements $s \in S$ correspond to subsets $s \subseteq X$, and we put $s \leqslant t$ if and only if $s \subseteq t$ as subsets of X. The meet of two elements is their intersection as subsets of X, $s \wedge t = s \cap t$, and the join of two elements is their union as subsets of X, $s \vee t = s \cup t$.

a. Is it possible to put a monoid structure on the set S in which the multiplication formula is given by meets? If so, what would the unit element be?

b. Is it possible to put a monoid structure on the set S in which the multiplication formula is given by joins? If so, what would the unit element be?

◊

---

[7]Use the displayed preorder, not any kind of completion of what is written there.

Solution 4.4.2.4.

a. Yes, this will work. The unit element is the subset $X \subseteq X$, because for any $s \in \mathbb{P}(X)$, we have $X \wedge s = X \cap s = s$. So the monoid is $(\mathbb{P}(X), X, \cap)$.

b. Yes, this will work. The unit element is the subset $\varnothing \subseteq X$, because for any $s \in \mathbb{P}(X)$, we have $\varnothing \vee s = \varnothing \cup s = s$. So the monoid is $(\mathbb{P}(X), \varnothing, \cup)$.

$\blacklozenge$

Example 4.4.2.5 (Trees). A *tree*, i.e., a system of nodes and branches, all of which emanate from a single node called the *root*, is a partial order but generally not a linear order. A tree $(T, \leqslant)$ can either be oriented toward the root (so the root is the largest element of the partial order) or away from the root (so the root is the smallest element); let's only consider the former.

A tree is pictured as a graph in (4.11). The root is labeled e.

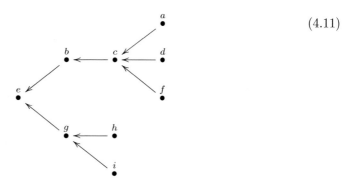

(4.11)

In a tree every pair of elements $s, t \in T$ has a join $s \wedge t$ (their closest mutual ancestor). On the other hand, if s and t have a join $c = s \vee t$, then either $c = s$ or $c = t$.

Exercise 4.4.2.6.

Consider the tree drawn in (4.11).

a. What is the join $i \vee h$?

b. What is the join $h \vee b$?

c. What is the meet $b \wedge a$?

d. What is the meet $b \wedge g$?

$\diamond$

Solution 4.4.2.6.

a. $i \vee h = g$.

b. $h \vee b = e$.

c. $b \wedge a = a$.

d. $b \wedge g$ does not exist.

♦

4.4.3 Opposite order

Definition 4.4.3.1. Let $\mathcal{S} := (S, \leqslant)$ be a preorder. The *opposite preorder*, denoted $\mathcal{S}^{\mathrm{op}}$, is the preorder $(S, \leqslant^{\mathrm{op}})$ having the same set of elements but where $s \leqslant^{\mathrm{op}} s'$ iff $s' \leqslant s$.

Example 4.4.3.2. Consider the preorder $\mathcal{N} := (\mathbb{N}, \texttt{divides})$, where a $\texttt{divides}$ b if "a goes into b evenly," i.e., if there exists $n \in \mathbb{N}$ such that $a * n = b$. So 5 $\texttt{divides}$ 35, and so on. Then $\mathcal{N}^{\mathrm{op}}$ is the set of natural numbers but where $m \leqslant n$ iff m is a multiple of n. So $6 \leqslant 2$ and $6 \leqslant 3$, but $6 \not\leqslant 4$.

Exercise 4.4.3.3.

Suppose that $\mathcal{S} := (S, \leqslant)$ is a preorder.

a. If $\mathcal{S}$ is a partial order, is $\mathcal{S}^{\mathrm{op}}$ also a partial order?

b. If $\mathcal{S}$ is a linear order, is $\mathcal{S}^{\mathrm{op}}$ a linear order?

◊

Solution 4.4.3.3.

a. Yes. If $\mathcal{S}$ has no nontrivial cliques, neither will $\mathcal{S}^{\mathrm{op}}$.

b. Yes. If every two elements in $\mathcal{S}$ are comparable, so are every two elements in $\mathcal{S}^{\mathrm{op}}$.

♦

Exercise 4.4.3.4.

Suppose that $\mathcal{S} := (S, \leqslant)$ is a preorder and that $s_1, s_2 \in S$ have join $s_1 \vee s_2 = t$ in $\mathcal{S}$. The preorder $\mathcal{S}^{\mathrm{op}}$ has the same elements as $\mathcal{S}$. Is t the join of s_1 and s_2 in $\mathcal{S}^{\mathrm{op}}$, or is it their meet, or is it not necessarily their meet or their join? ◊

Solution 4.4.3.4.

It is their meet.   ♦

4.4.4   Morphism of orders

An order $(S, \leqslant)$, be it a preorder, a partial order, or a linear order, involves a set and a binary relation. For two orders to be comparable, their sets and their relations should be appropriately comparable.

Definition 4.4.4.1. Let $\mathcal{S} := (S, \leqslant)$ and $\mathcal{S}' := (S', \leqslant')$ be preorders (resp. partial orders or linear orders). A *morphism of preorders* (resp. *partial orders* or *linear orders*) f *from* $\mathcal{S}$ *to* $\mathcal{S}'$, denoted $f\colon \mathcal{S} \to \mathcal{S}'$, is a function $f\colon S \to S'$ such that, for every pair of elements $s_1, s_2 \in S$, if $s_1 \leqslant s_2$, then $f(s_1) \leqslant' f(s_2)$.

Example 4.4.4.2. Let X and Y be sets, and let $f\colon X \to Y$ be a function. Then for every subset $X' \subseteq X$, its image $f(X') \subseteq Y$ is a subset (see Exercise 2.1.2.8). Thus we have a function $F\colon \mathbb{P}(X) \to \mathbb{P}(Y)$, given by taking images. This is a morphism of partial orders $(\mathbb{P}(X), \subseteq) \to (\mathbb{P}(Y), \subseteq)$. Indeed, if $a \subseteq b$ in $\mathbb{P}(X)$, then $f(a) \subseteq f(b)$ in $\mathbb{P}(Y)$.

Application 4.4.4.3. It is often said that a team is only as strong as its weakest member. Is this true for materials? The hypothesis that a material is only as strong as its weakest constituent can be understood as follows.

Recall from the beginning of Section 4.4 (page 162) that we can put several different orders on the set M of materials. One example is the order given by constituency ($m \leqslant_C m'$ if m is an ingredient or constituent of m'). Another order is given by strength: $m \leqslant_S m'$ if m' is stronger than m (in some fixed setting).

Is it true that if material m is a constituent of material m', then the strength of m' is less than or equal to the strength of m? Mathematically the question would be, Is there a morphism of preorders $(M, \leqslant_C) \longrightarrow (M, \leqslant_S)^{\mathrm{op}}$?

◊◊

Exercise 4.4.4.4.

Let X and Y be sets, and let $f\colon X \to Y$ be a function. Then for every subset $Y' \subseteq Y$, its preimage $f^{-1}(Y') \subseteq X$ is a subset (see Definition 3.2.1.12). Thus we have a function $F\colon \mathbb{P}(Y) \to \mathbb{P}(X)$, given by taking preimages. Is it a morphism of partial orders?   ◊

Solution 4.4.4.4.

Let's first ground the discussion with an olog

$$\boxed{\begin{array}{c} X \\ \text{a person} \end{array}} \xrightarrow{\text{lives in}} \boxed{\begin{array}{c} Y \\ \text{a country} \end{array}}$$

Now given any set of countries $Y' \subseteq Y$, we can consider the set $f^{-1}(Y')$ of persons living in (any one of) those countries—that is the preimage. This question is asking whether, if Mary chooses a set of countries $M \subseteq Y$ and John chooses all those and a few more $M \subseteq J$, does the set of persons living in John's countries include the set of persons living in Mary's countries, $f^{-1}(M) \subseteq f^{-1}(J)$? Well, clearly, yes. So $F = f^{-1}(-) \colon \mathbb{P}(Y) \to \mathbb{P}(X)$ is a morphism of partial orders. ♦

Example 4.4.4.5. Let S be a set. The smallest preorder structure that can be put on S is to say $a \leqslant b$ iff $a = b$. This is indeed reflexive and transitive, and it is called the *discrete preorder on S*.

The largest preorder structure that can be put on S is to say $a \leqslant b$ for all $a, b \in S$. This again is reflexive and transitive, and it is called the *indiscrete preorder on S*.

Exercise 4.4.4.6.

Let S be a set, and let $(T, \leqslant_T)$ be a preorder. Let $\leqslant_D$ be the discrete preorder on S.

a. A morphism of preorders $(S, \leqslant_D) \to (T, \leqslant_T)$ is a function $S \to T$ satisfying certain properties (see Definition 4.4.4.1). Which functions $S \to T$ arise in this way?

b. Given a morphism of preorders $(T, \leqslant_T) \to (S, \leqslant_D)$, we get a function $T \to S$. In terms of $\leqslant_T$, which functions $T \to S$ arise in this way?

◇

Solution 4.4.4.6.

a. All of them. Any function $S \to T$ will respect the discrete preorder.

b. We get exactly those functions $f \colon T \to S$ with the following property for all $t_1, t_2 \in T$: if $t_1 \leqslant_T t_2$, then $f(t_1) = f(t_2)$.

♦

Exercise 4.4.4.7.

Let S be a set, and let $(T, \leqslant_T)$ be a preorder. Let $\leqslant_I$ be the indiscrete preorder on S, as in Example 4.4.4.5.

a. Given a morphism of preorders $(S, \leqslant_I) \to (T, \leqslant_T)$, we get a function $S \to T$. In terms of $\leqslant_T$, which functions $S \to T$ arise in this way?

b. Given a morphism of preorders $(T, \leqslant_T) \to (S, \leqslant_I)$, we get a function $T \to S$. In terms of $\leqslant_T$, which functions $T \to S$ arise in this way?

◇

Solution 4.4.4.7.

a. We get exactly those functions $f\colon S \to T$ with the following property: there exists a clique $T' \subseteq T$ such that $f(S) \subseteq T'$.

b. All of them.

$\blacklozenge$

4.4.5   Other applications

4.4.5.1   Biological classification

Biological classification is a method for dividing the set of organisms into distinct classes, called taxa. In fact, it turns out that such a classification, say, a phylogenetic tree, can be understood as a partial order C on the set of taxa. The typical *ranking* of these taxa, including kingdom, phylum, and so on, can be understood as morphism of orders $f\colon C \to [n]$, for some $n \in \mathbb{N}$.

For example, we may have a tree (see Example 4.4.2.5) that looks like this:

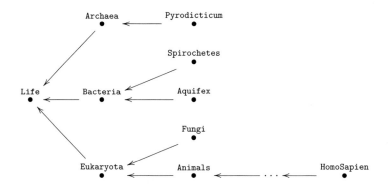

We also have a linear order that looks like this:

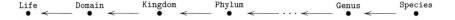

and the ranking system that puts Eukaryota at Domain and Homo Sapien at Species is an order-preserving function from the dots upstairs to the dots downstairs; that is, it is a morphism of preorders.

Exercise 4.4.5.2.

Since the phylogenetic tree is a tree, it has all joins.

a. Determine the join of dogs and humans.

b. If we did not require the phylogenetic partial order to be a tree, what would it mean if two taxa (nodes in the phylogenetic partial order), say, a and b, had meet c with $c \neq a$ and $c \neq b$?

◊

Solution 4.4.5.2.

a. Dogs and humans join in the class "mammal."

b. The requirements $a \wedge b \neq a$ and $a \wedge b \neq b$ mean that neither $a \leqslant b$ nor $b \leqslant a$. So this strange situation would mean that taxon c was classified as a subtaxon of both taxon a and taxon b, and that it was the largest such subtaxon.

♦

Exercise 4.4.5.3.

a. In your favorite scientific subject, are there any interesting classification systems that are actually orders?

b. Choose one such system; what would meets mean in that setting?

◊

Solution 4.4.5.3.

a. In geography, perhaps we can classify political regions by whether they are neighborhood associations, cities/counties, states/provinces, countries, or world.

b. The join of two political regions would be the smallest region containing both, e.g., Chicago $\vee$ North Carolina $=$ USA.

♦

4.4.5.4   Security

Security, say of sensitive information, is based on two things: a security clearance and need to know. Security clearance might consist of levels like confidential, secret, top secret. But maybe we can throw in "President's eyes only" and some others too, like "anyone."

Exercise 4.4.5.5.

Does it appear that security clearance is a preorder, a partial order, or a linear order?
◊

Solution 4.4.5.5.

It looks like a linear order.   ◆

"Need to know" is another classification of people. For each bit of information, we do not necessarily want everyone to know about it, even everyone with the specified clearance. It is only disseminated to those who need to know.

Exercise 4.4.5.6.

Let P be the set of all people, and let $\overline{I}$ be the set of all pieces of information known by the government. For each subset $I \subseteq \overline{I}$, let $K(I) \subseteq P$ be the set of people who need to know every piece of information in I. Let $S = \{K(I) \mid I \subseteq \overline{I}\}$ be the set of all "need to know" groups, with the subset relation denoted $\leqslant$.

a. Is $(S, \leqslant)$ a preorder? If not, find a nearby preorder.

b. If $I_1 \subseteq I_2$, do we always have $K(I_1) \leqslant K(I_2)$ or $K(I_2) \leqslant K(I_1)$ or possibly neither?

c. Should the preorder $(S, \leqslant)$ have all meets?

d. Should $(S, \leqslant)$ have all joins?

◊

Solution 4.4.5.6.

a. Yes, $(S, \leqslant)$ is a preorder.

b. If Alice needs to know every piece of information in I_1, and $I_1 \subseteq I_2$, this does not mean she needs to know everything in I_2; it is the other way around. So $K(I_2) \leqslant K(I_1)$. Another way to see this is, if $I_1 \subseteq I_2$, then "need to know I_2" is a higher kind of clearance, so fewer people have it.

c. If you need to know everything in I_1 and everything in I_2, then you need to know everything in $I_1 \cup I_2$. And, of course, if you need to know everything in $I_1 \cup I_2$, then you need to know everything in I_1 and everything in I_2. So $K(I_1 \cup I_2) = K(I_1) \wedge K(I_2)$. Yes, $(S, \leqslant)$ has all meets.

d. Take the people who need to know I_1 and the people who need to know I_2, and put them all in a room. Is there necessarily some set of information I_3 that this group, and only this group, needs to know? This does not seem necessary. So I would say 'no.

♦

4.4.5.7 Spaces and geography

Consider closed curves that can be drawn in the plane $\mathbb{R}^2$, e.g., circles, ellipses, and kidney-bean shaped curves. The interiors of these closed curves (not including the boundary itself) are called *basic open sets in* $\mathbb{R}^2$. The good thing about such an interior U is that any point $p \in U$ is not on the boundary, so no matter how close p is to the boundary of U, there will always be a tiny basic open set surrounding p and completely contained in U. In fact, the union of any collection of basic open sets still has this property. That is, an *open set in* $\mathbb{R}^2$ is any subset $U \subseteq \mathbb{R}^2$ that can be formed as the union of a collection of basic open sets.

Example 4.4.5.8. Let $U = \{(x, y) \in \mathbb{R}^2 \mid x > 0\}$. To see that U is open, define the following sets: for any $a, b \in \mathbb{R}$, let $S(a, b)$ be the square parallel to the axes, with side length 1, where the upper left corner is (a, b). Note that $S(a, b)$ is a closed curve, so if we let $S'(a, b)$ be the interior of $S(a, b)$, then each $S'(a, b)$ is a basic open set. Now U is the union of $S'(a, b)$ over the collection of all $a > 0$ and all b,

$$U = \bigcup_{\substack{a, b \in \mathbb{R}, \\ a > 0}} S'(a, b),$$

so U is open.

Example 4.4.5.9. The idea of open sets extends to spaces beyond $\mathbb{R}^2$. For example, on the earth one could define a basic open set to be the interior of any region one can draw a closed curve around (with a metaphorical pen), and define open sets to be unions of these basic open sets.

Exercise 4.4.5.10.

Let $(S, \subseteq)$ be the partial order of open subsets on earth as defined in Example 4.4.5.9.

a. If $\leqslant$ is the subset relation, is $(S, \leqslant)$ a partial order or just a preorder, or neither?

b. Does it have meets?

c. Does it have joins?

$\Diamond$

Solution 4.4.5.10.

a. It is a partial order.

b. It has meets (given by intersections).

c. It has joins (given by unions).

♦

Exercise 4.4.5.11.

Let S be the set of open subsets of earth as defined in Example 4.4.5.9. For each open subset of earth, suppose we know the range of recorded temperature throughout s (i.e., the low and high throughout the region). Thus to each element $s \in S$ we assign an interval $T(s) := \{x \in \mathbb{R} \mid a \leqslant x \leqslant b\}$. The set V of intervals of $\mathbb{R}$ can be partially ordered by the subset relation.

a. Does the assignment $T \colon S \to V$ amount to a morphism of orders?

b. If so, does it preserve meets or joins? Hint: It does not preserve both.

$\Diamond$

Solution 4.4.5.11.

a. Suppose s is a subregion of s', e.g., New Mexico as a subregion of North America. This question is asking whether the range of temperatures recorded throughout New Mexico is a subset of the range of temperatures recorded throughout North America, which, of course, it is.

b. The question on meets is, If we take two regions s and s' and intersect them, is the temperature range on $s \cap s'$ equal to the intersection $T(s) \cap T(s')$? Clearly, if a temperature t is recorded somewhere in $s \cap s'$, then it is recorded somewhere in s and somewhere in s', so $T(s \cap s') \subseteq T(s) \cap T(s')$. But is it true that if a temperature is recorded somewhere in s and somewhere in s', then it must be recorded somewhere in $s \cap s'$? No, that is false. So T does not preserve meets.

The question on joins is, If we take the union of two regions s and s', is the temperature range on $s \cup s'$ equal to the union $T(s) \cup T(s')$? If a temperature is recorded somewhere in $s \cup s'$, then it is either recorded somewhere in s or somewhere in s' (or both), so $T(s \cup s') \subseteq T(s) \cup T(s')$. And if a temperature is recorded somewhere in s, then it is recorded somewhere in $s \cup s'$, so $T(s) \subseteq T(s \cup s')$. Similarly, $T(s') \subseteq T(s \cup s')$, so in fact T does preserve joins: $T(s \cup s') = T(s) \cup T(s')$.

◆

Exercise 4.4.5.12.

a. Can you think of a space relevant to an area of science for which it makes sense to assign an interval of real numbers to each open set, analogously to Exercise 4.4.5.11? For example, for a sample of some material under stress, perhaps the strain on each open set is somehow an interval?

b. Check that your assignment, which you might denote as in Exercise 4.4.5.11 by $T \colon S \to V$, is a morphism of orders.

c. How does it act with respect to meets and/or joins?

◇

Solution 4.4.5.12.

a. Consider the roads in the United States, and let S denote the open sets within this space; its elements are the various regions of roads. With V again the set of intervals in $\mathbb{R}$, we could take the function $T \colon S \to V$ to be the weight range permissible throughout a given region of roadway. Note that this is qualitatively different than Exercise 4.4.5.11 in that $T(s)$ is not about weights that are permissible *somewhere* within s, it is about weights that are permissible *everywhere* within s.

b. This assignment T is not a morphism of orders, but it is a morphism of orders $T \colon S^{\mathrm{op}} \to V$. The reason is that if $s \subseteq s'$, then a weight permissible throughout s may not be permissible throughout the whole of s', but the reverse is true.

c. A weight permissible throughout $s \cup s'$ is permissible throughout s and throughout s', so we have $T(s \cup s') = T(s) \cap T(s')$. That is, T sends joins in S (which are meets in S^{op}) to meets in V.

◆

4.5   Databases: schemas and instances

So far this chapter has discussed classical objects from mathematics. The present section is about databases, which are classical objects from computer science. These are truly "categories and functors, without admitting it" (see Theorem 5.4.2.3).

4.5.1   What are databases?

Data, in particular, the set of observations made during experiment, plays a primary role in science of any kind. To be useful, data must be organized, often in a row-and-column display called a table. Columns existing in different tables can refer to the same data.

A database is a collection of tables, each table T of which consists of a set of columns and a set of rows. We roughly explain the role of tables, columns, and rows as follows. The existence of table T suggests the existence of a fixed methodology for observing objects or events of a certain type. Each column c in T prescribes a single kind or method of observation, so that the datum inhabiting any cell in column c refers to an observation of that kind. Each row r in T has a fixed sourcing event or object, which can be observed using the methods prescribed by the columns. The cell (r, c) refers to the observation of kind c made on event r. All of the rows in T should refer to uniquely identifiable objects or events of a single type, and the name of the table T should refer to that type.

Example 4.5.1.1. When graphene is strained (lengthened by a factor of $x \geqslant 1$), it becomes stressed (carries a force in the direction of the lengthening). The following is a madeup set of data:

Graphene Sample			
ID	**Source**	**Stress**	**Strain**
A118-1	C Smkt	0	0
A118-2	C Smkt	0.02	20
A118-3	C Smkt	0.05	40
A118-4	AC	0.04	37
A118-5	AC	0.1	80
A118-6	C Plat	0.1	82

Supplier		
ID	**Full Name**	**Phone**
C Smkt	Carbon Supermarket	(541) 781-6611
AC	Advanced Chemical	(410) 693-0818
C Plat	Carbon Platform	(510) 719-2857
McD	McDonard's Burgers	(617) 244-4400
APP	Acme Pen and Paper	(617) 823-5603

$$(4.12)$$

In the table in (4.12) titled "Graphene Sample," the rows refer to graphene samples, and the table is so named. Each graphene sample can be observed according to the source supplier from which it came, the strain that it was subjected to, and the stress that it carried. These observations are the columns. In the right-hand table the rows refer to suppliers of various things, and the table is so named. Each supplier can be observed according to its full name and its phone number; these are the columns.

In the left-hand table it appears either that each graphene sample was used only once, or that the person recording the data did not keep track of which samples were reused. If such details become important later, the lab may want to change the layout of the left-hand table by adding an appropriate column. This can be accomplished using morphisms of schemas (see Section 5.4.1).

4.5.1.2   Primary keys, foreign keys, and data columns

There is a bit more structure in the tables in (4.12) than first meets the eye. Each table has a *primary ID column*, on the left, as well as some *data columns* and some *foreign key columns*. The primary key column is tasked with uniquely identifying different rows. Each data column houses elementary data of a certain sort. Perhaps most interesting from a structural point of view are the foreign key columns, because they link one table to another, creating a connection pattern between tables. Each foreign key column houses data that needs to be further unpacked. It thus refers us to another *foreign* table, in particular, to the primary ID column of that table. In (4.12) the `Source` column is a foreign key to the `Supplier` table.

Here is another example, taken from Spivak [39].

Example 4.5.1.3. Consider the bookkeeping necessary to run a department store. We keep track of a set of employees and a set of departments. For each employee e, we keep track of

E.1   the **first** name of e, which is a `FirstNameString`,

E.2   the **last** name of e, which is a `LastNameString`,

E.3   the **manager** of e, which is an `Employee`,

E.4   the department that e **works in**, which is a `Department`.

For each department d, we keep track of

D.1   the **name** of d, which is a `DepartmentNameString`,

D.2   the **secretary** of d, which is an `Employee`.

We can suppose that E.1, E.2, and D.1 are data columns (referring to names of various sorts), and E.3, E.4, and D.2 are foreign key columns (referring to managers, secretaries, etc.).

The tables in (4.13) show how such a database might look at a particular moment in time.

Employee				
ID	**first**	**last**	**manager**	**worksIn**
101	David	Hilbert	103	q10
102	Bertrand	Russell	102	x02
103	Emmy	Noether	103	q10

Department		
ID	**name**	**secretary**
q10	Sales	101
x02	Production	102

$$(4.13)$$

4.5.1.4 Business rules

Looking at the tables in (4.13), one may notice a few patterns. First, every employee works in the same department as his or her manager. Second, every department's secretary works in that department. Perhaps the business counts on these rules for the way it structures itself. In that case the database should enforce those rules, i.e., it should check that whenever the data is updated, it conforms to the rules:

> Rule 1   For every employee e, the **manager** of e **works in** the same department that e **works in**.

> Rule 2   For every department d, the **secretary** of d **works in** department d.

$$(4.14)$$

Together, the statements E.1, E.2, E.3, E.4, D.1, and D.2 from Example 4.5.1.3 and Rule 1 and Rule 2 constitute the *schema* of the database. This is formalized in Section 4.5.2.

4.5.1.5 Data columns as foreign keys

To make everything consistent, we could even say that data columns are specific kinds of foreign keys. That is, each data column constitutes a foreign key to some non-branching *leaf table*, which has no additional data.

Example 4.5.1.6. Consider again Example 4.5.1.3. Note that first names and last names have a particular type, which we all but ignored. We could cease to ignore them by adding three tables, as follows:

(4.15)

FirstNameString
ID
Alan
Alice
Bertrand
Carl
David
Emmy
⋮

LastNameString
ID
Arden
Hilbert
Jones
Noether
Russell
⋮

DepartmentNameString
ID
Marketing
Production
Sales
⋮

In combination, (4.13) and (4.15) form a collection of five tables, each with the property that every column is either a primary key or a foreign key. The notion of data column is now subsumed under the notion of foreign key column. Each column is either a primary key (one per table, labeled ID) or a foreign key column (everything else).

4.5.2 Schemas

Pictures here, roughly graphs, should capture the *conceptual layout* to which the data conforms, without being concerned (yet) with the individual pieces of data that may populate the tables in this instant. We proceed at first by example; the precise definition of schema is given in Definition 4.5.2.7.

Example 4.5.2.1. In Examples 4.5.1.3 and 4.5.1.6, the conceptual layout for a department store was given, and some example tables were shown. We were instructed to keep track of employees, departments, and six types of data (E.1, E.2, E.3, E.4, D.1, and D.2), and to follow two rules (Rule 1, Rule 2). All of this is summarized in the following picture:

$\mathcal{C}$:= Schema for tables (4.13) and (4.15) conforming to (4.14)

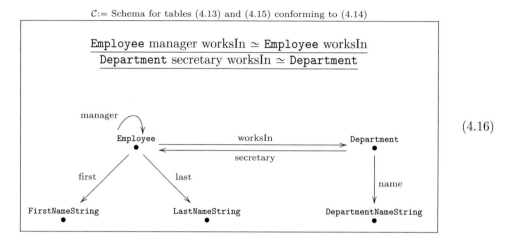

(4.16)

The five tables from (4.13) and (4.15) are seen as five vertices; this is also the number of primary ID columns. The six foreign key columns from (4.13) and (4.15) are seen as six arrows; each points from a table to a foreign table. The two rules from (4.14) are seen as declarations at the top of (4.16). These path equivalence declarations are explained in Definition 4.5.2.3.

Exercise 4.5.2.2.

Create a schema (consisting of dots and arrows) describing the conceptual layout of information presented in Example 4.5.1.1.   ◊

Solution 4.5.2.2.

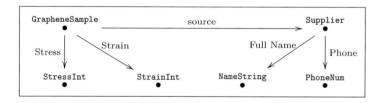

♦

In order to define schemas, we must first define the notion of *congruence* for an arbitrary graph G. Roughly a congruence is an equivalence relation that indicates how different paths in G are related (see Section 4.3.2). A notion of congruence for monoids was given in Definition 4.1.1.17, and the current notion is a generalization of that. A congruence (in addition to being reflexive, symmetric, and transitive) has two sorts of additional properties: congruent paths must have the same source and target, and the composition of congruent paths with other congruent paths must yield congruent paths. Formally we have Definition 4.5.2.3.

Definition 4.5.2.3. Let $G = (V, A, src, tgt)$ be a graph, and let Path_G denote the set of paths in G (see Definition 4.3.2.1). A *path equivalence declaration* (or PED) is an expression of the form $p \simeq q$, where $p, q \in \text{Path}_G$ have the same source and target, $src(p) = src(q)$ and $tgt(p) = tgt(q)$.

A *congruence* on G is a relation $\simeq$ on Path_G that has the following properties:

1. The relation $\simeq$ is an equivalence relation.

2. If $p \simeq q$, then $src(p) = src(q)$.

3. If $p \simeq q$, then $tgt(p) = tgt(q)$.

4. Suppose given paths $p, p' : a \to b$ and $q, q' : b \to c$. If $p \simeq p'$ and $q \simeq q'$, then $(p \mathbin{+\!\!+} q) \simeq (p' \mathbin{+\!\!+} q')$.

Remark 4.5.2.4. Any set of path equivalence declarations (PEDs) generates a congruence. The proof of this is analogous to that of Proposition 4.1.1.18. We tend to elide the difference between a congruence and a set of PEDs that generates it.

The basic idea for generating a congruence from a set R of PEDs is to proceed as follows. First find the equivalence relation generated by R. Then every time there are paths $p, p' : a \to b$ and $q, q' : b \to c$ with $p \simeq p'$ and $q \simeq q'$,

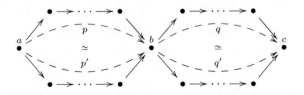

add to R the relation $(p \mathbin{+\!\!+} q) \simeq (p' \mathbin{+\!\!+} q')$.

Exercise 4.5.2.5.

Suppose given the following graph G, with the PED $_b[w, x] \simeq {}_b[y, z]$:

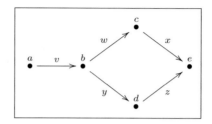

In the congruence generated by that PED, is it the case that $_a[v, w, x] \simeq {}_a[v, y, z]$?   ◇

Solution 4.5.2.5.

Yes. Since a congruence is an equivalence relation on $\mathrm{Paths}(G)$, it is in particular, reflexive, so we have $_a[v] \simeq {}_a[v]$. Thus we have

$$_a[v, w, x] = {}_a[v] \mathbin{++} {}_b[w, x] \simeq {}_a[v] \mathbin{++} {}_b[y, z] = {}_a[v, y, z].$$

◆

Exercise 4.5.2.6.

Consider the graph shown in (4.16) and the two declarations shown at the top. They generate a congruence.

a. Is it true that the following PED is an element of this congruence?

$$\texttt{Employee manager manager worksIn} \overset{?}{\simeq} \texttt{Employee worksIn}$$

b. What about this one?

$$\texttt{Employee worksIn secretary} \overset{?}{\simeq} \texttt{Employee}$$

c. What about this one?

$$\texttt{Department secretary manager worksIn name} \overset{?}{\simeq} \texttt{Department name}$$

◇

Solution 4.5.2.6.

a. Yes; every employee, even the manager of an employee, works in the same department as her manager.

$$\texttt{Employee} \text{ manager manager worksIn} \simeq \texttt{Employee} \text{ manager worksIn}$$
$$\simeq \texttt{Employee} \text{ worksIn}$$

b. No; this does not follow. An employee is not necessarily the secretary of the department he works in.

c. Yes; here is the derivation:

$$\texttt{Department} \text{ secretary manager worksIn name}$$
$$\simeq \texttt{Department} \text{ secretary worksIn name}$$
$$\simeq \texttt{Department} \text{ name.}$$

♦

Definition 4.5.2.7. A *database schema* (or simply *schema*) $\mathcal{C}$ consists of a pair $\mathcal{C} := (G, \simeq)$, where G is a graph and $\simeq$ is a congruence on G.

Example 4.5.2.8. Pictured in (4.16) is a graph with two PEDs; these generate a congruence, as discussed in Remark 4.5.2.4. Thus this constitutes a database schema.

A schema can be converted into a system of tables, each with a primary key and some number of foreign keys referring to other tables, as discussed in Section 4.5.1. Definition 4.5.2.7 gives a precise conceptual understanding of what a schema is, and the following rules describe how to convert it into a table layout.

Rules of good practice 4.5.2.9. Converting a schema $\mathcal{C} = (G, \simeq)$ into a table layout should be done as follows:

(i) There should be a table for every vertex in G, and if the vertex is named, the table should have that name.

(ii) Each table should have a leftmost column called ID, set apart from the other columns by a double vertical line.

(iii) To each arrow a in G having source vertex $s := src(a)$ and target vertex $t := tgt(a)$, there should be a foreign key column a in table s, referring to table t; if the arrow a is named, column a should have that name.

Example 4.5.2.10 (Discrete dynamical system). Consider the schema

$$\mathcal{Loop} := \boxed{\begin{array}{c} f \\ \circlearrowright \\ s \\ \bullet \end{array}} \tag{4.17}$$

in which the congruence is trivial (i.e., generated by the empty set of PEDs.) This schema is quite interesting. It encodes a set s and a function $f\colon s \to s$. Such a thing is called a *discrete dynamical system*. One imagines s as the set of states, and for any state $x \in s$, the function f encodes a notion of next state $f(x) \in s$. For example,

ID	f
A	B
B	C
C	C
D	B
E	C
F	G
G	H
H	G

... pictured ...

$$\boxed{\begin{array}{ccc} A & B & C \circlearrowright \\ \bullet \to \bullet \to \bullet \\ D \nearrow E \nearrow \\ \bullet \quad \bullet \\ F \quad G \quad H \\ \bullet \to \bullet \rightleftarrows \bullet \end{array}} \tag{4.18}$$

Application 4.5.2.11. Imagine a deterministic quantum-time universe in which there are discrete time steps. We model it as a discrete dynamical system, i.e., a table of the form (4.18). For every possible state of the universe we include a row in the table. The state in the next instant is recorded in the second column.[8]

◊◊

Example 4.5.2.12 (Finite hierarchy). The schema $\mathcal{Loop}$ can also be used to encode hierarchies, such as the manager relation from Examples 4.5.1.3 and 4.5.2.1,

One problem with this, however, is if a schema has even one loop, then it can have infinitely many paths (corresponding, e.g., to an employee's manager's manager's manager's ... manager).

---

[8]If we want nondeterminism, i.e., a probabilistic distribution as the next state, we can use monads. See Section 7.3.4.2.

Sometimes we know that in a given company that process eventually terminates, a famous example being that at Ben and Jerry's ice cream company, there were only seven levels. In that case we know that an employee's eighth-level manager is equal to his or her seventh-level manager. This can be encoded by the PED

$$_E[\mathrm{mgr, mgr, mgr, mgr, mgr, mgr, mgr, mgr}] \simeq {}_E[\mathrm{mgr, mgr, mgr, mgr, mgr, mgr, mgr}]$$

or more concisely, $_E[\mathrm{mgr}]^8 = {}_E[\mathrm{mgr}]^7$.

Exercise 4.5.2.13.

There is a nontrivial PED on $\mathcal{L}oop$ that holds for the data in Example 4.5.2.10.

a. What is it?

b. How many equivalence classes of paths in $\mathcal{L}oop$ are there after you impose that relation?

◊

Solution 4.5.2.13.

a. $f^4 = f^2$ (or to be pedantic, $_s[f, f, f, f] \simeq {}_s[f, f]$).

b. There are four: $_s[], {}_s[f], {}_s[f, f]$, and $_s[f, f, f]$. Any longer path is equivalent to one of these.

◆

Exercise 4.5.2.14.

Let P be a chess-playing program, playing against itself. Given any position (where a position includes the history of the game so far), P will make a move.

a. Is this an example of a discrete dynamical system?

b. How do the rules for ending the game in a win or draw play out in this model? (Look up online how chess games end if you do not know.)

◊

Solution 4.5.2.14.

a. Yes, as long as the program is deterministic (i.e., it plays the same move every time it is in the same position).

b. We need to make positions called "white win," "black win," and "draw." The only move from the position "white win" results in "white win," and similarly for "black win" and for "draw."

$\blacklozenge$

4.5.2.15   Ologging schemas

It should be clear that a database schema is nothing but an olog in disguise. The difference is basically the readability requirements for ologs. There is an important new addition in this section, namely, that schemas and ologs can be filled in with data. Conversely, we have seen that databases are not any harder to understand than ologs are.

Example 4.5.2.16. Consider the olog

$$\boxed{\text{a moon}} \xrightarrow{\;orbits\;} \boxed{\text{a planet}} \tag{4.19}$$

We can document some instances of this relationship using the following table:

orbits	
a moon	**a planet**
The Moon	Earth
Phobos	Mars
Deimos	Mars
Ganymede	Jupiter
Titan	Saturn

(4.20)

Clearly, this table of instances can be updated as more moons are discovered by the olog's owner (be it by telescope, conversation, or research).

Exercise 4.5.2.17.

In fact, Example 4.5.2.16 did not follow rules 4.5.2.9. Strictly following those rules, copy over the data from (4.20) into tables that are in accordance with schema (4.19). $\Diamond$

Solution 4.5.2.17.

a moon	
ID	**orbits**
The Moon	Earth
Phobos	Mars
Deimos	Mars
Ganymede	Jupiter
Titan	Saturn

a planet
ID
Earth
Mars
Jupiter
Saturn

Exercise 4.5.2.18.

a. Write a schema (olog) in terms of the boxes ⌜a thing I own⌝ and ⌜a place⌝ and one arrow that might help a person remember where she decided to put random things.

b. What is a good label for the arrow?

c. Fill in some rows of the corresponding set of tables for your own case.

Solution 4.5.2.18.

a.

b. I think "belongs in" is fine.

c.

a thing I own	
ID	**belongs in**
passport	file cabinet
spare keys	middle desk drawer
gloves	front closet
big umbrella	front closet

a place
ID
file cabinet
middle desk drawer
front closet

Exercise 4.5.2.19.

Consider the olog

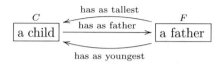

a. What path equivalence declarations would be appropriate for this olog? You can use
$y\colon F \to C$, $t\colon F \to C$, and $f\colon C \to F$ for "youngest," "tallest," and "father," if you
prefer.

b. How many PEDs are in the congruence?

$\Diamond$

Solution 4.5.2.19.

a. There are two: $F.t.f \simeq F$ and $F.y.f \simeq F$, meaning "a father F's tallest child has as
father F" and "a father F's youngest child has as father F."

b. There are infinitely many PEDs in this congruence, including $_F[t, f, t] \simeq {}_F[t]$ and
$_F[t, f, y] \simeq {}_F[y]$. But the congruence is *generated* by only two PEDs, those in part
(a).

♦

4.5.3   Instances

Given a database schema $(G, \simeq)$, an instance of it is just a bunch of tables whose data
conform to the specified layout. These can be seen throughout the previous section, most
explicitly in the relationship between schema (4.16) and tables (4.13) and (4.15), and
between schema (4.17) and table (4.18). Following is the mathematical definition.

Definition 4.5.3.1. Let $\mathcal{C} = (G, \simeq)$, where $G = (V, A, src, tgt)$. An *instance on* $\mathcal{C}$,
denoted $(\mathrm{PK}, \mathrm{FK})\colon \mathcal{C} \to \mathbf{Set}$, is defined as follows: One announces some constituents
(A. primary ID part, B. foreign key part) and shows that they conform to a law (1.
preservation of congruence). Specifically, one announces

   A. a function $\mathrm{PK}\colon V \to \mathbf{Set}$, i.e., to each vertex $v \in V$ one provides a set $\mathrm{PK}(v)$;[9]

---

[9]The elements of $\mathrm{PK}(v)$ are listed as the rows of table v, or more precisely, as the leftmost cells of
these rows.

B. for every arrow $a \in A$ with $v = src(a)$ and $w = tgt(a)$, a function $\mathrm{FK}(a) \colon \mathrm{PK}(v) \to \mathrm{PK}(w)$.[10]

One must then show that the following law holds for any vertices v, w and paths $p = {}_v[a_1, a_2, \ldots, a_m]$ and $q = {}_v[a'_1, a'_2, \ldots, a'_n]$ from v to w:

1. If $p \simeq q$, then for all $x \in \mathrm{PK}(v)$, we have

$$\mathrm{FK}(a_m) \circ \cdots \circ \mathrm{FK}(a_2) \circ \mathrm{FK}(a_1)(x) = \mathrm{FK}(a'_n) \circ \cdots \circ \mathrm{FK}(a'_2) \circ \mathrm{FK}(a'_1)(x)$$

in $\mathrm{PK}(w)$.

Exercise 4.5.3.2.

Consider the olog in (4.21):[11]

$$\mathcal{C} :=$$

$$(4.21)$$

It can be considered a schema of which the following is an instance:

a self-email			an email			a person
ID	**is**		**ID**	**is sent by**	**is sent to**	**ID**
SEm1207	Em1207		Em1206	Bob	Sue	Bob
SEm1210	Em1210		Em1207	Carl	Carl	Carl
SEm1211	Em1211		Em1208	Sue	Martha	Chris
			Em1209	Chris	Bob	Julia
			Em1210	Chris	Chris	Martha
			Em1211	Julia	Julia	Sue
			Em1212	Martha	Chris	

$$(4.22)$$

---

[10]The arrow a corresponds to a column, and to each row $r \in \mathrm{PK}(v)$ the (r, a) cell contains the datum $\mathrm{FK}(a)(r)$.

[11]The text at the bottom of the box in (4.21) is a summary of a fact, i.e., a path equivalence in the olog. Under the formal rules of Englishing a fact (see (2.20)), it would read as follows. Given x, a self-email, consider the following. We know that x is a self-email, which is an email, which is sent by a person who we call $P(x)$. We also know that x is a self-email, which is an email, which is sent to a person who we call $Q(x)$. Fact: Whenever x is a self-email, we have $P(x) = Q(x)$.

a. What is the set PK($\ulcorner$an email$\urcorner$)?

b. What is the set PK($\ulcorner$a person$\urcorner$)?

c. What is the function FK($\xrightarrow{\text{is sent by}}$): PK($\ulcorner$an email$\urcorner$) → PK($\ulcorner$a person$\urcorner$)?

d. Interpret the sentences at the bottom of $\mathcal{C}$ as the Englishing of a simple path equivalence declaration (PED).

e. Is your PED satisfied by the instance (4.22); that is, does law 1. from Definition 4.5.3.1 hold?

◇

Solution 4.5.3.2.

a. PK($\ulcorner$an email$\urcorner$) = {Em1206,Em1207,Em1208,Em1209,Em1210,Em1211,Em1212}.

b. PK($\ulcorner$a person$\urcorner$) = {Bob, Carl, Chris, Julia, Martha, Sue}.

c. It is the first two columns in the `an email` table in (4.22). For example, we have
FK($\xrightarrow{\text{is sent by}}$)(Em1206) = Bob and FK($\xrightarrow{\text{is sent by}}$)(Em1207) = Carl.

d. There are two paths from $\ulcorner$a self-email$\urcorner$ to $\ulcorner$a person$\urcorner$ (namely, the sender and receiver of the email that the self-email is). We declare them equivalent.

e. Yes. One can check that the emails Em1207, Em1210, and Em1211 have the same sender as receiver (Carl, Chris, and Julia respectively).

◆

Example 4.5.3.3 (Monoid action table). In Example 4.1.2.9 we saw how a monoid $\mathcal{M}$ could be captured as an olog with only one object. As a database schema, this means there is only one table. Every generator of $\mathcal{M}$ would be a column of the table. The notion of database instance for such a schema (see Definition 4.5.3.1) matches perfectly with the notion of action table from Section 4.1.3. Note that a monoid can act on itself, in which case this action table is the monoid's multiplication table, as in Example 4.1.3.2, but it can also act on any other set, as in Example 4.1.3.1. If $\mathcal{M}$ acts on a set S, then the set of rows in the action table will be S.

Exercise 4.5.3.4.

Draw (as a graph) a schema for which table (4.1), page 133, looks like an instance. ◇

Solution 4.5.3.4.

All three columns have the same type of data, so we can guess that there is only one box in the ologs, i.e., one vertex in the graph.

$$b \, \overset{S}{\curvearrowleft} \, \bullet \, \overset{}{\curvearrowright} \, a$$

One connection we can make is that (4.1) has something to do with an action, i.e., it corresponds to a function $\Sigma \times S \to S$. Currying, we have a function $\Sigma \to \mathrm{Hom}(S, S)$. Indeed, we see $\Sigma = \{a, b\}$ in the preceding picture as the maps from S to itself. ◆

Exercise 4.5.3.5.

Suppose that $\mathcal{M}$ is a monoid and some instance of it is written in table form, e.g., as in table (4.1). It is possible that $\mathcal{M}$ is a group. What evidence in an instance table for $\mathcal{M}$ might suggest that $\mathcal{M}$ is a group? ◊

Solution 4.5.3.5.

If there are no repeats in any column of the action table, it suggests that $\mathcal{M}$ might be a group. Why? First note that each column of the action table corresponds to an element of $\mathcal{M}$. Suppose $\mathcal{M} = (M, e, \star)$ is a group. Then for any element $m \in M$, there is some m^{-1} such that $m \star m^{-1} = m^{-1} \star m = e$. But then on foreign keys (i.e., in the columns of the table), we have

$$\mathrm{FK}(m) \circ \mathrm{FK}(m^{-1}) = \mathrm{FK}(m^{-1}) \circ \mathrm{FK}(m) = \mathrm{id}.$$

In other words, $\mathrm{FK}(m)$ is a bijection. If we assume the table is finite, this is tantamount to saying that there are no repeats in column m. ◆

4.5.3.6 Paths through a database

Let $\mathcal{C} := (G, \simeq)$ be a schema, and let $(\mathrm{PK}, \mathrm{FK}) \colon \mathcal{C} \to \mathbf{Set}$ be an instance on $\mathcal{C}$. Then for every arrow $a \colon v \to w$ in G we get a function $\mathrm{FK}(a) \colon \mathrm{PK}(v) \to \mathrm{PK}(w)$. Functions can be composed, so in fact for every path through G we get a function. Namely, if $p = {}_{v_0}[a_1, a_2, \ldots, a_n]$ is a path from v_0 to v_n, then the instance provides a function

$$\mathrm{FK}(p) := \mathrm{FK}(a_n) \circ \cdots \mathrm{FK}(a_2) \circ \mathrm{FK}(a_1) \colon \mathrm{PK}(v_0) \to \mathrm{PK}(v_n),$$

which first made an appearance as part of Law 1 in Definition 4.5.3.1.

Example 4.5.3.7. Consider the department store schema from Example 4.5.2.1. More specifically consider the path Employee[worksIn, secretary, last] in (4.16), which points from Employee to LastNameString. The instance lets us interpret this path as a function from the set of employees to the set of last names; this could be a useful function to have in real-life office settings. The instance from (4.13) would yield the following function:

Employee	
ID	**Secr. name**
101	Hilbert
102	Russell
103	Hilbert

Exercise 4.5.3.8.

Consider the path $p := {}_s[f, f]$ on the $\mathcal{L}oop$ schema in (4.17). Using the instance from (4.18), where $\mathrm{PK}(s) = \{A, B, C, D, E, F, G, H\}$, interpret p as a function $\mathrm{PK}(s) \rightarrow \mathrm{PK}(s)$, and write this as a two-column table, as in Example 4.5.3.7.   ◊

Solution 4.5.3.8.

The instance from (4.18) on the left is shown for convenience; the solution to the exercise is on the right.

s	
ID	f
A	B
B	C
C	C
D	B
E	C
F	G
G	H
H	G

s	
ID	$p = f \circ f$
A	C
B	C
C	C
D	C
E	C
F	H
G	G
H	H

♦

Exercise 4.5.3.9.

Given an instance $(\mathrm{PK}, \mathrm{FK})$ on a schema $\mathcal{C}$, and given a trivial path p (i.e., p has length 0; it starts at some vertex but does not go anywhere), what function does p yield as $\mathrm{FK}(p)$?   ◊

Solution 4.5.3.9.

Let c be the domain (and codomain) of the trivial path p. Then $FK(p): PK(c) \to PK(c)$ is the identity function, $\mathrm{id}_{PK(c)}$. ♦

Chapter 5

Basic Category Theory

"...We know only a very few—and, therefore, very precious—schemes whose unifying powers cross many realms."—Marvin Minsky.[1]

Categories, or an equivalent notion, have already been introduced as ologs, or equivalently, as database schemas. One can think of a category as a graph (as in Section 4.3) in which certain paths have been declared congruent. (Ologs demand an extra requirement that everything be readable in natural language, and this cannot be part of the mathematical definition of category.) The formal definition of category is given in Definition 5.1.1.1, but it will not appear obvious that it is equivalent to the graph + congruence notion of schema, found in Definition 4.5.2.7. Once we know how different categories can be compared using functors (Definition 5.1.2.1), and how different schemas can be compared using schema mappings (Definition 5.4.1.2), we prove that the two notions are indeed equivalent (Theorem 5.4.2.3).

5.1   Categories and functors

This section gives the standard definition of categories and functors. These, together with natural transformations (Section 5.3), form the backbone of category theory. It also gives several examples.

---

[1] In *Society of Mind* [32].

5.1.1 Categories

In everyday speech we think of a category as a kind of thing. A category consists of a collection of things, all of which are related in some way. In mathematics a category can also be construed as a collection of things and a type of relationship between pairs of such things. For this kind of thing-relationship duo to count as a category, we need to check two rules, which have the following flavor: every thing must be related to itself by simply being itself, and if one thing is related to another and the second is related to a third, then the first is related to the third. In a category the things are called *objects* and the relationships are called *morphisms*.

So far we have discussed things of various sorts, e.g., sets, monoids, graphs. In each case we discussed how such things should be appropriately compared as homomorphisms. In each case the things stand as the objects and the appropriate comparisons stand as the morphisms in the category. Here is the definition.

Definition 5.1.1.1. A *category* $\mathcal{C}$ is defined as follows: One announces some constituents (A. objects, B. morphisms, C. identities, D. compositions) and shows that they conform to some laws (1. identity law, 2. associativity law). Specifically, one announces

- A. a collection $\mathrm{Ob}(\mathcal{C})$, elements of which are called *objects*;

- B. for every pair $x, y \in \mathrm{Ob}(\mathcal{C})$, a set $\mathrm{Hom}_{\mathcal{C}}(x, y) \in \mathbf{Set}$; it is called the *hom-set from x to y*; its elements are called *morphisms from x to y*;[2]

- C. for every object $x \in \mathrm{Ob}(\mathcal{C})$, a specified morphism, denoted $\mathrm{id}_x \in \mathrm{Hom}_{\mathcal{C}}(x, x)$, and called *the identity morphism on x*;

- D. for every three objects $x, y, z \in \mathrm{Ob}(\mathcal{C})$, a function

$$\circ \colon \mathrm{Hom}_{\mathcal{C}}(y, z) \times \mathrm{Hom}_{\mathcal{C}}(x, y) \to \mathrm{Hom}_{\mathcal{C}}(x, z),$$

  called *the composition formula*.

Given objects $x, y \in \mathrm{Ob}(\mathcal{C})$, we can denote a morphism $f \in \mathrm{Hom}_{\mathcal{C}}(x, y)$ by $f \colon x \to y$; we say that x is the *domain* of f and that y is the *codomain* of f. Given also $g \colon y \to z$, the composition formula is written using infix notation, so $g \circ f \colon x \to z$ means $\circ(g, f) \in \mathrm{Hom}_{\mathcal{C}}(x, z)$.

One must then show that the following *category laws* hold:

1. For every $x, y \in \mathrm{Ob}(\mathcal{C})$ and every morphism $f \colon x \to y$, we have

$$f \circ \mathrm{id}_x = f \qquad \text{and} \qquad \mathrm{id}_y \circ f = f.$$

---

[2]The reason for the notation Hom and the word *hom-set* is that morphisms are often called *homomorphisms*, e.g., in group theory.

2. If $w, x, y, z \in \mathrm{Ob}(\mathcal{C})$ are any objects, and $f \colon w \to x$, $g \colon x \to y$, and $h \colon y \to z$ are any morphisms, then the two ways to compose yield the same element in $\mathrm{Hom}_{\mathcal{C}}(w, z)$:

$$(h \circ g) \circ f = h \circ (g \circ f) \in \mathrm{Hom}_{\mathcal{C}}(w, z).$$

Remark 5.1.1.2. There is perhaps much that is unfamiliar about Definition 5.1.1.1, but there is also one thing that is strange about it. The objects $\mathrm{Ob}(\mathcal{C})$ of $\mathcal{C}$ are said to be a collection rather than a set. This is because we sometimes want to talk about the category of all sets, in which every possible set is an object, and if we try to say that the collection of sets is itself a set, we run into Russell's paradox. Modeling this was a sticking point in the foundations of category theory, but it was eventually fixed by Grothendieck's notion of expanding universes. Roughly, the idea is to choose some huge set κ (with certain properties making it a *universe*), to work entirely inside of it when possible, and to call anything in that world κ-*small* (or just *small* if κ is clear from context). When we need to look at κ itself, we choose an even bigger universe κ' and work entirely within it.

A category in which the collection $\mathrm{Ob}(\mathcal{C})$ is a set (or a small set) is called a *small category*. From here on I do not take note of the difference; I refer to $\mathrm{Ob}(\mathcal{C})$ as a set. I do not think this will do any harm to scientists using category theory, at least not in the beginning phases of their learning.

Example 5.1.1.3 (The category **Set** of sets). Chapters 2 and 3 were about the category of sets, denoted **Set**. The objects are the sets and the morphisms are the functions; and the current notation $\mathrm{Hom}_{\mathbf{Set}}(X, Y)$ was used to refer to the set of functions $X \to Y$. The composition formula $\circ$ is given by function composition, and for every set X, the identity function $\mathrm{id}_X \colon X \to X$ serves as the identity morphism for $X \in \mathrm{Ob}(\mathbf{Set})$. The two laws clearly hold, so **Set** is indeed a category.

Example 5.1.1.4 (The category **Fin** of finite sets). Inside the category **Set** is a *subcategory* **Fin** $\subseteq$ **Set**, called the *category of finite sets*. Whereas an object $S \in \mathrm{Ob}(\mathbf{Set})$ is a set that can have arbitrary cardinality, **Fin** is defined such that $\mathrm{Ob}(\mathbf{Fin})$ includes all (and only) those sets S having finitely many elements, i.e., $|S| = n$ for some natural number $n \in \mathbb{N}$. Every object of **Fin** is an object of **Set**, but not vice versa.

Although **Fin** and **Set** have different collections of objects, their notions of morphism are in some sense the same. For any two finite sets $S, S' \in \mathrm{Ob}(\mathbf{Fin})$, we can also think of $S, S' \in \mathrm{Ob}(\mathbf{Set})$, and we have

$$\mathrm{Hom}_{\mathbf{Fin}}(S, S') = \mathrm{Hom}_{\mathbf{Set}}(S, S').$$

That is, a morphism in **Fin** between finite sets S and S' is simply a function $f \colon S \to S'$.

Example 5.1.1.5 (The category **Mon** of monoids). Monoids were defined in Definition 4.1.1.1, and monoid homomorphisms in Definition 4.1.4.1. Every monoid $\mathcal{M} := (M, e, \star_M)$

has an identity homomorphism $\text{id}_{\mathcal{M}} \colon \mathcal{M} \to \mathcal{M}$, given by the identity function $\text{id}_M \colon M \to M$. To compose two monoid homomorphisms $f \colon \mathcal{M} \to \mathcal{M}'$ and $g \colon \mathcal{M}' \to \mathcal{M}''$, we compose their underlying functions $f \colon M \to M'$ and $g \colon M' \to M''$, and check that the result $g \circ f$ is a monoid homomorphism. Indeed,

$$g \circ f(e) = g(e') = e'',$$

$$g \circ f(m_1 \star_M m_2) = g(f(m_1) \star_{M'} f(m_2)) = g \circ f(m_1) \star_{M''} g \circ f(m_2).$$

It is clear that the two category laws (unit and associativity) hold, because monoid morphisms are special kinds of functions, and functions compose unitally and associatively. So **Mon** is a category.

Remark 5.1.1.6. The following will be informal, but it can be formalized. Let's define a *questionable category* to be the specification of A, B, C, D from Definition 5.1.1.1, without enforcing either of the category laws (1, 2). Suppose that $\mathcal{Q}$ is a questionable category and $\mathcal{C}$ is a category. If $\mathcal{Q}$ sits somehow inside of $\mathcal{C}$, in the precise sense that

   A. there is a function $U \colon \text{Ob}(\mathcal{Q}) \to \text{Ob}(\mathcal{C})$,

   B. for all $a, b \in \text{Ob}(\mathcal{Q})$, we have an injection $U \colon \text{Hom}_{\mathcal{Q}}(a, b) \hookrightarrow \text{Hom}_{\mathcal{C}}(U(a), U(b))$,

   C. for all $a \in \text{Ob}(\mathcal{Q})$, both $\mathcal{Q}$ and $\mathcal{C}$ have the same version of the identity on a, i.e., $U(\text{id}_a) = \text{id}_{U(a)}$,

   D. for all $f \colon a \to b$ and $g \colon b \to c$ in $\mathcal{Q}$, both $\mathcal{Q}$ and $\mathcal{C}$ have the same version of composition $g \circ f$, i.e., $U(g \circ f) = U(g) \circ U(f)$,

then $\mathcal{Q}$ is a category (no longer questionable).

   This fact was used in Example 5.1.1.5 for **Mon** $\subseteq$ **Set**.

Exercise 5.1.1.7.

Suppose we set out to define a category **Grp**, having groups as objects and group homomorphisms as morphisms (see Definition 4.2.1.16). Show that the rest of the conditions for **Grp** to be a category are satisfied.   ◊

Solution 5.1.1.7.

Groups were defined in Definition 4.2.1.1 and group homomorphisms in Definition 4.2.1.16. Every group $\mathcal{G} := (G, e, \star_G)$ has an identity homomorphism $\text{id}_{\mathcal{G}} \colon \mathcal{G} \to \mathcal{G}$, given by the identity function $\text{id}_G \colon G \to G$. To compose two group homomorphisms $f \colon \mathcal{M} \to \mathcal{M}'$ and $g \colon \mathcal{M}' \to \mathcal{M}''$, we compose their underlying functions $f \colon M \to M'$ and $g \colon M' \to M''$, and check that the result $g \circ f$ is a group homomorphism. Because group homomorphisms are just monoid homomorphisms, we can apply Remark 5.1.1.6, so **Grp** is indeed a category.   ♦

Exercise 5.1.1.8.

Suppose we set out to define a category **PrO**, having preorders as objects and preorder homomorphisms as morphisms (see Definition 4.4.4.1). Show (to the level of detail of Example 5.1.1.5) that the rest of the conditions for **PrO** to be a category are satisfied. ◊

Solution 5.1.1.8.

Preorders were defined in Definition 4.4.1.1 and morphisms of preorders in Definition 4.4.4.1. Let $\mathcal{P} := (S, R)$ denote a preorder with underlying set S and relation $R \subseteq S \times S$. There is an identity morphism $\mathrm{id}_{\mathcal{P}} \colon \mathcal{P} \to \mathcal{P}$, given by the identity function $\mathrm{id}_S \colon S \to S$. To compose two preorder morphisms $f \colon \mathcal{P} \to \mathcal{P}'$ and $g \colon \mathcal{P}' \to \mathcal{P}''$, we compose their underlying functions $f \colon S \to S'$ and $g \colon S' \to S''$, and check that the result $g \circ f$ is a preorder morphism. For $x \leqslant y$ in S, we have $f(x) \leqslant f(y)$, so $g(f(x)) \leqslant g(f(y))$, proving that $g \circ f \colon S \to S'$ preserves the order. So we have the composition formula. The fact that this composition formula satisfies the category laws follows from Remark 5.1.1.6. Thus **PrO** is a category.   ♦

Example 5.1.1.9 (Noncategory 1). What is not a category? Two things can go wrong: either one fails to specify all the relevant constituents (A, B, C, D from Definition 5.1.1.1), or the constituents do not obey the category laws (1, 2).

Let G be the following graph:

$$G = \boxed{\begin{array}{ccccc} a & \xrightarrow{\,f\,} & b & \xrightarrow{\,g\,} & c \\ \bullet & & \bullet & & \bullet \end{array}}$$

Suppose we try to define a category $\mathcal{G}$ by faithfully recording vertices as objects and arrows as morphisms. Will that be a category?

Following that scheme, we put $\mathrm{Ob}(\mathcal{G}) = \{a, b, c\}$. For all nine pairs of objects we need a hom-set. Since the only things we are calling morphisms are the arrows of G, we put

$$\begin{array}{lll} \mathrm{Hom}_{\mathcal{G}}(a, a) = \varnothing & \mathrm{Hom}_{\mathcal{G}}(a, b) = \{f\} & \mathrm{Hom}_{\mathcal{G}}(a, c) = \varnothing \\ \mathrm{Hom}_{\mathcal{G}}(b, a) = \varnothing & \mathrm{Hom}_{\mathcal{G}}(b, b) = \varnothing & \mathrm{Hom}_{\mathcal{G}}(b, c) = \{g\} \\ \mathrm{Hom}_{\mathcal{G}}(c, a) = \varnothing & \mathrm{Hom}_{\mathcal{G}}(c, b) = \varnothing & \mathrm{Hom}_{\mathcal{G}}(c, c) = \varnothing \end{array} \qquad (5.1^*)$$

If we say we are done, the listener should object that we have given neither identities (C) nor a composition formula (D), and these are necessary constituents. Now we are at a loss: it is impossible to give identities under this scheme, because, e.g., $\mathrm{Hom}_{\mathcal{G}}(a, a) = \varnothing$. So what we have for $\mathcal{G}$ is not a category.

Suppose we fix that problem, adding an element to each of the diagonals so that

$$\mathrm{Hom}_{\mathcal{G}}(a, a) = \{\mathrm{id}_a\}, \qquad \mathrm{Hom}_{\mathcal{G}}(b, b) = \{\mathrm{id}_b\}, \qquad \text{and} \qquad \mathrm{Hom}_{\mathcal{G}}(c, c) = \{\mathrm{id}_c\}.$$

But the listener still demands a composition formula. In particular, we need a function

$$\text{Hom}_{\mathcal{G}}(b, c) \times \text{Hom}_{\mathcal{G}}(a, b) \to \text{Hom}_{\mathcal{G}}(a, c),$$

but the domain is nonempty (it is $\{(f, g)\}$) and the codomain $\text{Hom}_{\mathcal{G}}(a, c) = \varnothing$ is empty; there is no such function. In other words, to satisfy the listener we need to add a composite for the arrows f and g.

So again we must make a change, adding an element to make $\text{Hom}_{\mathcal{G}}(a, c) = \{h\}$. We can now say $g \circ f = h$. Finally, this does the trick and we have a category with the following morphisms:

$$
\begin{array}{lll}
\text{Hom}_{\mathcal{G}}(a, a) = \{\text{id}_a\} & \text{Hom}_{\mathcal{G}}(a, b) = \{f\} & \text{Hom}_{\mathcal{G}}(a, c) = \{h\} \\
\text{Hom}_{\mathcal{G}}(b, a) = \varnothing & \text{Hom}_{\mathcal{G}}(b, b) = \{\text{id}_b\} & \text{Hom}_{\mathcal{G}}(b, c) = \{g\} \\
\text{Hom}_{\mathcal{G}}(c, a) = \varnothing & \text{Hom}_{\mathcal{G}}(c, b) = \varnothing & \text{Hom}_{\mathcal{G}}(c, c) = \{\text{id}_c\}
\end{array}
$$

A computer could check this quickly, as can someone with good intuition for categories; for everyone else, it may be a painstaking process involving determining whether there is a unique composition formula for each of the 27 pairs of hom-sets and whether the associative law holds in the 81 necessary cases. Luckily this computation is sparse (lots of $\varnothing$'s).

If all the morphisms are drawn as arrows, the graph becomes:

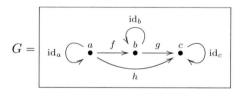

Example 5.1.1.10 (Noncategory 2). In this example, we make a faux category $\mathcal{F}$ with one object and many morphisms. The problem here is the composition formula.

Define $\mathcal{F}$ to have one object $\text{Ob}(\mathcal{F}) = \{\odot\}$, and $\text{Hom}_{\mathcal{F}}(\odot, \odot) = \mathbb{N}$. Define $\text{id}_{\odot} = 1 \in \mathbb{N}$. Define the composition formula $\circ \colon \mathbb{N} \times \mathbb{N} \to \mathbb{N}$ by the usual exponentiation function for natural numbers, $m \circ n = m^n$. This is a perfectly cromulent function, but it does not work right as a composition formula. Indeed, for the identity law to hold, we would need $m^1 = m = 1^m$, and one side of this is false. For the associativity law to hold, we would need $(m^n)^p = m^{(n^p)}$, but this is also not the case.

To fix this problem we must completely revamp the composition formula. It would work to use multiplication, $m \circ n = m * n$. Then the identity law would read $1 * m = m = m * 1$, and that holds; and the associativity law would read $(m * n) * p = m * (n * p)$, and that holds.

Example 5.1.1.11 (The category of preorders with joins). Suppose we are only interested in preorders $(X, \leqslant)$ for which every pair of elements has a join. We saw in Exercise 4.4.2.3 that not all preorders have this property. However, we can create a category $\mathcal{C}$ in which every object does have this property. To begin, let's put

$$C := \{(X, \leqslant) \in \mathrm{Ob}(\mathbf{PrO}) \mid (X, \leqslant) \text{ has all joins}\}$$

for the set of objects. What about morphisms?

One option would be to put in no morphisms (other than identities) and to just consider this collection of objects as having no structure other than a set. In other words, we can take $\mathcal{C}$ to be the discrete category on the preceding set $\mathrm{Ob}(\mathcal{C}) = C$.

Another option, say, $\mathcal{C}'$ with objects $\mathrm{Ob}(\mathcal{C}') := C$, would be to put in exactly the same morphisms as in $\mathbf{PrO}$: for any objects $a, b \in C$, we consider a and b as ordinary preorders and put $\mathrm{Hom}_{\mathcal{C}'}(a, b) := \mathrm{Hom}_{\mathbf{PrO}}(a, b)$. The resulting category $\mathcal{C}'$ of preorders with joins is called the *full subcategory of* $\mathbf{PrO}$ *spanned by the preorders with joins*.[3]

A third option, say, $\mathcal{C}''$ with objects $\mathrm{Ob}(\mathcal{C}'') := C$, would stand out to a category theorist. That is, the conscientious modeler takes the choice about how we define objects as a clue to how we should define morphisms.

Slogan 5.1.1.12.

> *If you like joins so much, why don't you marry them?*

Morphisms are often billed as preserving all the structure we care about, so it is worth asking whether we want to enforce that constraint on morphisms. That is, suppose $f \colon (X, \leqslant_X) \to (Y, \leqslant_Y)$ is a morphism of preorders. We might want to condition the decision of whether to include f as a morphism in $\mathcal{C}''$ on whether, for any join $w = x \vee x'$ in X, it is the case that $f(w) = f(x) \vee f(x')$ in Y. Concisely, we could define the morphisms in $\mathcal{C}''$ by

$$\mathrm{Hom}_{\mathcal{C}}(a, b) := \{f \in \mathrm{Hom}_{\mathbf{PrO}}(a, b) \mid f \text{ preserves joins}\}.$$

One can check easily that the identity morphisms preserve joins and that compositions of join-preserving morphisms are join-preserving, so this version of homomorphisms makes $\mathcal{C}''$ a well defined category.

These options are by no means comprehensive, and none of these options is better than any other. Which category to use is decided by whatever fits the situation being modeled.

---

[3]Full subcategory will be defined in Definition 6.2.3.1.

Example 5.1.1.13 (Category **FLin** of finite linear orders). We have a category **PrO** of preorders, and some of its objects are finite linear orders. Let **FLin** be the full subcategory of **PrO** spanned by the linear orders. That is, following Definition 4.4.4.1, given linear orders $X, Y \in \text{Ob}(\textbf{FLin})$, every morphism of preorders $X \to Y$ counts as a morphism in **FLin**:

$$\text{Hom}_{\textbf{FLin}}(X, Y) = \text{Hom}_{\textbf{PrO}}(X, Y).$$

Exercise 5.1.1.14.

Let **FLin** be the category of finite linear orders, defined in Example 5.1.1.13. For $n \in \mathbb{N}$, let $[n]$ be the linear order defined in Example 4.4.1.7. What are the cardinalities of the following sets?

a. $\text{Hom}_{\textbf{FLin}}([0], [3])$

b. $\text{Hom}_{\textbf{FLin}}([3], [0])$

c. $\text{Hom}_{\textbf{FLin}}([2], [3])$

d. $\text{Hom}_{\textbf{FLin}}([1], [n])$

e. (Challenge) $\text{Hom}_{\textbf{FLin}}([m], [n])$

It turns out that the category **FLin** of linear orders is sufficiently rich that much of algebraic topology (the study of arbitrary spaces, such as Mobius strips and seven-dimensional spheres) can be understood in its terms. See Example 6.2.1.7.   ◊

Solution 5.1.1.14.

a. $|\text{Hom}_{\textbf{FLin}}([0], [3])| = 4$.

b. $|\text{Hom}_{\textbf{FLin}}([3], [0])| = 1$.

c. $|\text{Hom}_{\textbf{FLin}}([2], [3])| = 20$. Finding a morphism $[2] \to [3]$ of linear orders is the same thing as writing a nondecreasing sequence of three numbers between 0 and 3:

0	0	0		1	1	1
0	0	1		1	1	2
0	0	2		1	1	3
0	0	3		1	2	2
0	1	1		1	2	3
0	1	2		1	3	3
0	1	3		2	2	2
0	2	2		2	2	3
0	2	3		2	3	3
0	3	3		3	3	3

d. $|\text{Hom}_{\textbf{FLin}}([1],[n])| = \frac{(n+1)(n+2)}{2}$, which may be reminiscent of triangle numbers. Perhaps the following arrangement of morphisms $[1] \to [n]$ (for $\leqslant n \leqslant 3$) will help:

$$(3,3)$$

$$(2,2) \quad (3,2)$$

$$(1,1) \quad (2,1) \quad (3,1)$$

$$(0,0) \quad (1,0) \quad (2,0) \quad (3,0)$$

e. These are generalized triangle numbers:

$$|\text{Hom}_{\textbf{FLin}}([m],[n])| = \frac{(n+1)(n+2)\cdots(n+m+1)}{(m+1)!} = \binom{n+m+1}{n}.$$

In the following table, each row is the running sum of the row above. For convenience, let $[-1]$ denote the empty linear order. Then is a table for $|\text{Hom}_{\textbf{FLin}}([m],[n])|$ would be:

n \ m	$[-1]$	$[0]$	$[1]$	$[2]$	$[3]$
$[-1]$	1	0	0	0	0
$[0]$	1	1	1	1	1
$[1]$	1	2	3	4	5
$[2]$	1	3	6	10	15
$[3]$	1	4	10	20	35
$[4]$	1	5	15	35	70

♦

Example 5.1.1.15 (Category of graphs). Graphs were defined in Definition 4.3.1.1 and graph homomorphisms in Definition 4.3.3.1. To see that these are sufficient to form a category is considered routine to a seasoned category theorist, so let's see why.

Since a morphism from $\mathcal{G} = (V, A, src, tgt)$ to $\mathcal{G}' = (V', A', src', tgt')$ involves two functions $f_0 \colon V \to V'$ and $f_1 \colon A \to A'$, the identity and composition formulas simply arise from the identity and composition formulas for sets. Associativity follow similarly. The only thing that needs to be checked is that the composition of two such morphisms, each satisfying (4.6), will itself satisfy (4.6). For completeness, we check that now.

Suppose that $f = (f_0, f_1) \colon \mathcal{G} \to \mathcal{G}'$ and $g = (g_0, g_1) \colon \mathcal{G}' \to \mathcal{G}''$ are graph homomorphisms, where $\mathcal{G}'' = (V'', A'', src'', tgt'')$. Then in each diagram in (5.2)

$$
\begin{array}{ccc}
A & \xrightarrow{f_1} & A' & \xrightarrow{g_1} & A'' \\
\downarrow{\scriptstyle src} & & \downarrow{\scriptstyle src'} & & \downarrow{\scriptstyle src''} \\
V & \xrightarrow{f_0} & V' & \xrightarrow{g_0} & V''
\end{array}
\qquad
\begin{array}{ccc}
A & \xrightarrow{f_1} & A' & \xrightarrow{g_1} & A'' \\
\downarrow{\scriptstyle tgt} & & \downarrow{\scriptstyle tgt'} & & \downarrow{\scriptstyle tgt''} \\
V & \xrightarrow{f_0} & V' & \xrightarrow{g_0} & V''
\end{array}
\tag{5.2}
$$

the left-hand square commutes because f is a graph homomorphism and the right-hand square commutes because g is a graph homomorphism. Thus the whole rectangle commutes, meaning that $g \circ f$ is a graph homomorphism, as desired.

We denote the category of graphs and graph homomorphisms **Grph**.

Remark 5.1.1.16. When one is struggling to understand basic definitions, notation, and style, a phase that naturally occurs when learning new mathematics (or any new language), the preceding example will probably appear long and tiring. I would say the reader has mastered the basics when the example seems straightforward. Around this time, I hope the reader will get a sense of the remarkable organizational potential of the categorical way of thinking.

Exercise 5.1.1.17.

Let F be a vector field defined on all of $\mathbb{R}^2$. Recall that for two points $x, x' \in \mathbb{R}^2$, any curve C with endpoints x and x', and any parameterization $r \colon [a, b] \to C$, the line integral $\int_C F(r) \cdot dr$ returns a real number. It does not depend on r, except its orientation (direction). Therefore, if we think of C has having an orientation, say, going from x to x', then $\int_C F$ is a well defined real number. If C goes from x to x', let's write $C \colon x \to x'$. Define an equivalence relation $\sim$ on the set of oriented curves in $\mathbb{R}^2$ by saying $C \sim C'$ if

- C and C' start at the same point;

- C and C' end at the same point;

- $\int_C F = \int_{C'} F$.

Suppose we try to make a category $\mathcal{C}_F$ as follows. Put $\mathrm{Ob}(\mathcal{C}_F) = \mathbb{R}^2$, and for every pair of points $x, x' \in \mathbb{R}^2$, let $\mathrm{Hom}_{\mathcal{C}_F}(x, x') = \{C \colon x \to x'\}/\sim$, where $C \colon x \to x'$ is an oriented curve and $\sim$ means "same line integral," as explained.

Is there an identity morphism and a composition formula that will make $\mathcal{C}_F$ into a category?   ◊

Solution 5.1.1.17.

Yes. For every object $x \in \mathbb{R}^2$, the constant curve at x serves as the identity on x. If $C: x \to y$ and $C': y \to z$ are curves, their composition is given by joining them to get a curve $x \to z$. ♦

5.1.1.18 Isomorphisms

In any category we have a notion of isomorphism between objects.

Definition 5.1.1.19. Let $\mathcal{C}$ be a category, and let $X, Y \in \mathrm{Ob}(\mathcal{C})$ be objects. An *isomorphism f from X to Y* is a morphism $f: X \to Y$ in $\mathcal{C}$ such that there exists a morphism $g: Y \to X$ in $\mathcal{C}$ with
$$g \circ f = \mathrm{id}_X \qquad \text{and} \qquad f \circ g = \mathrm{id}_Y.$$

In this case we say that the morphism f is *invertible* and that g is the *inverse* of f. We may also say that the objects X and Y are *isomorphic*.

Example 5.1.1.20. If $\mathcal{C} = \mathbf{Set}$ is the category of sets, then Definition 5.1.1.19 coincides precisely with the one given in Definition 2.1.2.14.

Exercise 5.1.1.21.

Let $\mathcal{C}$ be a category, and let $c \in \mathrm{Ob}(\mathcal{C})$ be an object. Show that id_c is an isomorphism. ◇

Solution 5.1.1.21.

We have a morphism $\mathrm{id}_c: c \to c$. To show it is an isomorphism we just need to find a morphism $f: c \to c$ such that $f \circ \mathrm{id}_c = \mathrm{id}_c$ and $\mathrm{id}_c \circ f = \mathrm{id}_c$. Taking $f = \mathrm{id}_c$ works. ♦

Exercise 5.1.1.22.

Let $\mathcal{C}$ be a category, and let $f: X \to Y$ be a morphism. Suppose that both $g: Y \to X$ and $g': Y \to X$ are inverses of f. Show that they are the same morphism, $g = g'$. ◇

Solution 5.1.1.22.

By definition, we have $g \circ f = \mathrm{id}_X$ and $f \circ g' = \mathrm{id}_Y$. We apply some category laws to $g \circ f \circ g'$ to obtain the result:
$$g = g \circ \mathrm{id}_Y = g \circ (f \circ g') = (g \circ f) \circ g' = \mathrm{id}_X \circ g' = g'.$$

♦

Exercise 5.1.1.23.

Suppose that $G = (V, A, src, tgt)$ and $G' = (V', A', src', tgt')$ are graphs and that $f = (f_0, f_1) \colon G \to G'$ is a graph homomorphism (as in Definition 4.3.3.1).

a. If f is an isomorphism in **Grph**, does this imply that $f_0 \colon V \to V'$ and $f_1 \colon A \to A'$ are isomorphisms in **Set**?

b. If so, why; if not, show a counterexample (where f is an isomorphism but either f_0 or f_1 is not).

$\Diamond$

Solution 5.1.1.23.

a. Yes.

b. If f is an isomorphism in **Grph**, then there is a graph homomorphism $g \colon G' \to G$ such that $g \circ f = \mathrm{id}_G$ and $f \circ g = \mathrm{id}_{G'}$. So we have the following diagrams:

$$
\begin{array}{ccccc}
A & \xrightarrow{f_1} & A' & \xrightarrow{g_1} & A \\
\scriptstyle src \downdownarrows \scriptstyle tgt & & \downdownarrows & & \downdownarrows \\
V & \xrightarrow{f_0} & V' & \xrightarrow{g_0} & V
\end{array}
\qquad\qquad
\begin{array}{ccccc}
A' & \xrightarrow{g_1} & A & \xrightarrow{f_1} & A' \\
\scriptstyle src' \downdownarrows \scriptstyle tgt' & & \downdownarrows & & \downdownarrows \\
V' & \xrightarrow{g_0} & V & \xrightarrow{f_0} & V'
\end{array}
$$

Because f and g are mutually inverse, their composite is assumed to be the identity morphism $\mathrm{id}_G \colon G \to G$, which by definition means that $g_1 \circ f_1 = \mathrm{id}_A$ and $g_0 \circ f_0 = \mathrm{id}_V$. Similarly, the other composite $f \circ g$ is identity on G' so $f_1 \circ g_1 = \mathrm{id}_{A'}$ and $f_0 \circ g_0 = \mathrm{id}_{V'}$. All together, these facts imply that f_1 and g_1 are mutually inverse functions, i.e., isomorphisms, and so are f_0 and g_0.

♦

Exercise 5.1.1.24.

Suppose that $G = (V, A, src, tgt)$ and $G' = (V', A', src', tgt')$ are graphs and that $f = (f_0, f_1) \colon G \to G'$ is a graph homomorphism (as in Definition 4.3.3.1).

a. If $f_0 \colon V \to V'$ and $f_1 \colon A \to A'$ are isomorphisms in **Set**, does this imply that f is an isomorphism in **Grph**?

b. If so, why; if not, show a counterexample (where f_0 and f_1 are isomorphisms but f is not).

$\Diamond$

Solution 5.1.1.24.

a. Yes.

b. Let $g_1 \colon A' \to A$ be the inverse of f_1, and let $g_0 \colon V' \to V$ be the inverse of f_0. We only need to check that (g_0, g_1) is an honest graph homomorphism, i.e., that the diagrams

commute. We use the following facts:

- $g_0 \circ f_0 = \mathrm{id}_V$.
- $g_1 \circ f_1 = \mathrm{id}_A$.
- $src' \circ f_1 = f_0 \circ src$.
- $tgt' \circ f_1 = f_0 \circ tgt$.

Now we write out the proof that the two diagrams above commute:

$$src \circ g_1 = g_0 \circ f_0 \circ src \circ g_1 = g_0 \circ src' \circ f_1 \circ g_1 = g_0 \circ src';$$
$$tgt \circ g_1 = g_0 \circ f_0 \circ tgt \circ g_1 = g_0 \circ tgt' \circ f_1 \circ g_1 = g_0 \circ tgt'.$$

$\blacklozenge$

Proposition 5.1.1.25. *Let $\mathcal{C}$ be a category, and let $\cong$ be the relation on $\mathrm{Ob}(\mathcal{C})$ given by saying $X \cong Y$ iff X and Y are isomorphic. Then $\cong$ is an equivalence relation.*

Proof. The proof of Proposition 2.1.2.18 can be mimicked in this more general setting.

$\square$

5.1.1.26 Another viewpoint on categories

Here is an alternative definition of category, using the work done in Chapter 2.

Exercise 5.1.1.27.

Suppose we begin our definition of category as follows.

A *category* $\mathcal{C}$ consists of a sequence $(\mathrm{Ob}(\mathcal{C}), \mathrm{Hom}_\mathcal{C}, dom, cod, ids, comp)$, where

- $\text{Ob}(\mathcal{C})$ is a set;[4]

- $\text{Hom}_{\mathcal{C}}$ is a set, and $dom, cod\colon \text{Hom}_{\mathcal{C}} \to \text{Ob}(\mathcal{C})$ are functions;

- $ids\colon \text{Ob}(\mathcal{C}) \to \text{Hom}_{\mathcal{C}}$ is a function;

- *comp* is a function as depicted in the commutative diagram (5.3)

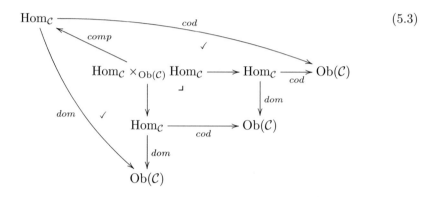

(5.3)

a. Add to diagram (5.3) to express the fact that for any $x \in \text{Ob}(\mathcal{C})$, the morphism id_x points from x to x.

b. Express the condition that composing a morphism f with an appropriate identity morphism yields f.

◇

Solution 5.1.1.27.

a. This is expressed by the equations: $dom \circ ids = \text{id}_{\text{Ob}(\mathcal{C})}$ and $cod \circ ids = \text{id}_{\text{Ob}(\mathcal{C})}$. One could express this with the diagram:

$$
\begin{array}{ccc}
\text{Hom}_{\mathcal{C}} \times_{\text{Ob}(\mathcal{C})} \text{Hom}_{\mathcal{C}} & \longrightarrow & \text{Hom}_{\mathcal{C}} \\
\downarrow & \lrcorner & \;\;\downarrow^{dom} \;\; \big) \, ids \\
\text{Hom}_{\mathcal{C}} & \xrightarrow{\;cod\;} & \text{Ob}(\mathcal{C}) \\
& \xleftarrow{\;ids\;} &
\end{array}
$$

---

[4]See Remark 5.1.1.2.

b. We have $\mathrm{id}_{\mathrm{Hom}_{\mathcal{C}}} \colon \mathrm{Hom}_{\mathcal{C}} \to \mathrm{Hom}_{\mathcal{C}}$ and $ids \circ cod \colon \mathrm{Hom}_{\mathcal{C}} \to \mathrm{Hom}_{\mathcal{C}}$, and these commute over $\mathrm{Ob}(\mathcal{C})$, meaning that for any morphism $f \colon A \to B$, its codomain is the domain of id_B. Thus a unique map

$$\langle \mathrm{id}_{\mathrm{Hom}_{\mathcal{C}}}, ids \circ cod \rangle_{\mathrm{Ob}(\mathcal{C})} \colon \mathrm{Hom}_{\mathcal{C}} \to \mathrm{Hom}_{\mathcal{C}} \times_{\mathrm{Ob}(\mathcal{C})} \mathrm{Hom}_{\mathcal{C}}$$

is induced (see Proposition 3.2.1.15). Similarly there is a function

$$\langle \mathrm{id}_{ids \circ dom} \mathrm{Hom}_{\mathcal{C}}, \mathrm{Ob}(\mathcal{C}) \rangle \colon \mathrm{Hom}_{\mathcal{C}} \to \mathrm{Hom}_{\mathcal{C}} \times_{\mathrm{Ob}(\mathcal{C})} \mathrm{Hom}_{\mathcal{C}}.$$

When we compose either of these morphisms with *comp*, we are taking the composition of a morphism and the identity (either on the domain or the codomain). Thus, the fact that composing any morphism with an identity morphism returns that morphism is expressed by asserting two path equivalences,

$$\begin{aligned} {}_{\mathrm{Hom}_{\mathcal{C}}}[\langle \mathrm{id}_{\mathrm{Hom}_{\mathcal{C}}}, ids \circ cod \rangle, comp] &\simeq {}_{\mathrm{Hom}_{\mathcal{C}}}[\,], \\ {}_{\mathrm{Hom}_{\mathcal{C}}}[\langle ids \circ dom, \mathrm{id}_{\mathrm{Hom}_{\mathcal{C}}} \rangle, comp] &\simeq {}_{\mathrm{Hom}_{\mathcal{C}}}[\,], \end{aligned}$$

in the following diagram:

Example 5.1.1.28 (Partial olog for a category). Diagram (5.4) is an olog that captures some of the essential structures of a category:

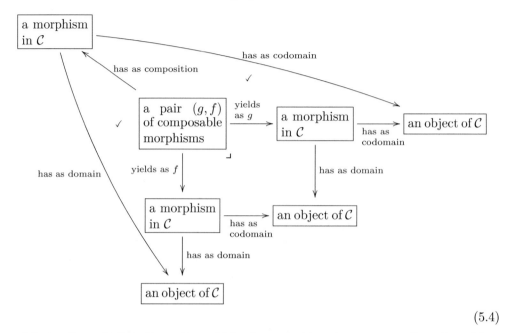

$$(5.4)$$

Missing from (5.4) is the notion of identity morphism (as an arrow from ⌜an object of $\mathcal{C}$⌝ to ⌜a morphism in $\mathcal{C}$⌝) and the associated path equivalences, as well as the identity and associativity laws. All of these can be added to the olog, at the expense of some clutter.

Remark 5.1.1.29. Perhaps it is already clear that category theory is very interconnected. It may feel like everything relates to everything, and this feeling may intensify as you go on. However, the relationships between different notions are rigorously defined, not random. Moreover, almost everything presented in this book can be formalized in a proof system like Coq (the most obvious exceptions being things like the readability requirement of ologs and the modeling of scientific applications).

Whenever you feel cognitive vertigo, use the interplay between examples and formal definitions to solidify your understanding. Go through each example, making sure it conforms to the definitions or theorems it purports to exemplify.

5.1.2   Functors

A category $\mathcal{C} = (\mathrm{Ob}(\mathcal{C}), \mathrm{Hom}_{\mathcal{C}}, dom, cod, ids, comp)$, involves a set of objects, a set of morphisms, a notion of domains and codomains, a notion of identity morphisms, and a composition formula. For two categories to be comparable, these various components should be appropriately comparable.

Definition 5.1.2.1. Let $\mathcal{C}$ and $\mathcal{C}'$ be categories. A *functor F from $\mathcal{C}$ to $\mathcal{C}'$*, denoted $F\colon \mathcal{C} \to \mathcal{C}'$, is defined as follows: One announces some constituents (A. on-objects part, B. on-morphisms part) and shows that they conform to some laws (1. preservation of identities, 2. preservation of composition). Specifically, one announces

A. a function $\mathrm{Ob}(F)\colon \mathrm{Ob}(\mathcal{C}) \to \mathrm{Ob}(\mathcal{C}')$, sometimes denoted simply $F\colon \mathrm{Ob}(\mathcal{C}) \to \mathrm{Ob}(\mathcal{C}')$;

B. for every pair of objects $c, d \in \mathrm{Ob}(\mathcal{C})$, a function

$$\mathrm{Hom}_F(c,d)\colon \mathrm{Hom}_{\mathcal{C}}(c,d) \to \mathrm{Hom}_{\mathcal{C}'}(F(c), F(d)),$$

sometimes denoted simply $F\colon \mathrm{Hom}_{\mathcal{C}}(c,d) \to \mathrm{Hom}_{\mathcal{C}'}(F(c), F(d))$.

One must then show that the following *functor laws* hold:

1. Identities are preserved by F, that is, for any object $c \in \mathrm{Ob}(\mathcal{C})$, we have $F(\mathrm{id}_c) = \mathrm{id}_{F(c)}$.

2. Composition is preserved by F, that is, for any objects $b, c, d \in \mathrm{Ob}(\mathcal{C})$ and morphisms $g\colon b \to c$ and $h\colon c \to d$, we have $F(h \circ g) = F(h) \circ F(g)$.

Example 5.1.2.2 (Monoids have underlying sets). Recall from Definition 4.1.1.1 that if $\mathcal{M} = (M, e, \star)$ is a monoid, then M is a set. And recall from Definition 4.1.4.1 that if $f\colon \mathcal{M} \to \mathcal{M}'$ is a monoid homomorphism, then $f\colon M \to M'$ is a function. Thus we can define a functor

$$U\colon \mathbf{Mon} \to \mathbf{Set}$$

The on-objects part of U sends every monoid to its underlying set, $U(\mathcal{M}) = M$, and sends every monoid homomorphism to its underlying function $U(f) = f$. It is easy to check that the functor laws hold, so U is indeed a functor.

Given two monoids $\mathcal{M} = (M, e, \star)$ and $\mathcal{M}' = (M', e', \star')$, there may be many functions from M to M' that do not arise from monoid homomorphisms. In other words, $U\colon \mathrm{Hom}_{\mathbf{Mon}}(\mathcal{M}, \mathcal{M}') \to \mathrm{Hom}_{\mathbf{Set}}(M, M')$ may not be surjective. It is often useful to speak of such functions. For example, one could assign to every command in one video game V a command in another video game V', but this may not work in accordance with the monoid laws when performing a sequence of commands. By being able to speak of M as a set or of $\mathcal{M}$ as a monoid, and understanding the relationship U between them, we can be clear about where we stand at all times in the discussion.

Example 5.1.2.3 (Groups have underlying monoids). Recall that a group is just a monoid $(M, e, \star)$ with the extra property that every element $m \in M$ has an inverse $m' \star m = e = m \star m'$. Thus to every group we can assign its *underlying monoid*. Similarly, a group homomorphism is just a monoid homomorphism of its underlying monoids. This means that there is a functor

$$U \colon \mathbf{Grp} \to \mathbf{Mon}$$

that sends every group or group homomorphism to its underlying monoid or monoid homomorphism. Identity and composition are preserved.

Application 5.1.2.4. Suppose you are a scientist working with symmetries. But then suppose that the symmetry breaks somewhere, or you add some extra observable that is not reversible under the symmetry. You want to seamlessly relax the requirement that every action be reversible without changing anything else. You want to know how you can proceed, or what is allowed. The answer is to simply pass from the category of groups (or group actions) to the category of monoids (or monoid actions).

We can also reverse this change of perspective. Recall that Example 4.1.2.9 discussed a monoid M controlling the actions of a video game character. The character position (P) could be moved up (u), moved down (d), or moved right (r). The path equivalences $P.u.d = P$ and $P.d.u = P$ imply that these two actions are mutually inverse, whereas moving right has no inverse. This, plus equivalences $P.r.u = P.u.r$ and $P.r.d = P.d.r$, defined a monoid M.

Inside M is a submonoid G, which includes just upward and downward movement. It has one object, just like M, i.e., $\mathrm{Ob}(M) = \{P\} = \mathrm{Ob}(G)$. But it has fewer morphisms. In fact, there is a monoid isomorphism $G \cong \mathbb{Z}$ because we can assign to any movement in G the number of ups, e.g., $_P[u, u, u, u, u]$ is assigned the integer 5, $_P[d, d, d]$ is assigned the integer -3, and $_P[d, u, u, d, d, u]$ is assigned the integer $0 \in \mathbb{Z}$. But $\mathbb{Z}$ is a group, because every integer has an inverse.

The upshot is that we can use functors to compare groups and monoids.

◇◇

Slogan 5.1.2.5.

    Out of all our available actions, some are reversible.

Example 5.1.2.6. Recall that we have a category **Set** of sets and a category **Fin** of finite sets. We said that **Fin** was a subcategory of **Set**. In fact, we can think of this subcategory relationship in terms of functors, just as we thought of the subset relationship in terms of functions in Example 2.1.2.4. Recall that if we have a subset $S \subseteq S'$, then every element $s \in S$ is an element of S', so we make a function $f \colon S \to S'$ such that $f(s) = s \in S'$.

To give a functor $i\colon \mathbf{Fin} \to \mathbf{Set}$, we have to announce how it works on objects and how it works on morphisms. We begin by announcing a function $i\colon \mathrm{Ob}(\mathbf{Fin}) \to \mathrm{Ob}(\mathbf{Set})$. By analogy with the preceding, we have a subset $\mathrm{Ob}(\mathbf{Fin}) \subseteq \mathrm{Ob}(\mathbf{Set})$. Hence every element $s \in \mathrm{Ob}(\mathbf{Fin})$ is an element of $\mathrm{Ob}(\mathbf{Set})$, so we put $i(s) = s$. We also have to announce, for each pair of objects $s, s' \in \mathrm{Ob}(\mathbf{Fin})$, a function

$$i\colon \mathrm{Hom}_{\mathbf{Fin}}(s, s') \to \mathrm{Hom}_{\mathbf{Set}}(s, s').$$

But again, that is easy because we know by definition (see Example 5.1.1.4) that these two sets are equal, $\mathrm{Hom}_{\mathbf{Fin}}(s, s') = \mathrm{Hom}_{\mathbf{Set}}(s, s')$. Hence we can simply take i to be the identity function on morphisms. It is evident that identities and compositions are preserved by i. Therefore, we have defined a functor i.

Remark 5.1.2.7. Recall that any group is just a monoid, except that it has an extra property: every element has an inverse. Thus one can start with a group, "forget" the fact that it is a group and remember only that it is a monoid. Doing this is functorial— Example 5.1.2.3 discussed it as a functor $U\colon \mathbf{Grp} \to \mathbf{Mon}$. We say that U is a *forgetful functor*. There is also a forgetful functor $\mathbf{Mon} \to \mathbf{Set}$ and so $\mathbf{Grp} \to \mathbf{Sct}$.

Slogan 5.1.2.8.

> *You can use a smartphone as a paperweight.*

Colloquially, people often say things like, "Carol wears many hats" to mean that Carol acts in different roles, even though substantively she is somehow the same. The *hat* Carol currently wears is the analogous to the category, or context of interaction, that she is currently in.

Exercise 5.1.2.9.

A partial order is just a preorder with a special property. A linear order is just a partial order with a special property.

a. Is there a useful functor $\mathbf{FLin} \to \mathbf{PrO}$?

b. Is there a useful functor $\mathbf{PrO} \to \mathbf{FLin}$?

$\Diamond$

Solution 5.1.2.9.

a. Yes, there is a forgetful functor $\mathbf{FLin} \to \mathbf{PrO}$. This functor takes a finite linear order $(X, \leqslant)$ and returns the preorder $(X, \leqslant)$. It takes a morphism $f\colon (X, \leqslant) \to (X', \leqslant')$ of finite linear orders and returns the preorder morphism $f\colon (X, \leqslant) \to (X', \leqslant')$. That is, it does nothing except allow us to place a finite linear order within a larger category. This is valuable if one wants to compare the linear order to other preorders (as opposed to only comparing it to other finite linear orders). It is like the situation in which a math graduate student $X \in$ Math goes to a university-wide graduate social event

$$\text{AtParty}\colon \text{Math} \longrightarrow \text{Univ.}$$

She is the same person at the party as she is when hanging out in the math department, and perhaps her interactions with a fellow math person $Y \in$ Math will be the same as they always are,

$$\text{Hom}_{\text{Univ}}(\text{AtParty}(X), \text{AtParty}(Y)) = \text{Hom}_{\text{Math}}(X, Y),$$

but she can also try her hand at interacting with a person $H \in$ Univ, say, from humanities, as well:

$$\text{Hom}_{\text{Univ}}(\text{AtParty}(X), H) = ??$$

This interaction cannot be discussed in the context of the Math category because H is not in Math, so for X to interact with H we have to forget that X is in the Math category using the functor AtParty.

b. No, not that I can think of. However, that is not to say that there are not *any* functors **PrO** $\to$ **FLin**. For example, there is a functor $c_{[6]}$ that sends every preorder X to the linear order [6], and sends every preorder morphism to $\text{id}_{[6]}$. But that is pretty arbitrary, and I would not consider it useful. As far as I know, there is no useful functor that extracts a linear order from a preorder, let alone to extract a *finite* linear order.

$\blacklozenge$

Proposition 5.1.2.10 (Preorders to graphs). *Let* **PrO** *be the category of preorders and* **Grph** *be the category of graphs. There is a functor* $P\colon$ **PrO** $\to$ **Grph** *such that for any preorder* $\mathcal{X} = (X, \leqslant)$, *the graph* $P(\mathcal{X})$ *has vertices* X.

Proof. Given a preorder $\mathcal{X} = (X, \leqslant_X)$, we can make a graph $F(\mathcal{X})$ with vertices X and an arrow $x \to x'$ whenever $x \leqslant_X x'$, as in Remark 4.4.1.10. More precisely, the preorder $\leqslant_X$ is a relation, i.e., a subset $R_{\mathcal{X}} \subseteq X \times X$, which we think of as a function $i\colon R_{\mathcal{X}} \to X \times X$. Composing with projections $\pi_1, \pi_2\colon X \times X \to X$ gives

$$src_{\mathcal{X}} := \pi_1 \circ i\colon R_{\mathcal{X}} \to X \qquad \text{and} \qquad tgt_{\mathcal{X}} := \pi_2 \circ i\colon R_{\mathcal{X}} \to X.$$

Then we put $F(\mathcal{X}) := (X, R_{\mathcal{X}}, src_{\mathcal{X}}, tgt_{\mathcal{X}})$. This gives us a function $F\colon \text{Ob}(\textbf{PrO}) \to \text{Ob}(\textbf{Grph})$.

Suppose now that $f\colon \mathcal{X} \to \mathcal{Y}$ is a preorder morphism, where $\mathcal{Y} = (Y, \leqslant_Y)$. This is a function $f\colon X \to Y$ such that for any $(x, x') \in X \times X$, if $x \leqslant_X x'$, then $f(x) \leqslant f(x')$. But that is the same as saying that there exists a dotted arrow making the following diagram of sets commute

$$
\begin{array}{ccc}
R_\mathcal{X} & \longrightarrow & X \times X \\
\Big\downarrow & & \Big\downarrow {\scriptstyle f \times f} \\
R_\mathcal{Y} & \longrightarrow & Y \times Y
\end{array}
$$

(Note that there cannot be two different dotted arrows making that diagram commute because $R_\mathcal{Y} \to Y \times Y$ is a monomorphism.) This commutative square is precisely what is needed for a graph homomorphism, as shown in Exercise 4.3.3.7. Thus, we have defined F on objects and on morphisms. It is clear that F preserves identity and composition.

□

Exercise 5.1.2.11.

Proposition 5.1.2.10 gave a functor $P\colon \mathbf{PrO} \to \mathbf{Grph}$.

a. Is every graph $G \in \mathrm{Ob}(\mathbf{Grph})$ in the image of P (or more precisely, is the function

$$
\mathrm{Ob}(P)\colon \mathrm{Ob}(\mathbf{PrO}) \to \mathrm{Ob}(\mathbf{Grph})
$$

surjective)?

b. If so, why; if not, name a graph not in the image.

c. Suppose that G' and H' are preorders with graph formats $P(G') = G$ and $P(H') = H$. Is every graph homomorphism $f\colon G \to H$ in the image of

$$
\mathrm{Hom}_P\colon \mathrm{Hom}_{\mathbf{PrO}}(G', H') \to \mathrm{Hom}_{\mathbf{Grph}}(G, H)?
$$

In other words, does every graph homomorphism between G and H come from a preorder homomorphism between G' and H'?

◊

Solution 5.1.2.11.

a. No. See, for example, Remark 4.4.1.10.

b. Neither of the following graphs are in the image of P:

The first does not work because there are too many arrows $a \to b$. The second does not work because there is a path $a \to a$ (namely, the trivial path), but no arrow $a \to a$.

c. Yes. Given a graph morphism $f = (f_0, f_1) \colon G \to H$, we take $f' \colon G' \to H'$ to be $f' = f_0$. That is, the elements of the preorder G' are just the vertices in graph G, so f' should do whatever f did on vertices. Now we must check that if $g_1 \leqslant g_2$ in G', then $f'(g_1) \leqslant f'(g_2)$. If $g_1 \leqslant g_2$, then there is a path from g_1 to g_2 in G, and graph morphisms preserve paths, so there is a path from $f(g_1)$ to $f(g_2)$ in H, so indeed $f'(g_1) \leqslant f'(g_2)$. Now it is easy to check that $P(f') = f$.

♦

Remark 5.1.2.12. There is a functor $W \colon \mathbf{PrO} \to \mathbf{Set}$ sending $(X, \leqslant)$ to X. There is a functor $T \colon \mathbf{Grph} \to \mathbf{Set}$ sending (V, A, src, tgt) to V. When we study the category of categories (see Section 5.1.2.30), it will be clear that Proposition 5.1.2.10 can be summarized as a commutative triangle in $\mathbf{Cat}$,

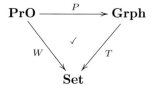

Exercise 5.1.2.13.

Recall from (2.3) that every function $f \colon A \to B$ has an image, $\mathrm{im}_f(A) \subseteq B$. Use this idea and Example 4.4.1.16 to construct a functor $Im \colon \mathbf{Grph} \to \mathbf{PrO}$ such that for any graph $G = (V, A, src, tgt)$, the vertices of G are the elements of $Im(G)$. That is, find some ordering $\leqslant_G$, such that we have $Im(G) = (V, \leqslant_G)$.   ◊

Solution 5.1.2.13.

Suppose given an object $G \in \mathrm{Ob}(\mathbf{Grph})$, i.e., a graph $G = (V, A, src, tgt)$. The source and target functions combine to give a function $\langle src, tgt \rangle \colon A \to V \times V$. Its image is a subset $R \subseteq V \times V$, i.e., a binary relation. But R is not necessarily a preorder. We can remedy that by using the preorder $\overline{R}$ generated by R, as in Example 4.4.1.16. On objects we put $Im(G) := \overline{R}$. One way to understand this preorder is that it has as elements V, the vertices of G, and it has $v \leqslant v'$ if and only if there exists a path from v to v' in G.

Given a morphism $f\colon G \to G'$, we need to provide a preorder morphism $Im(G) \to Im(G')$. The obvious choice is to use f_0 (what f does on vertices), but we need to check that it preserves the order. This is clear because graph morphisms send paths to paths—if there was a path from v to v' in G, there will be one from $f(v)$ to $f(v')$. We need to check that $Im(\mathrm{id}_G) = \mathrm{id}_{Im(G)}$, but this is straightforward.   ♦

Exercise 5.1.2.14.

In Exercise 5.1.2.13 you constructed a functor $Im\colon \mathbf{Grph} \to \mathbf{PrO}$. What is the preorder $Im(G)$ when $G \in \mathrm{Ob}(\mathbf{Grph})$ is the following graph?

$$G :=$$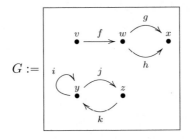

◊

Solution 5.1.2.14.

The easiest way to see it is that $v \leqslant v'$ in $Im(G)$ if there is a path from v to v' in G. But for completeness, we write out the relation, which we denote $\leqslant_{Im(G)}$:

$\leqslant_{Im(G)}$	
v	v
v	w
v	x
w	w
w	x
x	x
y	y
y	z
z	y
z	z

♦

Exercise 5.1.2.15.

Consider the functor $Im\colon \mathbf{Grph} \to \mathbf{PrO}$ constructed in Exercise 5.1.2.13.

a. Is every preorder $\mathcal{X} \in \mathrm{Ob}(\mathbf{PrO})$ in the image of Im (or more precisely, in the image of $\mathrm{Ob}(Im)\colon \mathrm{Ob}(\mathbf{Grph}) \to \mathrm{Ob}(\mathbf{PrO})$)?

b. If so, why; if not, name a preorder not in the image.

c. Suppose that $\mathcal{X}', \mathcal{Y}' \in \mathrm{Ob}(\mathbf{Grph})$ are graphs, with $\mathcal{X} := Im(\mathcal{X}')$ and $\mathcal{Y} := Im(\mathcal{Y}')$ in the preorder format. Is every preorder morphism $f\colon \mathcal{X} \to \mathcal{Y}$ in the image of

$$\mathrm{Hom}_{Im}\colon \mathrm{Hom}_{\mathbf{Grph}}(\mathcal{X}', \mathcal{Y}') \to \mathrm{Hom}_{\mathbf{PrO}}(\mathcal{X}, \mathcal{Y})?$$

In other words, does every preorder homomorphism between $\mathcal{X}$ and $\mathcal{Y}$ come from a graph homomorphism between $\mathcal{X}'$ and $\mathcal{Y}'$?

$\Diamond$

Solution 5.1.2.15.

a. Yes.

b. In Proposition 5.1.2.10 showed the construction of a functor $P\colon \mathbf{PrO} \to \mathbf{Grph}$. Given a preorder $\mathcal{X} := (X, \leqslant)$, we can make a graph $G = P(\mathcal{X})$ out of that, with vertices X and an edge $x \to x'$ whenever $x \leqslant x'$. The functor $Im\colon \mathbf{Grph} \to \mathbf{PrO}$ sends G back to $(X, \leqslant)$. Thus $\mathcal{X} = Im(P(\mathcal{X}))$ is in the image of Im.

c. No. There is no graph morphism

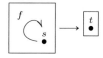

because the arrow f has nowhere to go. However, if we apply the functor Im, these two graphs become the same preorder, and so there is an identity morphism between them.

♦

Exercise 5.1.2.16.

We have functors $P\colon \mathbf{PrO} \to \mathbf{Grph}$ and $Im\colon \mathbf{Grph} \to \mathbf{PrO}$.

a. What can you say about $Im \circ P\colon \mathbf{PrO} \to \mathbf{PrO}$?

b. What can you say about $P \circ Im\colon \mathbf{Grph} \to \mathbf{Grph}$?

$\Diamond$

Solution 5.1.2.16.

a. It is the identity functor. That is, when a preorder is turned into a graph and then back into a preorder, it returns unchanged; similarly, when a preorder morphism is turned into a graph morphism and then back into a preorder morphism, it remains unchanged.

b. It is not the identity functor. When a graph G is turned into a preorder and then back into a graph, it has lost all redundancy (more than one edge $v \to v'$) and every path in G has become an edge in $P(Im(G))$.

◆

Exercise 5.1.2.17.

Consider the functors $P \colon \mathbf{PrO} \to \mathbf{Grph}$ and $Im \colon \mathbf{Grph} \to \mathbf{PrO}$. And consider the chain graph $[n]$ of length n from Example 4.3.1.8 and the linear order $[n]$ of length n from Example 4.4.1.7. To differentiate the two, let's rename them for this exercise as $[n]_{\mathbf{Grph}} \in \mathrm{Ob}(\mathbf{Grph})$ and $[n]_{\mathbf{PrO}} \in \mathrm{Ob}(\mathbf{PrO})$. We see a similarity between $[n]_{\mathbf{Grph}}$ and $[n]_{\mathbf{PrO}}$, and we might hope that the functors help formalize this similarity. That is, we might hope that one of the following hold:

$$P([n]_{\mathbf{PrO}}) \cong^? [n]_{\mathbf{Grph}} \qquad \text{or} \qquad Im([n]_{\mathbf{Grph}}) \cong^? [n]_{\mathbf{PrO}}.$$

Do either, both, or neither of these hold? ◇

Solution 5.1.2.17.

The first equation does not hold. For example, $P([0]_{\mathbf{PrO}})$ is the loop graph (with one vertex and one self-arrow), and that is different from $[0]_{\mathbf{Grph}}$, which has no arrows. The second equation does hold:

$$Im([n]_{\mathbf{Grph}}) \cong [n]_{\mathbf{PrO}}.$$

◆

Remark 5.1.2.18. In the course announcement for MIT's 18-S996 course, I wrote the following:

It is often useful to focus one's study by viewing an individual thing, or a group of things, as though it exists in isolation. However, the ability to rigorously change our point of view, seeing our object of study in a different context, often yields unexpected insights. Moreover, this ability to change perspective is indispensable for effectively communicating with and learning from others. It is the relationships between things, rather than the things in and by themselves, that are responsible for generating the rich variety of phenomena we observe in the physical, informational, and mathematical worlds.

This holds at many different levels. For example, one can study a group (in the sense of Definition 4.2.1.1) in isolation, trying to understand its subgroups or its automorphisms, and this is mathematically interesting. But one can also view it as a quotient of something else, or as a subgroup of something else. One can view the group as a monoid and look at monoid homomorphisms to or from it. One can look at the group in the context of symmetries by seeing how it acts on sets. These changes of viewpoint are all clearly and formally expressible within category theory. We know how the different changes of viewpoint compose and how they fit together in a larger context.

Exercise 5.1.2.19.

a. Is the preceding quotation also true in your scientific discipline of expertise? How so?

b. Can you imagine a way that category theory can help catalogue the kinds of relationships or changes of viewpoint that exist in your discipline?

c. What kinds of structures that you use often deserve to be better formalized?

                                                                                          ◊

Solution 5.1.2.19.

a. It is useful to study a person in isolation, e.g., in the context of anatomy. However, even psychology is so relational (about how a person relates with other people) that it does not make sense to consider psychology as the study of an individual in isolation. Being able to change one's point of view, e.g., helping a person see how others see him or how his past self or future self might see him, yields unexpected insights. And without an understanding of other points of view, it may be hard for a person to drive a car ("I was told that everyone is supposed to drive on the right side of the road. Why are those oncoming cars driving on the left side?") let alone cooperate with others. The rich variety of phenomena that exist in society cannot be reduced to the anatomy of an individual, even to the arrangement of cells in the brain.

b. If each person were assigned a database corresponding to her worldview and her acquired set of examples, then relationships between people might be formalizable as functors (or as some kind of morphisms) that relate these structures. Formalizing interaction in this way could allow us to produce a much more effective simulation of human behavior, or allow humans to interact with computers more seamlessly.

Rigorous communication with others (e.g., research papers) are written in prose. But they should be written in a rigorous way, so that different papers can be connected together in interesting ways to form a network of understanding. What are the connections? In precisely what sense is one paper an extension or a rebuttal of another? ♦

Example 5.1.2.20 (Free monoids). Let G be a set. Definition 4.1.1.15 defined a monoid $\mathrm{List}(G)$, called the free monoid on G. Given a function $f\colon G \to G'$, there is an induced function $\mathrm{List}(f)\colon \mathrm{List}(G) \to \mathrm{List}(G')$, and this preserves the identity element [ ] and concatenation of lists, so $\mathrm{List}(f)$ is a monoid homomorphism. It is easy to check that $\mathrm{List}\colon \mathbf{Set} \to \mathbf{Mon}$ is a functor.

Application 5.1.2.21. Application 2.1.2.16 discussed an isomorphism $\mathrm{Nuc_{DNA}} \cong \mathrm{Nuc_{RNA}}$ given by RNA transcription. Applying the functor List, we get a function

$$\mathrm{List}(\mathrm{Nuc_{DNA}}) \xrightarrow{\cong} \mathrm{List}(\mathrm{Nuc_{RNA}}),$$

which will send sequences of DNA nucleotides to sequences of RNA nucleotides, and vice versa. This is performed by polymerases.

◇◇

Exercise 5.1.2.22.

Let $G = \{1, 2, 3, 4, 5\}, G' = \{a, b, c\}$, and let $f\colon G \to G'$ be given by the sequence (a, c, b, a, c).[5] Then if $L = [1, 1, 3, 5, 4, 5, 3, 2, 4, 1]$, what is $\mathrm{List}(f)(L)$? ◇

Solution 5.1.2.22.

Use f to translate L, entry by entry:

$$\mathrm{List}(f)([1, 1, 3, 5, 4, 5, 3, 2, 4, 1]) = [a, a, b, c, a, c, b, c, a, a].$$

♦

Remark 5.1.2.23 (Questionable functor). Recall from Remark 5.1.1.6 that a questionable category is defined to be a structure that looks like a category (objects, morphisms, identities, composition formula), but which is not required to satisfy any laws. Similarly, given categories (or questionable categories) $\mathcal{C}$ and $\mathcal{D}$, we can define a questionable functor $F\colon \mathcal{C} \to \mathcal{D}$ to consist of

    A. a function $\mathrm{Ob}(F)\colon \mathrm{Ob}(\mathcal{C}) \to \mathrm{Ob}(\mathcal{C}')$, sometimes denoted simply $F\colon \mathrm{Ob}(\mathcal{C}) \to \mathrm{Ob}(\mathcal{C}')$;

---

[5]See Exercise 2.1.2.22 if there is any confusion about this.

B. for every pair of objects $c, d \in \mathrm{Ob}(\mathcal{C})$, a function

$$\mathrm{Hom}_F(c, d) \colon \mathrm{Hom}_{\mathcal{C}}(c, d) \to \mathrm{Hom}_{\mathcal{C}'}(F(c), F(d)),$$

sometimes denoted simply $F \colon \mathrm{Hom}_{\mathcal{C}}(c, d) \to \mathrm{Hom}_{\mathcal{C}'}(F(c), F(d))$.

Exercise 5.1.2.24.

We can rephrase the notion of functor in terms compatible with Exercise 5.1.1.27. We begin by saying that a functor $F \colon \mathcal{C} \to \mathcal{C}'$ consists of two functions,

$$\mathrm{Ob}(F) \colon \mathrm{Ob}(\mathcal{C}) \to \mathrm{Ob}(\mathcal{C}') \qquad \text{and} \qquad \mathrm{Hom}_F \colon \mathrm{Hom}_{\mathcal{C}} \to \mathrm{Hom}_{\mathcal{C}'},$$

called the *on-objects part* and the *on-morphisms part* respectively. They must follow some rules, expressed by the commutativity of the following squares in **Set**:

$$(5.5)$$

$$
\begin{array}{ccc}
\mathrm{Hom}_{\mathcal{C}} & \xrightarrow{\ dom\ } & \mathrm{Ob}(\mathcal{C}) \\
{\scriptstyle \mathrm{Hom}_F}\downarrow & & \downarrow{\scriptstyle \mathrm{Ob}(F)} \\
\mathrm{Hom}_{\mathcal{C}'} & \xrightarrow[\ dom'\ ]{} & \mathrm{Ob}(\mathcal{C}')
\end{array}
\qquad
\begin{array}{ccc}
\mathrm{Hom}_{\mathcal{C}} & \xrightarrow{\ cod\ } & \mathrm{Ob}(\mathcal{C}) \\
{\scriptstyle \mathrm{Hom}_F}\downarrow & & \downarrow{\scriptstyle \mathrm{Ob}(F)} \\
\mathrm{Hom}_{\mathcal{C}'} & \xrightarrow[\ cod'\ ]{} & \mathrm{Ob}(\mathcal{C}')
\end{array}
$$

$$
\begin{array}{ccc}
\mathrm{Ob}(\mathcal{C}) & \xrightarrow{\ ids\ } & \mathrm{Hom}_{\mathcal{C}} \\
{\scriptstyle \mathrm{Ob}(F)}\downarrow & & \downarrow{\scriptstyle \mathrm{Hom}_F} \\
\mathrm{Ob}(\mathcal{C}') & \xrightarrow[\ ids\ ]{} & \mathrm{Hom}_{\mathcal{C}'}
\end{array}
\qquad
\begin{array}{ccc}
\mathrm{Hom}_{\mathcal{C}} \times_{\mathrm{Ob}(\mathcal{C})} \mathrm{Hom}_{\mathcal{C}} & \xrightarrow{\ comp\ } & \mathrm{Hom}_{\mathcal{C}} \\
\downarrow & & \downarrow{\scriptstyle \mathrm{Hom}_F} \\
\mathrm{Hom}_{\mathcal{C}'} \times_{\mathrm{Ob}(\mathcal{C}')} \mathrm{Hom}_{\mathcal{C}'} & \xrightarrow[\ comp\ ]{} & \mathrm{Hom}_{\mathcal{C}'}
\end{array}
$$

$$(5.6)$$

a. In the right-hand diagram in (5.6), where does the (unlabeled) left-hand function come from? Hint: Use Exercise 3.2.1.20.

Consider diagram (5.3); imagine it as though it were contained in a pane of glass. Then imagine a parallel pane of glass involving $\mathcal{C}'$ in place of $\mathcal{C}$ everywhere.

b. Draw arrows from the $\mathcal{C}$ pane to the $\mathcal{C}'$ pane, each labeled $\mathrm{Ob}(F)$, Hom_F, and so on, as appropriate.

c. If F is a functor, i.e., it satisfies (5.5) and (5.6), do all the squares in your drawing commute?

d. Does the definition of functor involve anything not captured in this setup?

◊

Solution 5.1.2.24.

a. We have $\mathrm{Hom}_F \colon \mathrm{Hom}_C \to \mathrm{Hom}_{C'}$, and since it commutes with *dom* and *cod*, we have the desired function, by Exercise 3.2.1.20.

b. Let $CP_C = \mathrm{Hom}_C \times_{\mathrm{Ob}(C)} \mathrm{Hom}_C$ denote the set of composable pairs of arrows in C (and similarly define $CP_{C'}$ and $CP_F \colon CP_C \to CP_{C'}$). The two-pane diagram is a bit cluttered, but looks like this:

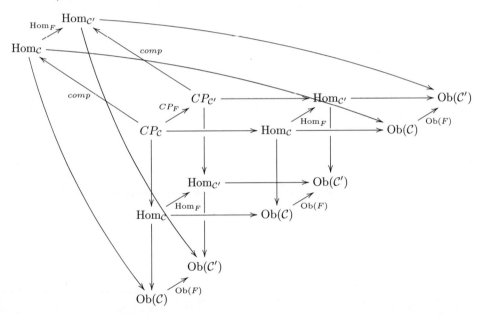

c. Yes.

d. No, this is all one needs: functions $\mathrm{Ob}(F) \colon \mathrm{Ob}(C) \to \mathrm{Ob}(C')$ and $\mathrm{Hom}_F \colon \mathrm{Hom}_C \to \mathrm{Hom}_{C'}$ such that all the squares commute.

♦

Example 5.1.2.25 (Paths-graph). Let $G = (V, A, src, tgt)$ be a graph. We have a set Path_G of paths in G, and functions $\overline{src}, \overline{tgt} \colon \mathrm{Path}_G \to V$. That information is enough to define a new graph,

$$\mathrm{Paths}(G) := (V, \mathrm{Path}_G, \overline{src}, \overline{tgt}).$$

Moreover, given a graph homomorphism $f: G \to G'$, every path in G is sent under f to a path in G'. So Paths: **Grph** $\to$ **Grph** is a functor.

Exercise 5.1.2.26.

a. Consider the graph G from Example 4.3.3.3. Draw the paths-graph Paths(G) for G.

b. Repeating part (a) for G' from the same example would be hard, because the paths-graph Paths(G') has infinitely many arrows. However, the graph homomorphism $f: G \to G'$ does induce a morphism of paths-graphs Paths(f): Paths(G) $\to$ Paths(G'). How does that act on the vertices and arrows of Paths(G)?

c. Given a graph homomorphism $f: G \to G'$ and two paths $p: v \to w$ and $q: w \to x$ in G, is it true that Paths(f) preserves the concatenation? Explain also what it means to say Paths(f) preserves the concatenation.

$\diamond$

Solution 5.1.2.26.

a. Here are G and Paths(G).

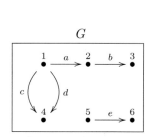

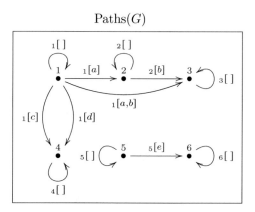

b. For the reader's convenience, here is a copy of $f\colon G \to G'$:

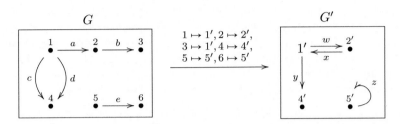

By definition Paths(f) acts like f on the vertices, and arrow by arrow on paths. Here is the formal answer:

$f_0\colon V \to V'$	
V	V'
1	$1'$
2	$2'$
3	$1'$
4	$4'$
5	$5'$
6	$5'$

$f_1\colon \mathrm{Path}_G \to \mathrm{Path}_{G'}$	
Path_G	$\mathrm{Path}_{G'}$
$_1[\,]$	$_{1'}[\,]$
$_1[a]$	$_{1'}[w]$
$_1[a,b]$	$_{1'}[w,x]$
$_1[c]$	$_{1'}[y]$
$_1[d]$	$_{1'}[y]$
$_2[\,]$	$_{2'}[\,]$
$_2[b]$	$_{2'}[x]$
$_3[\,]$	$_{1'}[\,]$
$_4[\,]$	$_{4'}[\,]$
$_5[\,]$	$_{5'}[\,]$
$_5[e]$	$_{5'}[z]$
$_6[\,]$	$_{5'}[\,]$

c. Yes, that is true. It means that $f(p) \mathbin{++} f(q) = f(p \mathbin{++} q)$, where $++$ denotes concatenation of paths.

♦

Exercise 5.1.2.27.

Suppose that $\mathcal{C}$ and $\mathcal{D}$ are categories, $c, c' \in \mathrm{Ob}(\mathcal{C})$ are objects, and $F\colon \mathcal{C} \to \mathcal{D}$ is a functor. Suppose that c and c' are isomorphic in $\mathcal{C}$. Show that this implies that $F(c)$ and $F(c')$ are isomorphic in $\mathcal{D}$.   ◊

Solution 5.1.2.27.

If c and c' are isomorphic, that means there exists a morphism $f\colon c \to c'$ and a morphism $f'\colon c' \to c$ in $\mathcal{C}$, such that $f' \circ f = \mathrm{id}_c$ and $f \circ f' = \mathrm{id}_{c'}$. But then $F(f)\colon F(c) \to F(c')$ and $F(f')\colon F(c') \to F(c)$ are mutually inverse morphisms between $F(c)$ and $F(c')$. Indeed, since F preserves composition and identities, we have $F(f') \circ F(f) = F(f' \circ f) = F(\mathrm{id}_c) = \mathrm{id}_{F(c)}$ and $F(f) \circ F(f') = F(f \circ f') = F(\mathrm{id}_{c'}) = \mathrm{id}_{F(c')}$. So $F(f)$ is an isomorphism, which means that $F(c)$ and $F(c')$ are isomorphic in $\mathcal{D}$.   ♦

Example 5.1.2.28. For any graph G, we can assign its set of length 1 loops $Eq(G)$ as in Exercise 4.3.1.12. This assignment is functorial in that given a graph homomorphism $G \to G'$, there is an induced function $Eq(G) \to Eq(G')$. Similarly, we can functorially assign the set of connected components of the graph, $Coeq(G)$. In other words, $Eq\colon \mathbf{Grph} \to \mathbf{Set}$ and $Coeq\colon \mathbf{Grph} \to \mathbf{Set}$ are functors. The assignment of vertex set and arrow set are two more functors $\mathbf{Grph} \to \mathbf{Set}$.

Suppose you want to decide whether two graphs G and G' are isomorphic. If the graphs have thousands of vertices and thousands of arrows, this could take a long time. However, the preceding functors, in combination with Exercise 5.1.2.27 give us some things to try.

The first thing to do is to count the number of loops of each, because these numbers are generally small. If the number of loops in G is different than the number of loops in G', then because functors preserve isomorphisms, G and G' cannot be isomorphic. Similarly, one can count the number of connected components, again generally a small number. If the number of components in G is different than the number of components in G', then $G \not\cong G'$. Similarly, one can simply count the number of vertices or the number of arrows in G and G'. These are all isomorphism invariants.

All this is a bit like trying to decide if a number is prime by checking if it is even, if its digits add up to a multiple of 3, or if it ends in a 5; these tests do not determine the answer, but they offer some level of discernment.

Remark 5.1.2.29. As mentioned, functors allow ideas in one domain to be rigorously imported to another. Example 5.1.2.28 is a first taste. Because functors preserve isomorphisms, we can tell graphs apart by looking at them in a simpler category, $\mathbf{Set}$, using various lenses (in that case, four). There is relatively simple theorem in $\mathbf{Set}$ that says that for different natural numbers m, n the sets $\underline{m}$ and $\underline{n}$ are never isomorphic. This theorem is transported via the four functors to four different theorems about telling graphs apart.

5.1.2.30   The category of categories

Recall from Remark 5.1.1.2 that a small category $\mathcal{C}$ is one in which $\mathrm{Ob}(\mathcal{C})$ is a set. But everything said so far works whether or not $\mathcal{C}$ is small. The following definition gives more precision.

Proposition 5.1.2.31. *There exists a category, called* the category of small categories *and denoted* **Cat**, *in which the objects are the small categories and the morphisms are the functors,*

$$\text{Hom}_{\textbf{Cat}}(\mathcal{C}, \mathcal{D}) = \{F \colon \mathcal{C} \to \mathcal{D} \mid F \text{ is a functor}\}.$$

That is, there are identity functors, functors can be composed, and the identity and associativity laws hold.

Proof. We follow Definition 5.1.1.1. We have already specified $\text{Ob}(\textbf{Cat})$ and $\text{Hom}_{\textbf{Cat}}$ in the statement of the proposition. Given a small category $\mathcal{C}$, there is an identity functor $\text{id}_{\mathcal{C}} \colon \mathcal{C} \to \mathcal{C}$ that is identity on the set of objects and the set of morphisms. And given a functor $F \colon \mathcal{C} \to \mathcal{D}$ and a functor $G \colon \mathcal{D} \to \mathcal{E}$, it is easy to check that $G \circ F \colon \mathcal{C} \to \mathcal{E}$, defined by composition of functions $\text{Ob}(G) \circ \text{Ob}(F) \colon \text{Ob}(\mathcal{C}) \to \text{Ob}(\mathcal{E})$ and $\text{Hom}_G \circ \text{Hom}_F \colon \text{Hom}_{\mathcal{C}} \to \text{Hom}_{\mathcal{E}}$ (see Exercise 5.1.2.24), is a functor; thus we have a composition formula. For the same reasons, one can show that functors, as morphisms, obey the identity law and the composition law. Therefore, this specification of **Cat** satisfies the definition of being a category.

$\square$

Example 5.1.2.32 (Categories have underlying graphs). Suppose given a category in the notation is as in Exercise 5.1.1.27, $\mathcal{C} = (\text{Ob}(\mathcal{C}), \text{Hom}_{\mathcal{C}}, dom, cod, ids, comp)$. Then $(\text{Ob}(\mathcal{C}), \text{Hom}_{\mathcal{C}}, dom, cod)$ is a graph, called the *graph underlying* $\mathcal{C}$ and denoted $U(\mathcal{C}) \in \text{Ob}(\textbf{Grph})$. A functor $F \colon \mathcal{C} \to \mathcal{D}$ induces a graph morphism $U(F) \colon U(\mathcal{C}) \to U(\mathcal{D})$, as seen in (5.5). So we have a functor,

$$U \colon \textbf{Cat} \to \textbf{Grph}.$$

Example 5.1.2.33 (Free category on a graph). Example 5.1.2.25 discussed a functor Paths: **Grph** $\to$ **Grph** that considered all the paths in a graph G as the arrows of a new graph $\text{Paths}(G)$. In fact, $\text{Paths}(G)$ could be construed as a category, denoted $F(G) \in \text{Ob}(\textbf{Cat})$ and called *the free category generated by* G.

The objects of the category $F(G)$ are the vertices of G. For any two vertices v, v', the hom-set $\text{Hom}_{F(G)}(v, v')$ is the set of paths in G from v to v'. The identity elements are given by the trivial paths, and the composition formula is given by concatenation of paths.

For the on-morphisms part of F, we need to see that a graph homomorphism $f \colon G \to G'$ induces a functor $F(f) \colon F(G) \to F(G')$. But this was shown in Exercise 5.1.2.26. Thus we have a functor

$$F \colon \textbf{Grph} \to \textbf{Cat}$$

called *the free category* functor.

Exercise 5.1.2.34.

Let G be the graph depicted

and let $[1] \in \mathrm{Ob}(\mathbf{Cat})$ denote the free category on G, i.e., $[1] := F(G)$, as in Example 5.1.2.33. We call $[1]$ the *free arrow category*.

a. What are the objects of $[1]$?

b. For every pair of objects in $[1]$, write the hom-set.

$\Diamond$

Solution 5.1.2.34.

a. $\mathrm{Ob}([1]) = \{v_0, v_1\}$.

b. There are four pairs of objects, so the four hom-sets are:

$$\mathrm{Hom}_{[1]}(v_0, v_0) = \{\mathrm{id}_{v_0}\}; \qquad \mathrm{Hom}_{[1]}(v_0, v_1) = \{e\};$$
$$\mathrm{Hom}_{[1]}(v_1, v_0) = \varnothing; \qquad \mathrm{Hom}_{[1]}(v_1, v_1) = \{\mathrm{id}_{v_1}\}.$$

$\blacklozenge$

Exercise 5.1.2.35.

Let G be the graph whose vertices are all U.S. cities and whose arrows are airplane flights connecting the cities. What idea is captured by the free category on G? $\Diamond$

Solution 5.1.2.35.

This captures the idea of flight itineraries. "I am leaving Boston, flying to Atlanta, then on to Chicago." You can compose itineraries if the arrival city of one itinerary equals the departure city of the next. And there is the identity "I am not going anywhere" itinerary for any U.S. city. $\blacklozenge$

Exercise 5.1.2.36.

Let $F \colon \mathbf{Grph} \to \mathbf{Cat}$ denote the free category functor from Example 5.1.2.33, and let $U \colon \mathbf{Cat} \to \mathbf{Grph}$ denote the underlying graph functor from Example 5.1.2.32. What is the composition $U \circ F \colon \mathbf{Grph} \to \mathbf{Grph}$ called? $\Diamond$

Solution 5.1.2.36.

Since $F\colon \mathbf{Grph} \to \mathbf{Cat}$ freely adds all paths, one can check that $U \circ F\colon \mathbf{Grph} \to \mathbf{Grph}$ is the construction that takes a graph and adds all paths; i.e., $U \circ F = \mathrm{Paths}$ (see Example 5.1.2.25). ◆

Exercise 5.1.2.37.

Recall the graph G from Example 4.3.1.2. Let $\mathcal{C} = F(G)$ be the free category on G.

a. What is $\mathrm{Hom}_{\mathcal{C}}(v, x)$?

b. What is $\mathrm{Hom}_{\mathcal{C}}(x, v)$?

◇

Solution 5.1.2.37.

a. The set $\mathrm{Hom}_{\mathcal{C}}(v, x)$ has two elements: $_v[f, g]$ and $_v[f, h]$.

b. $\mathrm{Hom}_{\mathcal{C}}(x, v) = \varnothing$.

◆

Example 5.1.2.38 (Discrete graphs, discrete categories). There is a functor $Disc\colon \mathbf{Set} \to \mathbf{Grph}$ that sends a set S to the graph

$$Disc(S) := (S, \varnothing, !, !),$$

where $!\colon \varnothing \to S$ is the unique function. We call $Disc(S)$ the *discrete graph on the set S*. It is clear that a function $S \to S'$ induces a morphism of discrete graphs. Now applying the free category functor $F\colon \mathbf{Grph} \to \mathbf{Cat}$, we get the *discrete category on the set S*. This composition is also denoted $Disc\colon \mathbf{Set} \to \mathbf{Cat}$.

Exercise 5.1.2.39.

Recall from (2.4) the definition of the set $\underline{n}$ for any natural number $n \in \mathbb{N}$, and let $D_n := Disc(\underline{n}) \in \mathrm{Ob}(\mathbf{Cat})$ be the discrete category on the set $\underline{n}$, as in Example 5.1.2.38.

a. List all the morphisms in D_4.

b. List all the functors $D_3 \to D_2$.

◇

Solution 5.1.2.39.

a. There are only identity morphisms, one for each object: $\{\mathrm{id}_1, \mathrm{id}_2, \mathrm{id}_3, \mathrm{id}_4\}$.

b. A functor $F \colon D_3 \to D_2$ consists of a function $F \colon \mathrm{Ob}(D_3) \to \mathrm{Ob}(D_2)$ on objects as well as a function on morphisms that respects identities and compositions. But since the only morphisms in D_3 are identities, there is no choice and no restriction in the morphism part. In other words, a functor $D_3 \to D_2$ is completely determined by a function $\underline{3} \to \underline{2}$. There are eight of these, which by (2.5), can be denoted by sequences:

$$(1,1,1); \quad (1,1,2); \quad (1,2,1); \quad (1,2,2);$$
$$(2,1,1); \quad (2,1,2); \quad (2,2,1); \quad (2,2,2).$$

◆

Exercise 5.1.2.40.

Let $\mathcal{C}$ be a category. How many functors are there $\mathcal{C} \to D_1$, where $D_1 := Disc(\underline{1})$ is the discrete category on one element?   ◊

Solution 5.1.2.40.

There is always one functor $\mathcal{C} \to D_1$. There is no choice about where to send objects (all go to the object 1), and there is no choice about where to send morphisms (all go to the morphism id_1).   ◆

We sometimes refer to $Disc(\underline{1})$ as the *terminal category* (see Section 6.1.3) and for simplicity denote it $\underline{1}$. Its unique object is denoted 1.

Exercise 5.1.2.41.

If someone said, "Ob is a functor from **Cat** to **Set**," what might they mean?   ◊

Solution 5.1.2.41.

They probably mean that there is a functor **Cat** $\to$ **Set** that sends a category $\mathcal{C}$ to its set of objects $\mathrm{Ob}(\mathcal{C})$. Since the speaker does not say what this functor, Ob, does on morphisms, he is suggesting it is obvious. A morphism in **Cat** is a functor $F \colon \mathcal{C} \to \mathcal{D}$, which includes an on-objects part by definition. In other words, it is indeed obvious what $\mathrm{Ob}(F) \colon \mathrm{Ob}(\mathcal{C}) \to \mathrm{Ob}(\mathcal{D})$ should mean because this is given in the specification of F (see Definition 5.1.2.1). It is not hard to check that Ob preserves identities and compositions, so it is indeed a functor.   ◆

Exercise 5.1.2.42.

If someone said, "Hom is a functor from **Cat** to **Set**, where by Hom I mean the mapping that takes $\mathcal{C}$ to the set $\mathrm{Hom}_{\mathcal{C}}$, as in Exercise 5.1.1.27," what might they mean?
◊

Solution 5.1.2.42.

They probably mean that there is a functor **Cat** $\to$ **Set** that sends a category $\mathcal{C}$ to its set of morphisms $\mathrm{Hom}_{\mathcal{C}}$. Since the speaker does not indicate what this functor, Hom, does on morphisms, she is suggesting it is obvious. A morphism in **Cat** is a functor $F\colon \mathcal{C} \to \mathcal{D}$, which includes an on-morphisms part by definition. In other words, it is indeed obvious what $\mathrm{Hom}(F)\colon \mathrm{Hom}(\mathcal{C}) \to \mathrm{Hom}(\mathcal{D})$ should mean because this is given in the specification of F (see Definition 5.1.2.1). It is easy to check that Hom preserves identities and compositions, so it is indeed a functor. ♦

5.2 Common categories and functors from pure math

5.2.1 Monoids, groups, preorders, and graphs

We saw in Section 5.1.1 that there is a category **Mon** of monoids, a category **Grp** of groups, a category **PrO** of preorders, and a category **Grph** of graphs. This section shows that each monoid $\mathcal{M}$, each group $\mathcal{G}$, and each preorder $\mathcal{P}$ can be considered as its own category. If each object in **Mon** is a category, we might hope that each morphism in **Mon** is just a functor, and this is true. The same holds for **Grp** and **PrO**. We saw in Example 5.1.2.33 how each graph can be regarded as giving a free category. Another perspective on graphs (i.e., graphs as functors) is discussed in Section 5.2.1.21.

5.2.1.1 Monoids as categories

Example 4.1.2.9 said that to olog a monoid, one should use only one box. And again Example 4.5.3.3 said that a monoid action could be captured by only one table. These ideas are encapsulated by the understanding that a monoid is perfectly modeled as a category with one object.

Each monoid as a category with one object Let $(M, e, \star)$ be a monoid. We consider it as a category $\mathcal{M}$ with one object, $\mathrm{Ob}(\mathcal{M}) = \{\blacktriangle\}$, and

$$\mathrm{Hom}_{\mathcal{M}}(\blacktriangle, \blacktriangle) := M.$$

The identity morphism $\mathrm{id}_{\blacktriangle}$ serves as the monoid identity e, and the composition formula

$$\circ\colon \mathrm{Hom}_{\mathcal{M}}(\blacktriangle, \blacktriangle) \times \mathrm{Hom}_{\mathcal{M}}(\blacktriangle, \blacktriangle) \to \mathrm{Hom}_{\mathcal{M}}(\blacktriangle, \blacktriangle)$$

is given by $\star\colon M \times M \to M$. The associativity and identity laws for the monoid match precisely with the associativity and identity laws for categories.

If a monoid is a category with one object, is there any categorical way of phrasing the notion of monoid homomorphism? Suppose that $\mathcal{M} = (M, e, \star)$ and $\mathcal{M}' = (M', e', \star')$. We know that a monoid homomorphism is a function $f\colon M \to M'$ such that $f(e) = e'$ and such that for every pair $m_0, m_1 \in M$, we have $f(m_0 \star m_1) = f(m_0) \star' f(m_1)$. What is a functor $\mathcal{M} \to \mathcal{M}'$?

Each monoid homomorphism as a functor between one-object categories   Say that $\mathrm{Ob}(\mathcal{M}) = \{\blacktriangle\}$ and $\mathrm{Ob}(\mathcal{M}') = \{\blacktriangle'\}$, and we know that $\mathrm{Hom}_{\mathcal{M}}(\blacktriangle, \blacktriangle) = M$ and $\mathrm{Hom}_{\mathcal{M}'}(\blacktriangle', \blacktriangle') = M'$. A functor $F\colon \mathcal{M} \to \mathcal{M}'$ consists first of a function $\mathrm{Ob}(\mathcal{M}) \to \mathrm{Ob}(\mathcal{M}')$, but these sets have only one element each, so there is nothing to say on that front: we must have $F(\blacktriangle) = \blacktriangle'$. It also consists of a function $\mathrm{Hom}_{\mathcal{M}} \to \mathrm{hom}_{\mathcal{M}'}$, but that is just a function $M \to M'$. The identity and composition formulas for functors match precisely with the identity and composition formula for monoid homomorphisms. Thus a monoid homomorphism is nothing more than a functor between one-object categories.

Slogan 5.2.1.2.

> *A monoid is a category with one object. A monoid homomorphism is just a functor between one-object categories.*

This is formalized in the following theorem.

Theorem 5.2.1.3. *There is a functor $i\colon \mathbf{Mon} \to \mathbf{Cat}$ with the following properties:*

- *For every monoid $\mathcal{M} \in \mathrm{Ob}(\mathbf{Mon})$, the category $i(\mathcal{M}) \in \mathrm{Ob}(\mathbf{Cat})$ itself has exactly one object,*

$$|\mathrm{Ob}(i(\mathcal{M}))| = 1.$$

- *For every pair of monoids $\mathcal{M}, \mathcal{M}' \in \mathrm{Ob}(\mathbf{Mon})$, the function*

$$\mathrm{Hom}_{\mathbf{Mon}}(\mathcal{M}, \mathcal{M}') \xrightarrow{\cong} \mathrm{Hom}_{\mathbf{Cat}}(i(\mathcal{M}), i(\mathcal{M}')),$$

induced by the functor i, is a bijection.

Proof. This is basically the content of the preceding paragraphs. The functor i sends a monoid to the corresponding category with one object and i sends a monoid homomorphism to the corresponding functor. One can check that i preserves identities and compositions.

□

Theorem 5.2.1.3 situates the theory of monoids very nicely within the world of categories. But we have other ways of thinking about monoids, namely, their actions on sets. It would greatly strengthen the story if we could subsume monoid actions within category theory also, and we can.

Each monoid action as a set-valued functor   Recall from Definition 4.1.2.1 that if $(M, e, \star)$ is a monoid, an action consists of a set S and a function $\circlearrowright \colon M \times S \to S$ such that $e \circlearrowright s = s$ and $m_0 \circlearrowright (m_1 \circlearrowright s) = (m_0 \star m_1) \circlearrowright s$ for all $s \in S$. How might we relate the notion of monoid actions to the notion of functors? Since monoids act on sets, one idea is to try asking what a functor $F \colon \mathcal{M} \to \mathbf{Set}$ is; this idea will work.

The monoid-as-category $\mathcal{M}$ has only one object, $\blacktriangle$, so F provides one set, $S := F(\blacktriangle) \in \mathrm{Ob}(\mathbf{Set})$. It also provides a function $\mathrm{Hom}_F \colon \mathrm{Hom}_{\mathcal{M}}(\blacktriangle, \blacktriangle) \to \mathrm{Hom}_{\mathbf{Set}}(F(\blacktriangle), F(\blacktriangle))$, or more concisely, a function

$$H_F \colon M \to \mathrm{Hom}_{\mathbf{Set}}(S, S).$$

By currying (see Proposition 3.4.2.3), this is the same as a function $\circlearrowright \colon M \times S \to S$. The first monoid action law, that $e \circlearrowright s = s$, becomes the law that functors preserve identities, $\mathrm{Hom}_F(\mathrm{id}_{\blacktriangle}) = \mathrm{id}_S$. The other monoid action law is equivalent to the composition law for functors.

5.2.1.4   Groups as categories

A group is just a monoid $(M, e, \star)$ in which every element $m \in M$ is invertible, meaning there exists some $m' \in M$ with $m \star m' = e = m' \star m$. If a monoid is the same thing as a category $\mathcal{M}$ with one object, then a group must be a category with one object and with an additional property having to do with invertibility. The elements of M are the morphisms of the category $\mathcal{M}$, so we need a notion of invertibility for morphisms. Luckily we have such a notion already, namely, isomorphism.

Slogan 5.2.1.5.

> *A group is a category $\mathcal{G}$ with one object, such that every morphism in $\mathcal{G}$ is an isomorphism. A group homomorphism is just a functor between such categories.*

Theorem 5.2.1.6. *There is a functor* $i\colon \mathbf{Grp} \to \mathbf{Cat}$ *with the following properties:*

- *For every group* $\mathcal{G} \in \mathrm{Ob}(\mathbf{Grp})$, *the category* $i(\mathcal{G}) \in \mathrm{Ob}(\mathbf{Cat})$ *itself has exactly one object, and every morphism* m *in* $i(\mathcal{G})$ *is an isomorphism.*

- *For every pair of groups* $\mathcal{G}, \mathcal{G}' \in \mathrm{Ob}(\mathbf{Grp})$, *the function*

$$\mathrm{Hom}_{\mathbf{Grp}}(\mathcal{G}, \mathcal{G}') \xrightarrow{\;\cong\;} \mathrm{Hom}_{\mathbf{Cat}}(i(\mathcal{G}), i(\mathcal{G}')),$$

induced by the functor i, *is a bijection.*

Just as with monoids, an action of some group $(G, e, \star)$ on a set $S \in \mathrm{Ob}(\mathbf{Set})$ is the same thing as a functor $\mathcal{G} \to \mathbf{Set}$ sending the unique object of $\mathcal{G}$ to the set S.

5.2.1.7   A monoid and a group stationed at each object in any category

If a monoid is just a category with one object, we can locate monoids in any category $\mathcal{C}$ by focusing on one object in $\mathcal{C}$. Similarly for groups.

Example 5.2.1.8 (Endomorphism monoid). Let $\mathcal{C}$ be a category and $x \in \mathrm{Ob}(\mathcal{C})$ an object. Let $M = \mathrm{Hom}_{\mathcal{C}}(x, x)$. Note that for any two elements $f, g \in M$, we have $f \circ g\colon x \to x$ in M. Let $\mathcal{M} = (M, \mathrm{id}_x, \circ)$. It is easy to check that $\mathcal{M}$ is a monoid; it is called the *endomorphism monoid of* x *in* $\mathcal{C}$, denoted $\mathrm{End}(x)$.

Example 5.2.1.9 (Automorphism group). Let $\mathcal{C}$ be a category and $x \in \mathrm{Ob}(\mathcal{C})$ an object. Let $G = \{f \in \mathrm{Hom}_{\mathcal{C}}(x, x) \mid f \text{ is an isomorphism}\}$. Let $\mathcal{G} = (G, \mathrm{id}_x, \circ)$. One can check that $\mathcal{G}$ is a group; it is called the *automorphism group of* x *in* $\mathcal{C}$ denoted $\mathrm{Aut}(x)$.

Exercise 5.2.1.10.

Let $S = \{1, 2, 3, 4\} \in \mathrm{Ob}(\mathbf{Set})$.

a. What is the automorphism group $\mathrm{Aut}(S)$ of S in $\mathbf{Set}$, and how many elements does this group have?

b. What is the endomorphism monoid $\mathrm{End}(S)$ of S in $\mathbf{Set}$, and how many elements does this monoid have?

c. Recall from Example 5.1.2.3 that every group has an underlying monoid $U(G)$. Is the endomorphism monoid of S the underlying monoid of the automorphism group of S? That is, is it the case that $\mathrm{End}(S) = U(\mathrm{Aut}(S))$?

$\diamond$

Solution 5.2.1.10.

a. It is the permutation group $\text{Aut}_S = (\text{Iso}(S), \text{id}_S, \circ)$, where $\text{Iso}(S)$ is as defined in Exercise 4.2.1.7. It has $4! = 24$ elements.

b. It is the monoid $(\text{Hom}_{\mathbf{Set}}(S, S), \text{id}_S, \circ)$ of all functions from S to S. It has $4^4 = 256$ elements.

c. No. The monoid underlying a group G has the same number of elements as G has, so it is not the case that the monoid underlying the permutation group of S is the endomorphism monoid of S.

The reader may note that there is a functor $\text{Core} \colon \mathbf{Mon} \to \mathbf{Grp}$ that takes a monoid M and returns its subset of invertible elements, which forms a group called the *core* of M. The core of the endomorphism monoid $\text{End}(S)$ is indeed the automorphism group $\text{Core}(\text{End}(S)) = \text{Aut}(S)$.

♦

Exercise 5.2.1.11.

Consider the following graph G, which has four vertices and eight arrows:

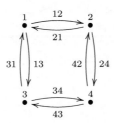

What is the automorphism group $\text{Aut}(G)$ of $G \in \text{Ob}(\mathbf{Grph})$ Hint: Every automorphism of G will induce an automorphism of the set $\{1, 2, 3, 4\}$; which ones will preserve the endpoints of arrows? ◊

Solution 5.2.1.11.

We use visual perception to guide us. The graph G has the shape of a square. Of the $4!$ different possible automorphisms of $\{1, 2, 3, 4\}$, only those preserving the square shape will be automorphisms of G. The group of automorphisms of G is called the dihedral group of order 8 (see Example 4.2.1.4). It has eight elements,

$$\{e, r, r^2, r^3, f, fr, fr^2, fr^3\},$$

where r means rotate the square clockwise $90°$, and f means flip the square horizontally. For example, flipping the square vertically can be obtained by flipping horizontally and then rotating twice: fr^2. ♦

5.2.1.12 Preorders as categories

A preorder $(X, \leqslant)$ consists of a set X and a binary relation $\leqslant$ that is reflexive and transitive. We can make from $(X, \leqslant) \in \mathrm{Ob}(\mathbf{PrO})$ a category $\mathcal{X} \in \mathrm{Ob}(\mathbf{Cat})$ as follows. Define $\mathrm{Ob}(\mathcal{X}) = X$ and for every two objects $x, y \in X$, define

$$\mathrm{Hom}_{\mathcal{X}}(x, y) = \begin{cases} \{\text{“}x \leqslant y\text{”}\} & \text{if } x \leqslant y, \\ \varnothing & \text{if } x \nleqslant y. \end{cases}$$

To clarify: if $x \leqslant y$, we assign $\mathrm{Hom}_{\mathcal{X}}(x, y)$ to be the set containing only one element, namely, the string "$x \leqslant y$."[6] If the pair (x, y) is not in relation $\leqslant$, then we assign $\mathrm{Hom}_{\mathcal{X}}(x, y)$ to be the empty set. The composition formula

$$\circ \colon \mathrm{Hom}_{\mathcal{X}}(x, y) \times \mathrm{Hom}_{\mathcal{X}}(y, z) \to \mathrm{Hom}_{\mathcal{X}}(x, z) \tag{5.7}$$

is completely determined because either one of two possibilities occurs. One possibility is that the left-hand side is empty (if either $x \nleqslant y$ or $y \nleqslant z$; in this case there is a unique function $\circ$ as in (5.7)). The other possibility is that the left-hand side is not empty in case $x \leqslant y$ and $y \leqslant z$, which implies $x \leqslant z$, so the right-hand side has exactly one element "$x \leqslant z$" in which case again there is a unique function $\circ$ as in (5.7).

On the other hand, if $\mathcal{C}$ is a category having the property that for every pair of objects $x, y \in \mathrm{Ob}(\mathcal{C})$, the set $\mathrm{Hom}_{\mathcal{C}}(x, y)$ is either empty or has one element, then we can form a preorder out of $\mathcal{C}$. Namely, take $X = \mathrm{Ob}(\mathcal{C})$ and say $x \leqslant y$ if there exists a morphism $x \to y$ in $\mathcal{C}$.

Proposition 5.2.1.13. *There is a functor $i \colon \mathbf{PrO} \to \mathbf{Cat}$ with the following properties for every preorder $(X, \leqslant)$:*

1. *the category $\mathcal{X} := i(X, \leqslant)$ has objects $\mathrm{Ob}(\mathcal{X}) = X$.*

2. *For each pair of elements $x, x' \in \mathrm{Ob}(\mathcal{X})$, the set $\mathrm{Hom}_{\mathcal{X}}(x, x')$ has at most one element.*

Moreover, any category with property 2 is in the image of the functor i.

---

[6]The name of this morphism is unimportant. What matters is that $\mathrm{Hom}_{\mathcal{X}}(x, y)$ has exactly one element iff $x \leqslant y$.

Proof. To specify a functor $i\colon \mathbf{PrO} \to \mathbf{Cat}$, we need to say what it does on objects and on morphisms. To an object $(X, \leqslant)$ in $\mathbf{PrO}$, we assign the category $\mathcal{X}$ with objects X and a unique morphism $x \to x'$ if $x \leqslant x'$. To a morphism $f\colon (X, \leqslant_X) \to (Y, \leqslant_Y)$ of preorders, we must assign a functor $i(f)\colon \mathcal{X} \to \mathcal{Y}$. Again, to specify a functor, we need to say what it does on objects and morphisms of $\mathcal{X}$. To an object $x \in \mathrm{Ob}(\mathcal{X}) = X$, we assign the object $f(x) \in Y = \mathrm{Ob}(\mathcal{Y})$. Given a morphism $f\colon x \to x'$ in $\mathcal{X}$, we know that $x \leqslant x'$, so by Definition 4.4.4.1 we have that $f(x) \leqslant f(x')$, and we assign to f the unique morphism $f(x) \to f(x')$ in $\mathcal{Y}$. To check that the rules of functors (preservation of identities and composition) are obeyed is routine.

$\square$

Slogan 5.2.1.14.

A preorder is a category in which every hom-set has either 0 elements or 1 element. A preorder morphism is just a functor between such categories.

Exercise 5.2.1.15.

Suppose that $\mathcal{C}$ is a preorder (considered as a category). Let $x, y \in \mathrm{Ob}(\mathcal{C})$ be objects such that $x \leqslant y$ and $y \leqslant x$. Prove that there is an isomorphism $x \to y$ in $\mathcal{C}$. ◊

Solution 5.2.1.15.

Categorically, $x \leqslant y$ means that there is a morphism $f\colon x \to y$ in $\mathcal{C}$, and similarly $g\colon y \leqslant x$ means that there is a morphism $y \to x$ in $\mathcal{C}$. We can compose these to get a morphism $g \circ f\colon x \to x$ and a morphism $f \circ g\colon y \to y$. But the condition that $\mathcal{C}$ is a preorder is that every hom-set has at most one element. We already have $\mathrm{id}_x\colon x \to x$ and $\mathrm{id}_y\colon y \to y$, so composites $g \circ f$ and $f \circ g$ must be the identity morphisms id_x and id_y respectively. By definition, then, f and g are isomorphisms. ◆

Exercise 5.2.1.16.

Proposition 5.2.1.13 stated that a preorder can be considered as a category $\mathcal{P}$. Recall from Definition 4.4.1.1 that a partial order is a preorder with an additional property. Phrase the defining property for partial orders in terms of isomorphisms in the category $\mathcal{P}$. ◊

Solution 5.2.1.16.

A preorder is a category $\mathcal{P}$ such that, for every pair of objects $x, y \in \mathrm{Ob}(\mathcal{P})$, the set $\mathrm{Hom}_{\mathcal{P}}(x, y)$ has at most one element. An element in $\mathrm{Hom}_{\mathcal{P}}(x, y)$ represents the fact that $x \leqslant y$ in $\mathcal{P}$. In a partial order, we never have $x \leqslant y$ and $y \leqslant x$, unless $x = y$. The situation $x \leqslant y$ and $y \leqslant x$ corresponds categorically to the situation when there is an isomorphism $x \cong y$. So the condition of $\mathcal{P}$ being a partial order can be phrased categorically as "whenever any two objects are isomorphic in $\mathcal{P}$, they must in fact be equal." ◆

Example 5.2.1.17. The olog from Example 4.4.1.3 depicted a partial order, call it $\mathcal{P}$. In it we have

$$\mathrm{Hom}_\mathcal{P}(\ulcorner\text{a diamond}\urcorner, \ulcorner\text{a red card}\urcorner) = \{\text{is}\}$$

and

$$\mathrm{Hom}_\mathcal{P}(\ulcorner\text{a black queen}\urcorner, \ulcorner\text{a card}\urcorner) \cong \{\text{is} \circ \text{is}\}.$$

Both of these sets contain exactly one element; the name is not important. The set $\mathrm{Hom}_\mathcal{P}(\ulcorner\text{a 4}\urcorner, \ulcorner\text{a 4 of diamonds}\urcorner) = \varnothing$.

Exercise 5.2.1.18.

Every linear order is a preorder with a special property. Using the categorical interpretation of preorders, can you phrase the property of being a linear order in terms of hom-sets? ◊

Solution 5.2.1.18.

A linear order is a preorder $\mathcal{P}$ such that for any two objects x, y either $\mathrm{Hom}_\mathcal{P}(x, y)$ has one element or $\mathrm{Hom}_\mathcal{P}(y, x)$ has one element, but not both (unless $x = y$). ◆

Exercise 5.2.1.19.

Recall the functor $P \colon \mathbf{PrO} \to \mathbf{Grph}$ from Proposition 5.1.2.10, the functors $F \colon \mathbf{Grph} \to \mathbf{Cat}$ and $U \colon \mathbf{Cat} \to \mathbf{Grph}$ from Example 5.1.2.36, and the functor $i \colon \mathbf{PrO} \to \mathbf{Cat}$ from Proposition 5.2.1.13.

a. Do either of the following diagrams of categories commute?

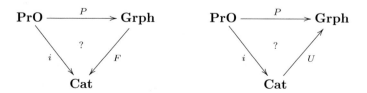

b. We also gave a functor $Im\colon \mathbf{Grph} \to \mathbf{PrO}$ in Exercise 5.1.2.13. Does the following diagram of categories commute?

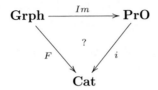

$\diamond$

Solution 5.2.1.19.

a. Only the second triangle commutes,

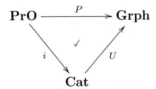

As an example of why the first triangle does not commute, let $X = \underline{1}$ be the unique preorder having one element. Then $P(X)$ is the loop graph, having an arrow $1 \to 1$ corresponding to the fact that $1 \leqslant 1$, and $i(X)$ is the terminal category, with one object 1 and one morphism $(\mathrm{Hom}_{i(X)}(1,1) = \{\mathrm{id}_1\})$. But the free category on the loop graph is the free monoid on one generator—it has $\mathrm{Hom}_{FP(X)}(1,1) \cong \mathbb{N}$. Since $\mathrm{Hom}_{i(X)}(1,1) \neq \mathrm{Hom}_{F \circ P(X)}(1,1)$, we must have $i \neq F \circ P$.

b. No, this does not commute. An easy example is the parallel arrows graph X drawn as follows:

$$X := \boxed{\; \overset{a}{\bullet} \underset{\;}{\overset{\;}{\rightrightarrows}} \overset{b}{\bullet} \;}$$

The free category $F(X)$ has four morphisms, whereas $i \circ Im(X)$ has only three morphisms.

$\blacklozenge$

Proposition 5.2.1.20. *There is a unique functor $R\colon \mathbf{Cat} \to \mathbf{PrO}$ with the following properties:*

1. *For each category $\mathcal{C}$, the preorder $(X, \leqslant) := R(\mathcal{C})$ has the same set of objects, $X = \mathrm{Ob}(\mathcal{C})$.*

2. *For each pair of objects $x, y \in \mathrm{Ob}(\mathcal{C})$, we have $x \leqslant y$ in $R(C)$ if and only if the hom-set $\mathrm{Hom}_{\mathcal{C}}(x, y) \neq \varnothing$ is nonempty.*

Furthermore, if $i \colon \mathbf{PrO} \to \mathbf{Cat}$ is the inclusion from Proposition 5.2.1.13, we have $R \circ i = \mathrm{id}_{\mathbf{PrO}}$.

Proof. Given a category $\mathcal{C}$, we define a preorder $R(\mathcal{C}) := (\mathrm{Ob}(\mathcal{C}), \leqslant)$, where $x \leqslant y$ if and only if $\mathrm{Hom}_{\mathcal{C}}(x, y) \neq \varnothing$. This is indeed a preorder because the identity law and composition law for a category ensure the reflexivity and transitivity properties of preorders hold. Given a functor $F \colon \mathcal{C} \to \mathcal{D}$ (i.e., a morphism in $\mathbf{Cat}$), we get $\mathrm{Ob}(F) \colon \mathrm{Ob}(\mathcal{C}) \to \mathrm{Ob}(\mathcal{C}')$, and for R to be defined on morphisms, we need to check that this function preserves order. If $x \leqslant y$ in $R(\mathcal{C})$, then there is a morphism $g \colon x \to y$ in $\mathcal{C}$, so there is a morphism $F(g) \colon F(x) \to F(y)$, which means $F(x) \leqslant F(y)$ in $\mathcal{C}'$. It is straightforward to see now that R is a functor, and there was no other way to construct R satisfying the desired properties. It is also easy to see that $R \circ i = \mathrm{id}_{\mathbf{PrO}}$. $\square$

5.2.1.21   Graphs as functors

Let $\mathcal{C}$ denote the category depicted as follows:

$$
\mathbf{GrIn} := \boxed{\begin{array}{c} Ar \quad \xrightarrow[tgt]{src} \quad Ve \\ \bullet \qquad\qquad \bullet \end{array}}
\tag{5.8}
$$

Then a functor $G \colon \mathbf{GrIn} \to \mathbf{Set}$ is the same thing as two sets $G(Ar), G(Ve)$ and two functions $G(src) \colon G(Ar) \to G(Ve)$ and $G(tgt) \colon G(Ar) \to G(Ve)$. This is precisely what is needed for a graph; see Definition 4.3.1.1. We call $\mathbf{GrIn}$ the *graph-indexing category*.

Exercise 5.2.1.22.

   Consider the terminal category, $\underline{1}$, also known as the discrete category on one element (see Exercise 5.1.2.40). Let $\mathbf{GrIn}$ be as in (5.8) and consider the functor $i_0 \colon \underline{1} \to \mathbf{GrIn}$ sending the unique object of $\underline{1}$ to the object $Ve \in \mathrm{Ob}(\mathbf{GrIn})$.

a. If $G \colon \mathbf{GrIn} \to \mathbf{Set}$ is a graph, what is the composite $G \circ i_0$? It consists of only one set; in terms of the graph G, what set is it?

b. As an example, what set is it when G is the graph from Example 4.3.3.3?

                                                                              ◇

Solution 5.2.1.22.

a. The composite $\underline{1} \xrightarrow{i_0} \mathbf{GrIn} \xrightarrow{G} \mathbf{Set}$ is the functor that sends the unique object of $\underline{1}$ to the set of vertices of G.

b. The set of vertices in this graph is $\{1, 2, 3, 4, 5, 6\}$.

$\blacklozenge$

If a graph is a functor $\mathbf{GrIn} \to \mathbf{Set}$, what is a graph homomorphism? Example 5.3.1.20 shows that graph homomorphisms are homomorphisms between functors, which are called natural transformations. (Natural transformations are the highest-level structure in ordinary category theory.)

Example 5.2.1.23. Let $\mathbf{SGrIn}$ be the category depicted as follows:

$$\mathbf{SGrIn} := \boxed{\rho \circlearrowleft \overset{A}{\bullet} \underset{tgt}{\overset{src}{\rightrightarrows}} \overset{V}{\bullet}} \qquad (5.9)$$

with the following composition formula:

$$\rho \circ \rho = \mathrm{id}_A; \qquad src \circ \rho = tgt; \qquad \text{and} \qquad tgt \circ \rho = src.$$

The idea here is that the morphism $\rho \colon A \to A$ reverses arrows. The PED $_A[\rho, \rho] = _A[\,]$ forces the fact that the reverse of the reverse of an arrow yields the original arrow. The PEDs $_A[\rho, src] = _A[tgt]$ and $_A[\rho, tgt] = _A[src]$ force the fact that when we reverse an arrow, its source and target switch roles.

This category $\mathbf{SGrIn}$ is the *symmetric graph-indexing category*. Just as any graph can be understood as a functor $\mathbf{GrIn} \to \mathbf{Set}$, where $\mathbf{GrIn}$ is the graph-indexing category displayed in (5.8), any symmetric graph can be understood as a functor $\mathbf{SGrIn} \to \mathbf{Set}$, where $\mathbf{SGrIn}$ is the category drawn in (5.9). Given a functor $G \colon \mathbf{SGrIn} \to \mathbf{Set}$, we will have a set of arrows, a set of vertices, a source operation, a target operation, and a reverse-direction operation (ρ) that all behave as expected.

It is customary to draw the connections in a symmetric graph G as line segments rather than arrows between vertices. However, a better heuristic is to think that each connection between vertices in G consists of two arrows, one pointing in each direction.

Slogan 5.2.1.24.

In a symmetric graph, every arrow has an equal and opposite arrow.

Exercise 5.2.1.25.

Which of the following graphs are symmetric:

a. The graph G from (4.4)?

b. The graph G from Exercise 4.3.1.10?

c. The graph G' from (4.7)?

d. The graph $\mathcal{L}oop$ from (4.17), i.e., the graph having exactly one vertex and one arrow?

e. The graph G from Exercise 5.2.1.11?

$\diamond$

Solution 5.2.1.25.

a. No, f has no inverse.

b. Yes.

c. No, y has no inverse (but close).

d. Yes.

e. Yes.

$\blacklozenge$

Exercise 5.2.1.26.

Let **GrIn** be the graph-indexing category shown in (5.8), and let **SGrIn** be the symmetric graph-indexing category displayed in (5.9).

a. How many functors are there of the form **GrIn** → **SGrIn**?

b. Is one more reasonable than the others? If so, call it i: **GrIn** → **SGrIn**, and write how it acts on objects and morphisms.

c. Choose a functor i: **GrIn** → **SGrIn**, the most reasonable one, if such a thing exists. seems most reasonable and call it i: **GrIn** → **SGrIn**. If a symmetric graph is a functor S: **SGrIn** → **Set**, you can compose with i to get a functor $S \circ i$: **GrIn** → **Set**. This is a graph; what graph is it? What has changed?

$\diamond$

Solution 5.2.1.26.

a. There are 9. We could send both objects of **GrIn** to $A \in \mathrm{Ob}(\mathbf{SGrIn})$: there are four of these. We could send both objects of **GrIn** to $V \in \mathrm{Ob}(\mathbf{SGrIn})$: there is one of these. Or we could send $Ar \mapsto A$ and $Ve \mapsto V$: there are four of these.

b. The most reasonable one is the inclusion that preserves names as well as possible:

$$Ar \mapsto A; \qquad Ve \mapsto V; \qquad src \mapsto src; \qquad tgt \mapsto tgt.$$

Let's call this $i \colon \mathbf{GrIn} \to \mathbf{SGrIn}$.

c. It is basically S again, except now we are considering it as a graph rather than as a symmetric graph. It is still symmetric, but it is interacting in the context of mere graphs. The inclusion $i \colon \mathbf{GrIn} \to \mathbf{SGrIn}$ of the graph-indexing category into the symmetric-graph-indexing category has something to do with forgetful functors. (See Remark 5.1.2.7 and Section 7.1.4.2.)

♦

Example 5.2.1.27. Let $\mathcal{C}$ be a category, and consider the set of isomorphisms in $\mathcal{C}$. Each isomorphism $f \colon c \to c'$ in $\mathcal{C}$ has an inverse as well as a domain (c) and a codomain (c'). Thus we can build a symmetric graph $I(\mathcal{C}) \colon \mathbf{SGrIn} \to \mathbf{Set}$. Its vertices are the objects in $\mathcal{C}$, and its arrows are the isomorphisms in $\mathcal{C}$.

5.2.2 Database schemas present categories

Recall from Definition 4.5.2.7 that a database schema (or schema, for short) consists of a graph together with a certain kind of equivalence relation, namely a congruence, on its paths. Section 5.4.1 defines a category **Sch** that has schemas as objects and appropriately modified graph homomorphisms as morphisms. Section 5.4.2 proves that the category of schemas is equivalent (in the sense of Definition 5.3.4.1) to the category of categories,

$$\mathbf{Sch} \simeq \mathbf{Cat}.$$

The difference between schemas and categories is like the difference between monoid presentations, given by generators and relations as in Definition 4.1.1.19, and the monoids themselves. The same monoid has (infinitely) many different presentations, and so it is for categories: many different schemas can *present* the same category. Computer scientists may think of the schema as *syntax* and the category it presents as the corresponding *semantics*. A schema is a compact form and can be specified in finite space and time, whereas the category it generates can be infinite.

Slogan 5.2.2.1.

A database schema is a category presentation.

Section 5.4.2 formally shows how to turn a schema into a category (the category it *presents*). For now, it seems better not to be so formal, because the idea is fairly straightforward. Suppose given a schema $\mathcal{S}$, which consists of a graph $G = (V, A, src, tgt)$ equipped with a congruence $\sim$ (see Definition 4.5.2.3). It presents a category $\mathcal{C}$ defined as follows. The set of objects in $\mathcal{C}$ is defined to be the vertices V; the set of morphisms in $\mathcal{C}$ is defined to be the quotient Paths$(G)/\sim$; and the composition formula is given by concatenation of paths. The path equivalences making up $\sim$ become commutative diagrams in $\mathcal{C}$.

Example 5.2.2.2. The following schema $\mathcal{L}oop$ has no path equivalence declarations. As a graph it has one vertex and one arrow.

$$\mathcal{L}oop := \boxed{\begin{array}{c} f \\ \circlearrowright \ s \\ \bullet \end{array}}$$

The category it generates, however, is the free monoid on one generator, $\mathbb{N}$. It has one object s, but a morphism $f^n \colon s \to s$ for every natural number $n \in \mathbb{N}$, thought of as "how many times to go around the loop f." Clearly, the schema is more compact than the infinite category it generates.

Exercise 5.2.2.3.

Consider the olog from Exercise 4.5.2.19, which says that for any father x, his youngest child's father is x and his tallest child's father is x. It is redrawn here as a schema $\mathcal{S}$, which includes the desired path equivalence declarations, $_F[t, f] = {}_F[\,]$ and $_F[y, f] = {}_F[\,]$.

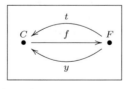

How many morphisms are there (total) in the category presented by $\mathcal{S}$?   ◊

Solution 5.2.2.3.

There are seven. Let $\overline{\mathcal{S}}$ be the category presented by $\mathcal{S}$. We have

$$\mathrm{Hom}_{\overline{\mathcal{S}}}(F, F) = \{{}_F[\,]\}; \qquad \mathrm{Hom}_{\overline{\mathcal{S}}}(F, C) = \{{}_F[t], \ {}_F[y]\};$$
$$\mathrm{Hom}_{\overline{\mathcal{S}}}(C, F) = \{{}_C[f]\}; \qquad \mathrm{Hom}_{\overline{\mathcal{S}}}(C, C) = \{{}_C[\,], \ {}_C[f, t], \ {}_C[f, y]\}.$$

Given a child, the three morphisms $C \to C$ respectively return the child herself, her tallest sibling (technically, her father's tallest child), and her youngest sibling (technically, her father's youngest child). ♦

Exercise 5.2.2.4.

Suppose that G is a graph and that $\mathcal{G}$ is the schema generated by G with no PEDs. What is the relationship between the category generated by $\mathcal{G}$ and the free category $F(G) \in \mathrm{Ob}(\mathbf{Cat})$, as defined in Example 5.1.2.33? ◊

Solution 5.2.2.4.

These are the same category. ♦

Exercise 5.2.2.5.

Let $\mathcal{C} = (G, \simeq)$ be a schema. A leaf table is an object $c \in \mathrm{Ob}(\mathcal{C})$ with no outgoing arrows.

a. Express the condition of being a leaf table mathematically in three different languages: that of graphs (using symbols V, A, src, tgt), that of categories (using $\mathrm{Hom}_\mathcal{C}$, etc.), and that of tables (in terms of columns, tables, rows, etc.).

b. In the language of categories, is there a difference between a terminal object and a leaf table? Explain.

◊

Solution 5.2.2.5.

a. **Graphs:** A leaf vertex is a vertex $v \in V$ with no outgoing arrows, that is, one for which there is no arrow $a \in A$ with $src(a) = v$.

   Categories: A leaf object is an object $c \in \mathrm{Ob}(\mathcal{C})$ such that for all $d \in \mathrm{Ob}(\mathcal{C})$, we have

$$\mathrm{Hom}_\mathcal{C}(c, d) = \begin{cases} \varnothing & \text{if } c \neq d, \\ \{\mathrm{id}_c\} & \text{if } c = d. \end{cases}$$

   Tables: A leaf table is a table with only one column, namely, the ID column.

b. There is a big difference. An object c can be a leaf object and not a terminal object, e.g., in the discrete category $\mathrm{Disc}(\{c, d\})$. And an object c can be a terminal object and not a leaf object, e.g., in the father-child category of Exercise 5.2.2.3 the object F is terminal.

♦

5.2.2.6   Instances on a schema $\mathcal{C}$

If schemas are like categories, what are instances? Recall that an instance I on a schema $\mathcal{S} = (G, \simeq)$ assigns to each vertex v in G a set of rows, say, $I(v) \in \mathrm{Ob}(\mathbf{Set})$. And to every arrow $a\colon v \to v'$ in G the instance assigns a function $I(a)\colon I(v) \to I(v')$. The rule is that given two equivalent paths, their compositions must give the same function. Concisely, an instance is a functor $I\colon \mathcal{S} \to \mathbf{Set}$.

Example 5.2.2.7. We have seen that a monoid is just a category $\mathcal{M}$ with one object and that a monoid action is a functor $\mathcal{M} \to \mathbf{Set}$. With database schemas as categories, $\mathcal{M}$ is a schema, and so an action becomes an instance of that schema. The monoid action table from Example 4.1.3.1 was simply a manifestation of the database instance according to the Rules 4.5.2.9.

Exercise 5.2.2.8.

Section 5.2.1.21 discussed how each graph is a functor $\mathbf{GrIn} \to \mathbf{Set}$ for the graph-indexing category depicted here:

$$\mathbf{GrIn} := \boxed{\begin{array}{c} Ar \xrightarrow[tgt]{src} Ve \\ \bullet \qquad\quad \bullet \end{array}}$$

But now we know that if a graph is a set-valued functor, then we can consider $\mathbf{GrIn}$ as a database schema.

a. How many tables, and how many foreign key columns of each should there be (if unsure, consult Rules 4.5.2.9)?

b. Write the table view of graph G from Example 4.3.3.3.

◊

Solution 5.2.2.8.

a. Two tables, with two and zero foreign key columns respectively; see part (b).

b.

	Ar				Ve
ID	*src*	*tgt*		**ID**	
a	1	2		1	
b	2	3		2	
c	1	4		3	
d	1	4		4	
e	5	6		5	
				6	

◆

5.2.3 Spaces

Category theory was invented for use in algebraic topology, and in particular, to discuss natural transformations between certain functors. Section 5.3 discusses natural transformations more formally. It suffices now to say a natural transformation is some kind of morphism between functors. In the original use, Eilenberg and Mac Lane were interested in functors that connect topological spaces (e.g., shapes such as spheres) to algebraic systems (e.g., groups).

For example, there is a functor that assigns to each space X its group $\pi_1(X)$ of round-trip voyages (starting and ending at some chosen point $x \in X$), modulo some equivalence relation. There is another functor that assigns to every space its group $H\mathbb{Z}_1(X)$ of ways to drop some (positive or negative) number of circles on X.

These two functors, π_1 and $H\mathbb{Z}_1$ are related, but they are not equal. For example, when X is the figure-8 space (two circles joined at a point) the group $\pi_1(X)$ is much bigger than the group $H\mathbb{Z}_1(X)$. Indeed, $\pi_1(X)$ includes information about the order and direction of loops traveled during the voyage, whereas the group $H\mathbb{Z}_1(X)$ includes only information about how many times one goes around each loop. However, there is a natural transformation of functors $\pi_1 \to H\mathbb{Z}_1$, called the Hurewicz transformation, which takes π_1's voyage, counts how many times it went around each loop, and delivers that information to $H\mathbb{Z}_1$.

Example 5.2.3.1. Given a set X, recall that $\mathbb{P}(X)$ denotes the preorder of subsets of X. A *topology* on X is a choice of which subsets $U \in \mathbb{P}(X)$ will be called *open sets*. To be a topology, these open sets must follow two rules. Namely, the union of any number of open sets must be considered to be an open set, and the intersection of any finite number of open sets must be considered open. One could say succinctly that a topology on X is a suborder $\mathrm{Open}(X) \subseteq \mathbb{P}(X)$ that is closed under taking finite meets and infinite joins.

A *topological space* is a pair $(X, \mathrm{Open}(X))$, where X is a set and $\mathrm{Open}(X)$ is a topology on X. The elements of the set X are called *points*. A *morphism of topological spaces* (also called a *continuous map*) is a function $f\colon X \to Y$ such that for every $V \in \mathrm{Open}(Y)$, the preimage $f^{-1}(V) \in \mathbb{P}(X)$ is actually in $\mathrm{Open}(X)$, that is, such that there exists a dashed arrow making the following diagram commute:

$$
\begin{array}{ccc}
\mathrm{Open}(Y) & \dashrightarrow & \mathrm{Open}(X) \\
\downarrow & & \downarrow \\
\mathbb{P}(Y) & \xrightarrow[f^{-1}]{} & \mathbb{P}(X).
\end{array}
$$

The *category of topological spaces*, denoted **Top**, is the category having the preceding objects and morphisms.

Exercise 5.2.3.2.

a. Explain how looking at points gives a functor **Top** $\to$ **Set**.

b. Does looking at open sets give a functor **Top** $\to$ **PrO**?

$\Diamond$

Solution 5.2.3.2.

a. A topological space $(X, \mathrm{Open}(X))$ includes a set $X \in \mathrm{Ob}(\mathbf{Set})$ of points. A morphism $(X, \mathrm{Open}(X)) \to (Y, \mathrm{Open}(Y))$ of spaces includes a function $X \to Y$. Thus we have a functor **Top** $\to$ **Set**, because the identity morphisms and compositions of morphisms in **Top** are sent to their counterparts in **Set**.

b. No. A morphism $(X, \mathrm{Open}(X)) \to (Y, \mathrm{Open}(Y))$ includes a preorder morphism in the direction $\mathrm{Open}(Y) \to \mathrm{Open}(X)$, not the other way around. Definition 6.2.1.1 shows that every category $\mathcal{C}$ has an opposite category $\mathcal{C}^{\mathrm{op}}$. Looking at open sets does give a functor $\mathrm{Open}\colon \mathbf{Top}^{\mathrm{op}} \to \mathbf{PrO}$.

$\blacklozenge$

Example 5.2.3.3 (Continuous dynamical systems). The set $\mathbb{R}$ can be given a topology in a standard way.[7] But $(\mathbb{R}, 0, +)$ is also a monoid. Moreover, for every $x \in \mathbb{R}$, the monoid

---

[7]The topology is given by saying that $U \subseteq \mathbb{R}$ is open iff for every $x \in U$, there exists $\epsilon > 0$ such that $\{y \in \mathbb{R} \mid |y - x| < \epsilon\} \subseteq U$. One says, "$U \subseteq \mathbb{R}$ is open if every point in U has an epsilon-neighborhood fully contained in U."

operation $+\colon \mathbb{R} \times \mathbb{R} \to \mathbb{R}$ is continuous.[8] So we say that $\mathcal{R} := (\mathbb{R}, 0, +)$ is a *topological monoid*, or that it is a monoid *enriched in topological spaces*.

Recall from Section 5.2.1.1 that an action of $\mathcal{R}$ is a functor $\mathcal{R} \to \mathbf{Set}$. Imagine a functor $a\colon \mathcal{R} \to \mathbf{Top}$. Since $\mathcal{R}$ is a category with one object, this amounts to an object $X \in \mathrm{Ob}(\mathbf{Top})$, a space. And for every real number $t \in \mathbb{R}$, we obtain a continuous map $a(t)\colon X \to X$. Further we can ask this $a(t)$ to vary continuously as t moves around in $\mathbb{R}$. If we consider X as the set of states of some system and $\mathbb{R}$ as the time line, we have modeled what is called a *continuous dynamical system*.

Example 5.2.3.4. Recall (see Axler [3]) that a *real vector space* is a set X, elements of which are called *vectors*, which is closed under addition and scalar multiplication. For example, $\mathbb{R}^3$ is a vector space. A *linear transformation f from X to Y* is a function $f\colon X \to Y$ that appropriately preserves addition and scalar multiplication. The *category of real vector spaces*, denoted $\mathbf{Vect}_{\mathbb{R}}$, has as objects the real vector spaces and as morphisms the linear transformations.

There is a functor $\mathbf{Vect}_{\mathbb{R}} \to \mathbf{Grp}$ sending a vector space to its underlying group of vectors, where the group operation is addition of vectors and the group identity is the 0-vector.

Exercise 5.2.3.5.

Every vector space has vector subspaces, ordered by inclusion (the origin is inside of any line that is inside of certain planes, and all are inside of the whole space V). If you know about this topic, answer the following questions.

a. Does a linear transformation $V \to V'$ induce a morphism of these orders? In other words, is there a functor subspaces: $\mathbf{Vect}_{\mathbb{R}} \to \mathbf{PrO}$?

b. Would you guess that there is a nice functor $\mathbf{Vect}_{\mathbb{R}} \to \mathbf{Top}$? By "nice functor" I mean a substantive one. For example, there is a functor $\mathbf{Vect}_{\mathbb{R}} \to \mathbf{Top}$ that sends every vector space to the empty topological space; if someone asked for a functor $\mathbf{Vect}_{\mathbb{R}} \to \mathbf{Top}$ for their birthday, this functor would make them sad. Give a functor $\mathbf{Vect}_{\mathbb{R}} \to \mathbf{Top}$ that would make them happy.

$\Diamond$

Solution 5.2.3.5.

a. Yes. If $A \subseteq B$ are subspaces of V and $f\colon V \to V'$ is a linear transformation, then $T(A) \subseteq T(B)$.

---

[8]The topology on $\mathbb{R} \times \mathbb{R}$ is similar; a subset $U \subseteq \mathbb{R} \times \mathbb{R}$ is open if every point $x \in U$ has an epsilon-neighborhood (a disk around x of some positive radius) fully contained in U.

b. Yes. Send a vector space $\mathbb{R}^n$ to the topological space $\mathbb{R}^n$. Every linear transformation will induce a continuous map.

♦

There is a functor $|\cdot|\colon \mathbf{Vect}_\mathbb{R} \to \mathbf{Set}$ sending every vector space X to its set $|X|$ of vectors. A categorically nice way to understand this functor is as $\mathrm{Hom}_{\mathbf{Vect}_\mathbb{R}}(\mathbb{R}, -)$, which sends X to the set of linear transformations $\mathbb{R} \to X$. Each linear transformation $\mathbb{R} \to X$ is completely determined by where it sends $1 \in \mathbb{R}$, which can be any vector in X. Thus we get the bijection $|X| \cong \mathrm{Hom}_{\mathbf{Vect}_\mathbb{R}}(\mathbb{R}, X)$.

Exercise 5.2.3.6.

Suppose we think of $\mathbf{Vect}_\mathbb{R}$ as a database schema, and we think of $|\cdot|\colon \mathbf{Vect}_\mathbb{R} \to \mathbf{Set}$ as an instance (see Section 4.5). Of course, the schema and the instance are both infinite, but let's not worry about that.

a. Pick two objects x, y and two morphisms $f, g\colon x \to y$ from $\mathbf{Vect}_\mathbb{R}$, actual vector spaces and linear transformations, and call this your subschema. Draw it as dots and arrows.

b. Write four rows in each table of the instance $|\cdot|$ on your subschema.

◇

Solution 5.2.3.6.

a.

$$
\mathbb{R}^3 \bullet
\begin{array}{c}
\left(\begin{array}{ccc} 2 & 2 & 0 \\ 1 & 1 & 2 \end{array} \right) \\
\rule{3cm}{0.4pt} \\
\left(\begin{array}{ccc} -2 & 1 & 3 \\ -1 & 2 & 2 \end{array} \right)
\end{array}
\rightrightarrows \mathbb{R}^2 \bullet
$$

b.

$\mathbb{R}^3$		
ID	$\left(\begin{smallmatrix} 2 & 2 & 0 \\ 1 & 1 & 2 \end{smallmatrix} \right)$	$\left(\begin{smallmatrix} -2 & 1 & 3 \\ -1 & 2 & 2 \end{smallmatrix} \right)$
$(0,0,0)$	$(0,0)$	$(0,0)$
$(0,2,3)$	$(4,8)$	$(11,10)$
$(-1,2,1)$	$(2,3)$	$(7,7)$
$(2,0,1.5)$	$(4,5)$	$(.5,1)$
$\vdots$	$\vdots$	$\vdots$

$\mathbb{R}^2$
ID
$(0,0)$
$(4,8)$
$(2,3)$
$(4,5)$
$(.5,1)$
$\vdots$

♦

5.2.3.7 Groupoids

Groupoids are like groups except a groupoid can have more than one object.

Definition 5.2.3.8. A *groupoid* is a category $\mathcal{C}$ such that every morphism is an isomorphism. If $\mathcal{C}$ and $\mathcal{D}$ are groupoids, a *morphism of groupoids*, denoted $F\colon \mathcal{C} \to \mathcal{D}$, is simply a functor. The category of groupoids is denoted **Grpd**.

Example 5.2.3.9. There is a functor **Grpd** $\to$ **Cat**, sending a groupoid to its underlying category. There is also a functor **Grp** $\to$ **Grpd** sending a group to itself as a groupoid with one object.

There is also a functor Core: **Cat** $\to$ **Grpd**, sending a category $\mathcal{C}$ to the largest groupoid inside $\mathcal{C}$, called its *core*. That is, $\mathrm{Ob}(\mathrm{Core}(\mathcal{C})) = \mathrm{Ob}(\mathcal{C})$ and

$$\mathrm{Hom}_{\mathrm{Core}(\mathcal{C})}(x, y) = \{f \in \mathrm{Hom}_{\mathcal{C}}(x, y) \mid f \text{ is an isomorphism}\}.$$

Application 5.2.3.10. Let M be a material in some original state s_0.[9] Construct a category $\mathcal{S}_M$ whose objects are the states of M (which are obtained by pulling on M in different ways, heating it up, and so on). Include a morphism from state s to state s' for every physical transformation from s to s'. Physical transformations can be performed one after another, so we can compose morphisms, and perhaps we can agree this composition is associative. Note that there is a morphism $i_s\colon s_0 \to s$ representing any physical transformation that can bring M from its initial state s_0 to s.

The elastic deformation region of the material is the set of states s such that there exists an inverse $s \to s_0$ to the morphism i_s. A transformation is irreversible if its representing morphism has no inverse. If a state s_1 is not in the elastic deformation region, we can still talk about the region that is (inventing a term) elastically equivalent to s_1. It is all the objects in $\mathcal{S}_M$ that are isomorphic to s_1. If we consider only elastic equivalences in $\mathcal{S}_M$, we are looking at a groupoid inside it, namely, the core $\mathrm{Core}(\mathcal{S}_M)$, as in Example 5.2.3.9.

$$\Diamond\Diamond$$

Example 5.2.3.11. Alan Weinstein [45] explains groupoids in terms of tiling patterns on a bathroom floor. This is worth reading.

Example 5.2.3.12. Let $I = \{x \in \mathbb{R} \mid 0 \leqslant x \leqslant 1\}$ denote the unit interval. It can be given a topology in a standard way, as a subset of $\mathbb{R}$ (see Example 5.2.3.3).

For any topological space X, a *path in X* is a continuous map $I \to X$. Two paths are called *homotopic* if one can be continuously deformed to the other, where the deformation

---

[9]This example may be somewhat crude, in accordance with the crudeness of my understanding of materials science.

occurs completely within X.[10] One can prove that being homotopic is an equivalence relation on paths.

Paths in X can be composed, one after the other, and the composition is associative (up to homotopy). Moreover, for any point $x \in X$, there is a trivial path (that stays at x). Finally every path is invertible (by traversing it backward) up to homotopy.

This all means that to any space $X \in \mathrm{Ob}(\mathbf{Top})$ we can associate a groupoid, called the *fundamental groupoid of X* and denoted $\Pi_1(X) \in \mathrm{Ob}(\mathbf{Grpd})$. The objects of $\Pi_1(X)$ are the points of X; the morphisms in $\Pi_1(X)$ are the paths in X (up to homotopy). A continuous map $f \colon X \to Y$ can be composed with any path $I \to X$ to give a path $I \to Y$, and this preserves homotopy. So, in fact, $\Pi_1 \colon \mathbf{Top} \to \mathbf{Grpd}$ is a functor.

Exercise 5.2.3.13.

Let T denote the surface of a doughnut, i.e., a torus. Choose two points $p, q \in T$. Since $\Pi_1(T)$ is a groupoid, it is also a category. What would the hom-set $\mathrm{Hom}_{\Pi_1(T)}(p, q)$ represent?   ◊

Solution 5.2.3.13.

The set $\mathrm{Hom}_{\Pi_1(T)}(p, q)$ represents the set of (equivalence classes of) paths from p to q, where two are considered equivalent if one can be deformed to the other.

In a course in algebraic topology, one proves that there is a bijection $\mathrm{Hom}_{\Pi_1(T)}(p, q) \cong \mathbb{Z} \times \mathbb{Z}$. This means that in drawing a line from p to q in T, one can wrap around the tight circle or the center hole any number of times (clockwise or counterclockwise), and the order does not matter.   ♦

Exercise 5.2.3.14.

Let $U \subseteq \mathbb{R}^2$ be an open subset of the plane, and let F be an irrotational vector field on U (i.e., one with $\mathrm{curl}(F) = 0$). Following Exercise 5.1.1.17, we have a category $\mathcal{C}_F$. If two curves C, C' in U are homotopic, then they have the same line integral, $\int_C F = \int_{C'} F$.

We also have a category $\Pi_1 U$, given by the fundamental groupoid, as in Example 5.2.3.12. Both categories have the same objects, $\mathrm{Ob}(\mathcal{C}_F) = |U| = \mathrm{Ob}(\Pi_1 U)$, the set of points in U.

---

[10] Let $I \times I = \{(x, y) \in \mathbb{R}^2 \mid 0 \leqslant x \leqslant 1 \text{ and } 0 \leqslant y \leqslant 1\}$ denote the square. There are two inclusions $i_0, i_1 \colon I \to S$ that put the interval inside the square at the left and right sides. Two paths $f_0, f_1 \colon I \to X$ are homotopic if there exists a continuous map $f \colon I \times I \to X$ such that $f_0 = f \circ i_0$ and $f_1 = f \circ i_1$,

$$I \underset{i_1}{\overset{i_0}{\rightrightarrows}} I \times I \xrightarrow{\ f\ } X$$

a. Is there a functor $\mathcal{C}_F \xrightarrow{?} \Pi_1 U$ or a functor $\Pi_1 U \xrightarrow{?} \mathcal{C}_F$ that is identity on the underlying objects?

b. Let $\mathcal{C}'_F \subseteq \mathcal{C}_F$ denote the subcategory with the same objects but only those morphisms corresponding to curves C with $\int_C F = 0$. Is $\mathcal{C}'_F$ a groupoid?

c. If F is a conservative vector field, what is $\mathcal{C}_F$?

d. If F is a conservative vector field, how does $\mathcal{C}_F$ compare with $\Pi_1 U$?

$\diamond$

Solution 5.2.3.14.

a. There is not a functor $\mathcal{C}_F \xrightarrow{?} \Pi_1 U$ that is identity on objects, in general; the rough idea is that two nonhomotopic paths may have the same line integral. But there is a functor $\Pi_1 U \to \mathcal{C}_F$ that is identity on objects, and that sends a homotopy class of paths to its equivalence class modulo line integral.

b. Yes. Let C be a curve, and let $-C$ denote its negative, i.e., if $C \colon [0,1] \to U$, then $-C$ is the result of composing C with the function $(x \mapsto 1 - x) \colon [0,1] \to [0,1]$. It is easy to check that concatenating C and $-C$ is homotopic to the constant path at $C(0)$. Therefore, since $\int_{C \star -C} F = 0$, it follows that $\int_C F = 0$ if and only if $\int_{-C} F = 0$. Now we see that every morphism in $\mathcal{C}'_F$ is invertible, i.e., $\mathcal{C}'_F$ is a groupoid.

c. If F is conservative, then the line integral $\int_C F$ is independent of path, so for every $x, y \in U = \mathrm{Ob}(\mathcal{C})$, there is exactly one morphism $x \to y$ if there is a path from x to y, and no morphisms $x \to y$ if there is no path between them.

d. One somewhat roundabout way to describe the relationship is via preorders. Recall the functors $R \colon \mathbf{Cat} \to \mathbf{PrO}$ and $i \colon \mathbf{PrO} \to \mathbf{Cat}$ from Propositions 5.2.1.20 and 5.2.1.13. If F is conservative, we have $\mathcal{C}_F = i \circ R(\Pi_1 U)$. In other words, since every path (morphism in $\Pi_1 U$) induces the same value for its line-integral, $\mathcal{C}_F$ is obtained by destroying all information in $\Pi_1 U$ except whether or not a path exists between two points.

$\blacklozenge$

Exercise 5.2.3.15.

Consider the set A of all (well-formed) arithmetic expressions that can be written with the symbols

$$\{0, 1, 2, 3, 4, 5, 6, 7, 8, 9, +, -, *, (,)\}.$$

For example, here are four different elements of A:

$$52, \qquad 52 - 7, \qquad 45 + 0, \qquad 50 + 3 * (6 - 2).$$

We can say that an equivalence between two arithmetic expressions is a justification that they give the same final answer, e.g., $52 + 60$ is equivalent to $10 * (5 + 6) + (2 + 0)$, which is equivalent to $10 * 11 + 2$.

a. I have basically described a category G. What are its objects, and what are its morphisms?

b. Is G a groupoid?

◇

Solution 5.2.3.15.

a. Its objects are the well-formed arithmetic expressions, $\text{Ob}(G) = A$. A morphism $f : a \to a'$ is a justification that they give the same final answer. The identity morphism id_a for a is the empty justification that $a = a$, and composition of morphisms is given by concatenating justifications. Now G has been defined as a category.

b. It does not appear to be a groupoid. The idea that it should be a groupoid would come from the belief that every justification for $a = a'$ would have an inverse justification for $a' = a$. But when we concatenate these justifications, is the result the empty justification? No, at least not as defined so far. Perhaps there is a way to make such a forward-then-backward justification equivalent to the empty justification, but this would take some thinking, and G as defined is not a groupoid.

♦

5.2.4   Logic, set theory, and computer science

5.2.4.1   The category of propositions

Given a domain of discourse, a logical proposition is a statement that is evaluated in any model of that domain as either true or not always true, which the black-and-white thinker might dub "false." For example, in the domain of real numbers we might have the proposition

For any real number $x \in \mathbb{R}$, there exists a real number $y \in \mathbb{R}$ such that $y > 3x$.

That is true: for $x = 22$, we can offer $y = 100$. But the following proposition is not true:

Every integer $x \in \mathbb{Z}$ is divisible by 2 or 3.

It is true for the majority of integers, but not for all integers; thus it is dubbed false.

We say that one logical proposition P *implies* another proposition Q, denoted $P \Rightarrow Q$, if for every model in which P is true, so is Q. There is a category **Prop** whose objects are logical propositions and whose morphisms are proofs that one statement implies another. Crudely, one might say that B *holds at least as often as* A if there is a morphism $A \to B$ (meaning in any model for which A holds, so does B). So the proposition "$x \neq x$" holds very seldom, and the proposition "$x = x$" holds very often.

Example 5.2.4.2. We can repeat this idea for nonmathematical statements. Take the set of all possible statements that are verifiable by experiment as the objects of a category. Given two such statements, it may be that one implies the other (e.g., "If the speed of light is fixed, then there are relativistic effects"). Every statement implies itself (identity) and implication is transitive, so we have a category.

Let's consider differences in proofs to be irrelevant, in which case the category **Prop** is simply a preorder $(\mathbf{Prop}, \Rightarrow)$: either A implies B or it does not. Then it makes sense to discuss meets and joins. It turns out that meets are "and's," and joins are "or's." That is, given propositions A, B, the meet $A \wedge B$ is defined to be a proposition that holds as often as possible subject to the constraint that it implies both A and B; the proposition "A holds and B holds" fits the bill. Similarly, the join $A \vee B$ is given by "A holds or B holds."

Exercise 5.2.4.3.

Consider the set of possible laws (most likely an infinite set) that can be dictated to hold throughout a jurisdiction. Consider each law as a proposition ("such and such is the case"), i.e., as an object of the preorder **Prop**. Given a jurisdiction V, and a set of laws $\{\ell_1, \ell_2, \ldots, \ell_n\}$ that are dictated to hold throughout V, we take their meet $L(V) := \ell_1 \wedge \ell_2 \wedge \cdots \wedge \ell_n$ and consider it to be the single law of the land V. Suppose that V is a jurisdiction and U is a subjurisdiction (e.g., U is a county and V is a state); write $U \subseteq V$. Then any law dictated by the large jurisdiction (the state) must also hold throughout the small jurisdiction (the county). Let J be the set of jurisdictions, so that $(J, \subseteq)$ is a preorder.

a. If $V \subseteq U$ are jurisdictions, what is the relation in **Prop** between $L(U)$ and $L(V)$?

b. Consider the preorder $(J, \subseteq)$ of jurisdictions. Is the law of the land a morphism of preorders $J \to \mathbf{Prop}$? That is, considering both J and **Prop** to be categories (by Proposition 5.2.1.13), we have a function $L \colon \mathrm{Ob}(J) \to \mathrm{Ob}(\mathbf{Prop})$; does L extend to a functor $J \to \mathbf{Prop}$.

◊

Solution 5.2.4.3.

   This exercise is strangely tricky, so we go through it slowly.

a. Suppose that the proposition $L(V)$ is true, i.e., we are in a model where all V's laws are being followed. Does this imply that $L(U)$ is true? Since $V \subseteq U$, every law of U is a law of V (e.g., if one may not own slaves anywhere in the United States, one may not own slaves in Maine). So indeed $L(U)$ is true; thus we have $L(V) \Rightarrow L(U)$.

b. Yes, L extends to a preorder morphism $L\colon J \to \textbf{Prop}$ because if $V \subseteq U$, then $L(V) \Rightarrow L(U)$.

$\blacklozenge$

Exercise 5.2.4.4.

   Take again the preorder $(J, \subseteq)$ of jurisdictions from Exercise 5.2.4.3 and the idea that laws are propositions. But this time, let $R(V)$ be the set of all possible laws (not just those dictated to hold) that are, in actuality, being respected, i.e., followed, by all people in V. This assigns to each jurisdiction a set. Does the "set of respected laws" function $R\colon \text{Ob}(J) \to \text{Ob}(\textbf{Set})$ extend to a functor $J \to \textbf{Set}$? $\diamond$

Solution 5.2.4.4.

   If $V \subseteq U$, then any law respected throughout U is respected throughout V, i.e., $R(U) \subseteq R(V)$. In other words, R is *contravariant* (see Section 6.2.1), meaning it constitutes a functor $R\colon J^{\text{op}} \to \textbf{Set}$. (Every law is being respected throughout the jurisdiction $\varnothing$, and physicists want to know what laws are being respected throughout the universe-as-jurisdiction.) $\blacklozenge$

5.2.4.5   A categorical characterization of Set

The category $\textbf{Set}$ of sets is fundamental in mathematics, but instead of thinking of it as something given or somehow special, it can be shown to merely be a category with certain properties, each of which can be phrased purely categorically. This was shown by Lawvere [23]. A very readable account is given in [26].

5.2.4.6   Categories in computer science

Computer science makes heavy use of trees, graphs, orders, lists, and monoids. All of these can be understood in the context of category theory, although it seems the categorical interpretation is rarely mentioned explicitly in computer science textbooks. However, categories are used explicitly in the theory of programming languages (PL). Researchers in that field attempt to understand the connection between what programs are supposed to do (their denotation) and what they actually cause to occur (their operation). Category theory provides a useful mathematical formalism in which to study this.

The kind of category most often considered by a PL researcher is known as a *Cartesian closed category*, or CCC, which means a category $\mathcal{T}$ that has products (like $A \times B$ in **Set**) and exponential objects (like B^A in **Set**). So **Set** is an example of a CCC, but there are others that are more appropriate for actual computation. The objects in a PL person's CCC represent the *types* of the programming language, types such as `integers`, `strings`, `floats`. The morphisms represent computable functions, e.g., `length:` `strings`$\longrightarrow$`integers`. The products allow one to discuss pairs (a, b), where a is of one type and b is of another type. Exponential objects allow one to consider computable functions as things that can be input to a function (e.g., given any computable function `floats`$\rightarrow$`integers`, one can consistently multiply its results by 2 and get a new computable function `floats`$\rightarrow$`integers`). Products are studied in Section 6.1.1.8 and exponential objects in Section 5.3.2.

But category theory does not only offer a language for thinking about programs, it offers an unexpected tool called monads. The CCC model for types allows researchers only to discuss functions, leading to the notion of functional programming languages; however, not all things that a computer does are functions. For example, reading input and output, changing internal state, and so on, are operations that can be performed on a computer but that ruin the functional aspect of programs. Monads were found in 1991 by Moggi [33] to provide a powerful abstraction that opens the doors to such nonfunction operations without forcing the developer to leave the category-theoretic paradise. Monads are discussed in Section 7.3.

Section 5.2.2 showed that databases are well captured by the language of categories (this is formalized in Section 5.4). Databases are used in this book to bring clarity to concepts within standard category theory.

5.2.5 Categories applied in science

Categories are used throughout mathematics to relate various subjects as well as to draw out the essential structures within these subjects. For example, there is active research in categorifying classical theories like that of knots, links, and braids (Khovanov [21]). It is similarly applied in science to clarify complex subjects. Here are some very brief descriptions of scientific disciplines to which category theory is applied.

Quantum field theory was categorified by Atiyah [2] in the late 1980s, with much success (at least in producing interesting mathematics). In this domain, one takes a category in which an object is a reasonable space, called a manifold, and a morphism is a manifold connecting two manifolds, like a cylinder connecting two circles. Such connecting manifolds are called cobordisms and the category of manifolds and cobordisms is denoted **Cob**. Topological quantum field theory is the study of functors **Cob** → **Vect** that assign a vector space to each manifold and a linear transformation of vector spaces to each cobordism.

Samson Abramsky [1] showed a relationship between database theory, category theory, and quantum physics. He used the notion of sheaves on a database (see Section 7.2.3) and the sheaf cohomology thereof, to derive Bell's theorem, which roughly states that certain variables that can be observed locally do not extend to globally observable variables.

Information theory, invented in 1948 by Claude Shannon, is the study of how to ideally compress messages so that they can be sent quickly and accurately across a noisy channel.[11] Its main quantity of interest is the number of bits necessary to encode a piece of information. For example, the amount of information in an English sentence can be greatly reduced. The fact that t's are often followed by h's, or that e's are much more common than z's, implies that letters are not being used as efficiently as possible. The amount of bits necessary to encode a message is called its *entropy* and has been linked to the commonly used notion of the same name in physics.

Baez, Fritz, and Leinster [7] show that entropy can be captured quite cleanly using category theory. They make a category `FinProb` whose objects are finite sets equipped with a probability measure, and whose morphisms are probability-preserving functions. They characterize *information loss* as a way to assign numbers to such morphisms, subject to certain explicit constraints. They then show that the entropy of an object in `FinProb` is the amount of information lost under the unique map to the singleton set {☺}. This approach explicates (by way of the explicit constraints for information loss functions) the essential idea of Shannon's information theory, allowing it to be generalized to categories other than `FinProb`. Thus Baez and colleagues effectively *categorified* information theory.

Robert Rosen proposed in the 1970s that category theory could play a major role in biology. That is only now starting to be fleshed out. There is a categorical account of

---

[11]The discipline called *information theory*, invented by Claude Shannon, is concerned only with ideal compression schemes. It does not pay attention to the content of the messages—what they mean—as Shannon says specifically in his seminal paper: "Frequently the messages have meaning; that is they refer to or are correlated according to some system with certain physical or conceptual entities. These semantic aspects of communication are irrelevant to the engineering problem." Thus I think the subject is badly named. It should be called compression theory or redundancy theory.

Information is inherently meaningful—that is its purpose—so a theory unconcerned with meaning is not really studying information per se. (The people who decide on speed limits for roads and highways may care about human health, but a study limited to understanding ideal speed limit schemes would not be called "human health theory.")

Information theory is extremely important in a diverse array of fields, including computer science [28], neuroscience [5], [27], and physics [16]. I am not trying to denigrate the field; I only disagree with its name.

evolution and memory, called *Memory Evolutive Systems* [15]. There is also a paper [10] by Brown and Porter with applications to neuroscience.

5.3 Natural transformations

The Big 3 of category theory are categories, functors, and natural transformations. This section introduces the last of these, natural transformations. Category theory was originally invented to discuss natural transformations. These were sufficiently conceptually challenging that they required formalization and thus the invention of category theory. If we think of categories as domains (e.g., of discourse, interaction, comparability) and functors as translations between different domains, the natural transformations compare different translations.

Natural transformations can seem a bit abstruse at first, but hopefully some examples and exercises may help.

5.3.1 Definition and examples

Let's begin with an example. There is a functor List: **Set** → **Set**, which sends a set X to the set List(X) consisting of all lists whose entries are elements of X. Given a morphism $f\colon X \to Y$, we can transform a list with entries in X into a list with entries in Y by applying f to each entry (see Exercise 5.1.2.22). Call this process translating the list.

It may seem a strange thing to contemplate, but there is also a functor List∘List: **Set** → **Set** that sends a set X to the set of lists of lists in X. If $X = \{a, b, c\}$, then List ∘ List(X) contains elements like $\big[[a, b], [a, c, a, b, c], [c]\big]$ and $\big[[\ ]\big]$ and $\big[[a], [\ ], [a, a, a]\big]$. We can *naturally transform* a list of lists into a list by concatenation. In other words, for any set X there is a function $\mu_X\colon$ List ∘ List$(X) \to$ List(X), which sends that list of lists to $[a, b, a, c, a, b, c, c]$ and $[\ ]$ and $[a, a, a, a]$ respectively. In fact, even if we use a function $f\colon X \to Y$ to translate a list of X's into a list of Y's (or a list of lists of X's into a list of lists of Y's), the concatenation works correctly.

Slogan 5.3.1.1.

> *What does it mean to say that concatenation of lists is natural with respect to translation? It means that concatenating then translating is the same thing as translating then concatenating.*

Let's make this concrete. Let $X = \{a, b, c\}$, let $Y = \{1, 2, 3\}$, and let $f\colon X \to Y$ assign $f(a) = 1, f(b) = 1, f(c) = 2$. The naturality condition says the following for any list of lists of X's, in particular, for $\big[[a,b],[a,c,a,b,c],[c]\big] \in \text{List} \circ \text{List}(X)$:

$$
\begin{array}{ccc}
\big[[a,b],[a,c,a,b,c],[c]\big] & \xmapsto{\ \ \mu_X\ \ } & [a,b,a,c,a,b,c,c] \\[2pt]
\Big\downarrow{\scriptstyle \text{List}\circ\text{List}(f)} & \checkmark & \Big\downarrow{\scriptstyle \text{List}(f)} \\[2pt]
\big[[1,1],[1,2,1,1,2],[2]\big] & \xmapsto[\ \ \mu_Y\ \ ]{} & [1,1,1,2,1,1,2,2]
\end{array}
$$

The top right path is concatenating then translating, and the left bottom path is translating then concatenating, and one sees here that they do the same thing.

Here is how the preceding example fits with the terminology of Definition 5.3.1.2. The categories $\mathcal{C}$ and $\mathcal{D}$ are both **Set**, the functor $F\colon \mathcal{C} \to \mathcal{D}$ is $\text{List} \circ \text{List}$, and the functor $G\colon \mathcal{C} \to \mathcal{D}$ is List. The natural transformation is $\mu\colon \text{List}\circ\text{List} \to \text{List}$. It can be depicted:

Definition 5.3.1.2. Let $\mathcal{C}$ and $\mathcal{D}$ be categories, and let $F\colon \mathcal{C} \to \mathcal{D}$ and $G\colon \mathcal{C} \to \mathcal{D}$ be functors. A *natural transformation α from F to G*, denoted $\alpha\colon F \to G$ and depicted

$$
\mathcal{C}\ \overset{F}{\underset{G}{\Longrightarrow\ \Downarrow\alpha}}\ \mathcal{D},
$$

is defined as follows. One announces some constituents (A. components) and shows that they conform to a law (1. naturality squares). Specifically, one announces

    A. for each object $X \in \text{Ob}(\mathcal{C})$, a morphism $\alpha_X\colon F(X) \to G(X)$ in $\mathcal{D}$, called *the X-component of α*.

One must then show that the following *natural transformation law* holds:

    1. For every morphism $f\colon X \to Y$ in $\mathcal{C}$, the square (5.10), called the *naturality square for f*, must commute:

$$
\begin{array}{ccc}
F(X) & \xrightarrow{\ \ \alpha_X\ \ } & G(X) \\[2pt]
{\scriptstyle F(f)}\Big\downarrow & \checkmark & \Big\downarrow{\scriptstyle G(f)} \\[2pt]
F(Y) & \xrightarrow[\ \ \alpha_Y\ \ ]{} & G(Y)
\end{array}
\qquad\qquad (5.10)
$$

The set of natural transformations $F \to G$ is denoted $\mathrm{Nat}(F, G)$.

Remark 5.3.1.3. If we have two functors $F, G \colon \mathcal{C} \to \mathcal{D}$, providing a morphism $\alpha_X \colon F(X) \to G(X)$ for every object $X \in \mathrm{Ob}(\mathcal{C})$ is called a *questionably natural transformation*. Once we check the commutativity of all the naturality squares, i.e., once we know it satisfies Definition 5.3.1.2, we drop the "questionably" part.

Example 5.3.1.4. Consider the following categories $\mathcal{C} \cong [1]$ and $\mathcal{D} \cong [2]$:

$$\mathcal{C} := \boxed{\overset{0}{\bullet} \xrightarrow{\ p\ } \overset{1}{\bullet}} \qquad \mathcal{D} := \boxed{\overset{A}{\bullet} \xrightarrow{\ f\ } \overset{B}{\bullet} \xrightarrow{\ g\ } \overset{C}{\bullet}.}$$

Consider the functors $F, G \colon [1] \to [2]$, where $F(0) = A$, $F(1) = B$, $G(0) = A$, and $G(1) = C$. It turns out that there is only one possible natural transformation $F \to G$; we call it α and explore its naturality square. The components of $\alpha \colon F \to G$ are shown in green. These components are $\alpha_0 = \mathrm{id}_A \colon F(0) \to G(0)$ and $\alpha_1 = g \colon F(1) \to G(1)$. The naturality square for $p \colon 0 \to 1$ is shown twice below, once with notation following that in (5.10) and once in local notation:

$$
\begin{array}{ccc}
F(0) & \xrightarrow{\ \alpha_0\ } & G(0) \\
{\scriptstyle F(p)}\downarrow & & \downarrow{\scriptstyle G(p)} \\
F(1) & \xrightarrow[\ \alpha_1\ ]{} & G(1)
\end{array}
\qquad\qquad
\begin{array}{ccc}
A & \xrightarrow{\ \mathrm{id}_A\ } & A \\
{\scriptstyle f}\downarrow & & \downarrow{\scriptstyle g \circ f} \\
B & \xrightarrow[\ g\ ]{} & C
\end{array}
$$

It is clear that this diagram commutes, so the components α_0 and α_1 satisfy the law of Definition 5.3.1.2, making α a natural transformation.

Proposition 5.3.1.5. *Let $\mathcal{C}$ and $\mathcal{D}$ be categories, let $F, G \colon \mathcal{C} \to \mathcal{D}$ be functors, and for every object $c \in \mathrm{Ob}(\mathcal{C})$, let $\alpha_c \colon F(c) \to G(c)$ be a morphism in $\mathcal{D}$. Suppose given a path $c_0 \xrightarrow{f_1} c_1 \xrightarrow{f_2} \cdots \xrightarrow{f_n} c_n$ such that for each arrow f_i in it, the following naturality square commutes:*

$$
\begin{array}{ccc}
F(c_{i-1}) & \xrightarrow{\ \alpha_{c_{i-1}}\ } & G(c_{i-1}) \\
{\scriptstyle F(f_i)}\downarrow & & \downarrow{\scriptstyle G(f_i)} \\
F(c_i) & \xrightarrow[\ \alpha_{c_i}\ ]{} & G(c_i)
\end{array}
$$

Then the naturality square for the composite $p := f_n \circ \cdots \circ f_2 \circ f_1 \colon c_0 \to c_n$

$$
\begin{array}{ccc}
F(c_0) & \xrightarrow{\ \alpha_{c_0}\ } & G(c_0) \\
{\scriptstyle F(p)}\Big\downarrow & & \Big\downarrow{\scriptstyle G(p)} \\
F(c_n) & \xrightarrow[\ \alpha_{c_n}\ ]{} & G(c_n)
\end{array}
$$

also commutes. In particular, the naturality square commutes for every identity morphism id_c.

Proof. When $n = 0$, we have a path of length 0 starting at each $c \in \mathrm{Ob}(\mathcal{C})$. It vacuously satisfies the condition, so we need to see that its naturality square

$$
\begin{array}{ccc}
F(c) & \xrightarrow{\ \alpha_c\ } & G(c) \\
{\scriptstyle F(\mathrm{id}_c)}\Big\downarrow & & \Big\downarrow{\scriptstyle G(\mathrm{id}_c)} \\
F(c) & \xrightarrow[\ \alpha_c\ ]{} & G(c)
\end{array}
$$

commutes. But this is clear because functors preserve identities.

The rest of the proof follows by induction on n. Suppose $q = f_{n-1} \circ \cdots \circ f_2 \circ f_1 \colon c_0 \to c_{n-1}$ and $p = f_n \circ q$ and that the naturality squares for q and for f_n commute; we need only show that the naturality square for p commutes. That is, we assume the two small squares commute; it follows that the large rectangle does too, completing the proof.

$$
\begin{array}{ccc}
F(c_0) & \xrightarrow{\ \alpha_{c_0}\ } & G(c_0) \\
{\scriptstyle F(q)}\Big\downarrow & & \Big\downarrow{\scriptstyle G(q)} \\
F(c_{n-1}) & \xrightarrow{\ \alpha_{c_{n-1}}\ } & G(c_{n-1}) \\
{\scriptstyle F(f_n)}\Big\downarrow & & \Big\downarrow{\scriptstyle G(f_n)} \\
F(c_n) & \xrightarrow{\ \alpha_{c_n}\ } & G(c_n)
\end{array}
$$

$\square$

Example 5.3.1.6. Let $\mathcal{C} = \mathcal{D} = [1]$ be the linear order of length 1, thought of as a category (by Proposition 5.2.1.13). There are three functors $\mathcal{C} \to \mathcal{D}$, which we can write as $(0,0), (0,1)$, and $(1,1)$; these are depicted left to right as follows:

These are just functors so far. What are the natural transformations say, $\alpha\colon (0,0) \to (0,1)$? To specify a natural transformation, we must specify a component for each object in $\mathcal{C}$. In this case $\alpha_0\colon 0 \to 0$ and $\alpha_1\colon 0 \to 1$. There is only one possible choice: $\alpha_0 = \mathrm{id}_0$ and $\alpha_1 = f$. Now that we have chosen components, we need to check the naturality squares.

There are three morphisms in $\mathcal{C}$, namely, $\mathrm{id}_0, f, \mathrm{id}_1$. By Proposition 5.3.1.5, we need only check the naturality square for f. We write it twice, once in abstract notation and once in concrete notation:

$$
\begin{array}{ccc}
F(0) & \xrightarrow{\ \alpha_0\ } & G(0) \\
{\scriptstyle F(f)}\big\downarrow & & \big\downarrow{\scriptstyle G(f)} \\
F(1) & \xrightarrow[\ \alpha_1\ ]{} & G(1)
\end{array}
\qquad\qquad
\begin{array}{ccc}
0 & \xrightarrow{\ \mathrm{id}_0\ } & 0 \\
{\scriptstyle \mathrm{id}_0}\big\downarrow & & \big\downarrow{\scriptstyle f} \\
0 & \xrightarrow[\ f\ ]{} & 1
\end{array}
$$

This commutes, so α is indeed a natural transformation.

Exercise 5.3.1.7.

With notation as in Example 5.3.1.6, we have three functors $\mathcal{C} \to \mathcal{D}$, namely, $(0,0), (0,1)$, and $(1,1)$. How many natural transformations are there from F to G, i.e., what is the cardinality of $\mathrm{Nat}(F, G)$

a. when $F = (0,0)$ and $G = (1,1)$?

b. when $F = (0,0)$ and $G = (0,0)$?

c. when $F = (0,1)$ and $G = (0,0)$?

d. when $F = (0,1)$ and $G = (1,1)$?

$\diamond$

Solution 5.3.1.7.

What is a natural transformation $\alpha\colon F \to G$? The idea is that for every object in $\mathcal{C}$, we need to move in $\mathcal{D}$. This object needs to go from where F points it to where G points it. In other words, for every object $c \in \mathcal{C}$, we need to pick a morphism $\alpha_c \colon F(c) \to G(c)$ called the c-component of α.

There are two objects in $\mathcal{C}$, namely, 0 and 1, so we need to choose two components. Once we have chosen all these components, we need to check naturality. If it works, we will have found a natural transformation.

a. We need morphisms $\alpha_0 \colon 0 \to 1$ and $\alpha_1 \colon 0 \to 1$. There is only one choice, $\alpha_0 = \alpha_1 = f$. Is this natural? We now must go through every morphism in $\mathcal{C}$ and check that a certain square commutes. There are three morphisms in $\mathcal{C}$, namely, $\mathrm{id}_0, \mathrm{id}_1$, and f. But by Proposition 5.3.1.5, the naturality squares for identities automatically commute. So we just need to check the commutativity of the following left-hand square, which after substituting the choices becomes the right-hand square:

It commutes; hence there is one natural transformation $(0,0) \to (1,1)$.

b. We need morphisms $\alpha_0 \colon 0 \to 0$ and $\alpha_1 \colon 0 \to 0$. There is only one choice, $\alpha_0 = \alpha_1 = \mathrm{id}_0$. Again by Proposition 5.3.1.5, we only need to check the commutativity of the following square, which is obvious. Thus there is one natural transformation $(0,0) \to (0,0)$.

$$
\begin{array}{ccc}
0 & \xrightarrow{\mathrm{id}_0} & 0 \\
{\scriptstyle \mathrm{id}_0}\downarrow & & \downarrow{\scriptstyle \mathrm{id}_0} \\
0 & \xrightarrow{\mathrm{id}_0} & 0
\end{array}
$$

c. We need morphisms $\alpha_0 \colon 0 \to 0$ and $\alpha_1 \colon 1 \to 0$. The latter is impossible because $\mathrm{Hom}_{\mathcal{D}}(1,0) = \varnothing$. Hence there are no natural transformations $(0,1) \to (0,0)$.

d. We need morphisms $\alpha_0 \colon 0 \to 1$ and $\alpha_1 \colon 1 \to 1$. There is only one choice, $\alpha_0 = f$ and $\alpha_1 = \mathrm{id}_1$. Again by Proposition 5.3.1.5, we only need to check the commutativity of the following square, which is clear.

Thus there is one natural transformation $(0,1) \to (1,1)$.

♦

Exercise 5.3.1.8.

Let $\underline{1}$ denote the discrete category on one object, $\mathrm{Ob}(\underline{1}) = \{1\}$, and let $\mathcal{L}oop$ denote the category with one object $\mathrm{Ob}(\mathcal{L}oop) = \{s\}$ and $\mathrm{Hom}_{\mathcal{L}oop}(s, s) = \mathbb{N}$ (see Example 5.2.2.2). There is exactly one functor $S: \underline{1} \to \mathcal{L}oop$. Characterize the natural transformations $\alpha: S \to S$. ◊

Solution 5.3.1.8.

Note that $S(1) = s$. We need to give just one component $\alpha_1: s \to s$. Since we have an isomorphism $\mathrm{Hom}_{\mathcal{L}oop}(s, s) \xrightarrow{\cong} \mathbb{N}$, we just need to check for each natural number $n \in \mathbb{N}$ that the naturality square commutes. But since the only morphism in $\underline{1}$ is an identity, it commutes regardless of n by Proposition 5.3.1.5. Thus there is a bijection $\mathrm{Nat}(S, S) \xrightarrow{\cong} \mathbb{N}$. ♦

Exercise 5.3.1.9.

Let $[1]$ denote the free arrow category,

$$[1] = \boxed{\overset{0}{\bullet} \overset{f}{\longrightarrow} \overset{1}{\bullet}}$$

as in Exercise 5.1.2.34, and let $\mathcal{L}oop$ be as in Example 5.2.2.2.

a. What are all the functors $[1] \to \mathcal{L}oop$?

b. For any two functors $F, G: [1] \to \mathcal{L}oop$, characterize the set $\mathrm{Nat}(F, G)$ of natural transformations $F \to G$.

◊

Solution 5.3.1.9.

a. To give a functor $F: [1] \to \mathcal{L}oop$ we provide $F(0), F(1) \in \mathrm{Ob}(\mathcal{L}oop)$ and $F(f): F(0) \to F(1)$. The functor laws will invariably hold, basically because $[1]$ is so simple, or more precisely because it is the free category on a graph (see Example 5.1.2.33). And since $\mathcal{L}oop$ has one object, we must have $F(0) = F(1) = s$, so choosing a functor $[1] \to \mathcal{L}oop$ amounts simply to choosing an element $F(f) \in \mathrm{Hom}_{\mathcal{L}oop}(s, s) \cong \mathbb{N}$. In other words, we have a bijection

$$\mathrm{Hom}_{\mathbf{Cat}}([1], \mathcal{L}oop) \xrightarrow{\cong} \mathbb{N},$$

"how many times should I wind around?"

b. Suppose given two functors $F, G: [1] \to \mathcal{Loop}$, which we can think of as winding numbers $n_F, n_G \in \mathbb{N}$. To give a natural transformation $\alpha: F \to G$, we need to give components $\alpha_0, \alpha_1 \in \mathrm{Hom}_{\mathcal{Loop}}(s, s) \cong \mathbb{N}$. We can think of these as winding numbers too; let's call them $n_0, n_1 \in \mathbb{N}$ respectively. For the choices of components to be natural, we require that this diagram commutes:

$$
\begin{array}{ccc}
s & \xrightarrow{\;n_0\;} & s \\
{\scriptstyle n_F}\big\downarrow & & \big\downarrow{\scriptstyle n_G} \\
s & \xrightarrow[\;n_1\;]{} & s
\end{array}
$$

This square commutes in $\mathcal{Loop}$ if and only if $n_0 + n_G = n_F + n_1$ in $\mathbb{N}$. Since n_F and n_G are given, let $N = n_G - n_F$. There is a bijection

$$\mathrm{Nat}(F, G) \xrightarrow{\;\cong\;} \{(n_0, n_1) \in \mathbb{N} \times \mathbb{N} \mid n_1 - n_0 = N\}.$$

◆

Exercise 5.3.1.10.

Consider the functor List: **Set** → **Set** sending a set X to the set List(X) of lists with entries in X. There is a natural transformation List ◦ List → List given by concatenation.

a. If someone said, "Singleton lists give a natural transformation σ from id$_{\mathbf{Set}}$ to List," what might she mean? That is, for a set X, what component σ_X might she be suggesting?

b. Do these components satisfy the necessary naturality squares for functions $f: X \to Y$? In other words, given your interpretation of what the person is saying, is she correct?

◇

Solution 5.3.1.10.

a. She is certainly telling us about a natural transformation $\sigma: \mathrm{id}_{\mathbf{Set}} \to \mathrm{List}$, and she seems to be telling us about how its components work. Since this is a natural transformation of functors **Set** → **Set**, to give components of σ is to provide, for each set X, a function $\sigma_X: \mathrm{id}_{\mathbf{Set}}(X) \to \mathrm{List}(X)$, i.e., a function $X \to \mathrm{List}(X)$. The person is telling us that this function is given by singleton lists. We know what a singleton list is: it looks something like [56] or [a]. To give a function $X \to \mathrm{List}(X)$, we need to provide, for each $x \in X$, a list of X's. Then we see that singleton lists work: to each $x \in X$, let $\sigma_X(x) = [x]$. We now have component σ_X.

b. We need to check that for every function $f \colon X \to Y$, the following square commutes:

$$
\begin{array}{ccc}
X & \xrightarrow{\ \sigma_X\ } & \mathrm{List}(X) \\
{\scriptstyle f}\downarrow & & \downarrow{\scriptstyle \mathrm{List}(f)} \\
Y & \xrightarrow[\ \sigma_Y\ ]{} & \mathrm{List}(Y)
\end{array}
$$

This is easy to check, once we recall the function $\mathrm{List}(f) \colon \mathrm{List}(X) \to \mathrm{List}(Y)$ (see beginning of Section 5.3.1).

♦

Exercise 5.3.1.11.

Let $\mathcal{C}$ and $\mathcal{D}$ be categories, and suppose that $d \in \mathrm{Ob}(\mathcal{D})$ is a terminal object. Consider the constant functor $\{d\}^{\mathcal{C}} \colon \mathcal{C} \to \mathcal{D}$, which sends each object $c \in \mathrm{Ob}(\mathcal{C})$ to d and each morphism in $\mathcal{C}$ to the identity morphism id_d on d.

a. For any other functor $F \colon \mathcal{C} \to \mathcal{D}$, how many natural transformations are there $F \to \{d\}^{\mathcal{C}}$?

b. Let $\mathcal{D} = \mathbf{Set}$, and let $d = \{\odot\}$, which is a terminal object in $\mathbf{Set}$ (see Exercise 3.2.3.5 or Warning 6.1.3.14). If $\mathcal{C} = [1]$ is the linear order of length 1, and $F \colon \mathcal{C} \to \mathbf{Set}$ is any functor, what does it mean to give a natural transformation $\{d\}^{\mathcal{C}} \to F$?

◊

Solution 5.3.1.11.

a. To give a natural transformation $\alpha \colon F \to \{d\}^{\mathcal{C}}$, one needs to provide a component $\alpha_c \colon F(c) \to d$ for every $c \in \mathrm{Ob}(\mathcal{C})$. But since $d \in \mathrm{Ob}(\mathcal{D})$ is terminal, there is exactly one such morphism, denoted $!_{F(c)} \colon F(c) \to d$. Thus there is exactly one questionably natural transformation α, and we must check that it is natural. For any $f \colon c \to c'$ in $\mathcal{C}$, the square looks like this

$$
\begin{array}{ccc}
F(c) & \xrightarrow{\ !_{F(c)}\ } & d \\
{\scriptstyle F(f)}\downarrow & & \downarrow{\scriptstyle \mathrm{id}_d} \\
F(c') & \xrightarrow[\ !_{F(c')}\ ]{} & d
\end{array}
$$

This may look complicated but in fact we are just asking whether two morphisms $F(c) \to d$ are the same; they surely are because d is terminal. Thus there is exactly one natural transformation $\alpha \colon F \to \{d\}^{\mathcal{C}}$.

b. A functor $F\colon [1] \to \mathbf{Set}$ can be identified with a morphism in $\mathbf{Set}$, i.e., with a function $F\colon X_0 \to X_1$, for arbitrary sets $X_0, X_1 \in \mathrm{Ob}(\mathbf{Set})$. To give a natural transformation $\{d\}^{\mathcal{C}} \to F$ is to give components $\alpha_0\colon \{\odot\} \to X_0$ and $\alpha_1\colon \{\odot\} \to X_1$ such that this diagram commutes:

$$
\begin{array}{ccc}
\{\odot\} & \xrightarrow{\ \alpha_0\ } & X_0 \\
{\scriptstyle \mathrm{id}_{\{\odot\}}}\Big\downarrow & & \Big\downarrow {\scriptstyle F} \\
\{\odot\} & \xrightarrow[\ \alpha_1\ ]{} & X_1
\end{array}
$$

A function $\alpha_0\colon \{\odot\} \to X_0$ can be identified with an element $\alpha_0 \in X_0$, and similarly for X_1. Hence to give a natural transformation $\{d\}^{\mathcal{C}} \to F$ is to give an element $\alpha_0 \in X_0$, an element $\alpha_1 \in X_1$, such that $F(\alpha_0) = \alpha_1$.

Going a bit further, if we pick α_0, we find that α_1 is forced on us. So it turns out that there is a bijection $\mathrm{Nat}(\{d\}, F) \xrightarrow{\ \cong\ } X_0$, i.e., giving a natural transformation $\{d\}^{\mathcal{C}} \to F$ is the same as picking an element in the domain of the function corresponding to F.

$\blacklozenge$

Application 5.3.1.12. Figure 4.2 showed a finite state machine on alphabet $\Sigma = \{a, b\}$, and Example 4.1.3.1 shows its associated action table. Imagine this was your model for understanding the behavior of some system when acted on by commands a and b. Suppose a colleague tells you he has a more refined model that fits with the same data. His model has six states rather than three, but it is compatible. What might that mean?

Both the original state machine, X, the proposed model, Y, and their associated action tables are shown in Figure 5.1 (see page 314).

How are these models compatible? In the table for Y, if one removes the distinction between states 1A, 1B, 1C and between states 2A and 2B, then one returns with the table for X. The table for Y is more specific, but it is fully compatible with the table for X. The sense in which it is compatible is precisely the sense defined by there being a natural transformation.

Recall that $\mathcal{M} = (\mathrm{List}(\Sigma), [\,], +\!+\,)$ is a monoid, and that a monoid is simply a category with one object, say, $\mathrm{Ob}(\mathcal{M}) = \{\blacktriangle\}$ (see Section 5.2.1). With $\Sigma = \{a, b\}$, the monoid $\mathcal{M}$ can be visualized as follows:

$$
\mathcal{M} = \boxed{\ a\ \circlearrowright\ \overset{\blacktriangle}{}\ \circlearrowleft\ b\ }
$$

Recall also that a state machine on $\mathcal{M}$ is simply a functor $\mathcal{M} \to \mathbf{Set}$. We thus have two such functors, X and Y. A natural transformation $\alpha\colon Y \to X$ would consist of a component α_m for every object $m \in \mathrm{Ob}(\mathcal{M})$ such that certain diagrams commute. But $\mathcal{M}$ having only one object, we need only one function $\alpha_{\blacktriangle}\colon Y(\blacktriangle) \to X(\blacktriangle)$, where $Y(\blacktriangle)$ is the set of (6) states of Y and $X(\blacktriangle)$ is the set of (3) states of X.

The states of Y have been named so as to make the function $\alpha_{\blacktriangle}$ particularly easy to guess.[12] We need to check that two squares commute:

$$
\begin{array}{ccc}
Y(\blacktriangle) & \xrightarrow{\alpha_{\blacktriangle}} & X(\blacktriangle) \\
{\scriptstyle Y(a)}\downarrow & & \downarrow{\scriptstyle X(a)} \\
Y(\blacktriangle) & \xrightarrow{\alpha_{\blacktriangle}} & X(\blacktriangle)
\end{array}
\qquad\qquad
\begin{array}{ccc}
Y(\blacktriangle) & \xrightarrow{\alpha_{\blacktriangle}} & X(\blacktriangle) \\
{\scriptstyle Y(b)}\downarrow & & \downarrow{\scriptstyle X(b)} \\
Y(\blacktriangle) & \xrightarrow{\alpha_{\blacktriangle}} & X(\blacktriangle)
\end{array}
\tag{5.11}
$$

This can only be checked by going through and making sure that certain things match, as specified by (5.11); this is spelled out in detail. The columns that should match are those whose entries are written in blue. These correspond to the left bottom composites being matched with the top right composites in the naturality squares of (5.11).

Naturality square for $a\colon \blacktriangle \to \blacktriangle$					
$Y(\blacktriangle)$ **[ID]**	$Y(a)$	$\alpha_{\blacktriangle} \circ Y(a)$	$\alpha_{\blacktriangle}$	$X(a) \circ \alpha_{\blacktriangle}$	
State 0	State 1A	State 1	State 0	State 1	
State 1A	State 2A	State 2	State 1	State 2	
State 1B	State 2B	State 2	State 1	State 2	(5.12)
State 1C	State 2B	State 2	State 1	State 2	
State 2A	State 0	State 0	State 2	State 0	
State 2B	State 0	State 0	State 2	State 0	

Naturality square for $b\colon \blacktriangle \to \blacktriangle$					
$Y(\blacktriangle)$ **[ID]**	$Y(b)$	$\alpha_{\blacktriangle} \circ Y(b)$	$\alpha_{\blacktriangle}$	$X(b) \circ \alpha_{\blacktriangle}$	
State 0	State 2A	State 2	State 0	State 2	
State 1A	State 1B	State 1	State 1	State 1	
State 1B	State 1C	State 1	State 1	State 1	(5.13)
State 1C	State 1B	State 1	State 1	State 1	
State 2A	State 0	State 0	State 2	State 0	
State 2B	State 0	State 0	State 2	State 0	

To recap, scientists may often have the idea that two models Y and X are compatible, and such notions of compatibility may be broadly agreed upon. However, these notions can at the same time be challenging to explain to an outsider, e.g., a regulatory body or auditor, especially in more complex situations. On the other hand, it is unambiguous to

---

[12]The function $\alpha_{\blacktriangle}\colon Y(\blacktriangle) \to X(\blacktriangle)$ makes the following assignments: State 0 $\mapsto$ State 0, State 1A $\mapsto$ State 1, State 1B $\mapsto$ State 1, State 1C $\mapsto$ State 1, State 2A $\mapsto$ State 2, State 2B $\mapsto$ State 2.

simply claim "there is a natural transformation from Y to X." If, in a given domain, the notion of natural transformation captures the essence of compatible models, it may bring clarity.

◇◇

Exercise 5.3.1.13.

   Let $F: \mathcal{C} \to \mathcal{D}$ be a functor. Suppose someone said, "The identity on F is a natural transformation from F to itself."

a. What might he mean?

b. What components is he suggesting?

c. Are the components natural?

◇

Solution 5.3.1.13.

a. He is certainly telling us about a natural transformation $\alpha: F \to F$, and he seems to be telling us that it will somehow act like an identity.

b. To give a questionably natural transformation, we need to provide, for every $c \in \mathrm{Ob}(\mathcal{C})$ a morphism $\alpha_c: F(c) \to F(c)$ in $\mathcal{D}$. Since we have in mind the word *identity*, we could take $\alpha_c := \mathrm{id}_{F(c)}$ for all c. This is probably what the person means.

c. For α to be natural we need to check that the following square commutes for any $f: c \to c'$ in $\mathcal{C}$:

$$\begin{array}{ccc} F(c) & \xrightarrow{\ \mathrm{id}_{F(c)}\ } & F(c) \\ {\scriptstyle F(f)}\downarrow & & \downarrow{\scriptstyle F(f)} \\ F(c') & \xrightarrow[\ \mathrm{id}_{F(c')}\ ]{} & F(c') \end{array}$$

   It clearly does commute, so α is natural. This natural transformation α is usually denoted $\mathrm{id}_F: F \to F$.

♦

Example 5.3.1.14. Let $[1] \in \mathrm{Ob}(\mathbf{Cat})$ be the free arrow category described in Exercise 5.1.2.34, and let $\mathcal{D}$ be any category. To specify a functor $F: [1] \to \mathcal{D}$ requires the specification of two objects, $F(v_1), F(v_2) \in \mathrm{Ob}(\mathcal{D})$ and a morphism $F(e): F(v_1) \to F(v_2)$

in $\mathcal{D}$. The identity and composition formulas are taken care of once that much is specified. To recap, a functor $F\colon [1] \to \mathcal{D}$ is the same thing as a morphism in $\mathcal{D}$.

Thus, choosing two functors $F, G\colon [1] \to \mathcal{D}$ is precisely the same thing as choosing two morphisms in $\mathcal{D}$. Let us call them $f\colon a_0 \to a_1$ and $g\colon b_0 \to b_1$, where we have $f = F(e), a_0 = F(v_0), a_1 = F(v_1)$ and $g = G(e), b_0 = G(v_0), b_1 = G(v_1)$.

A natural transformation $\alpha\colon F \to G$ consists of two components, i.e., morphisms $\alpha_{v_0}\colon a_0 \to b_0$ and $\alpha_{v_1}\colon a_1 \to b_1$, drawn as dashed lines:

$$
\begin{array}{ccc}
a_0 & \overset{\alpha_{v_0}}{\dashrightarrow} & b_0 \\
{\scriptstyle f}\downarrow & & \downarrow{\scriptstyle g} \\
a_1 & \underset{\alpha_{v_1}}{\dashrightarrow} & b_1
\end{array}
$$

The condition for α to be a natural transformation is that this square commutes.

In other words, a functor $[1] \to \mathcal{D}$ is a morphism in $\mathcal{D}$ and a natural transformation between two such functors is just a commutative square in $\mathcal{D}$.

Example 5.3.1.15. Recall that to any graph G we can associate the paths-graph $\mathrm{Paths}(G)$ (see Example 5.1.2.25). This is a functor $\mathrm{Paths}\colon \mathbf{Grph} \to \mathbf{Grph}$. There is also an identity functor $\mathrm{id}_{\mathbf{Grph}}\colon \mathbf{Grph} \to \mathbf{Grph}$. A natural transformation $\eta\colon \mathrm{id}_{\mathbf{Grph}} \to \mathrm{Paths}$ would consist of a graph homomorphism $\eta_G\colon \mathrm{id}_{\mathbf{Grph}}(G) \to \mathrm{Paths}(G)$ for every graph G. But $\mathrm{id}_{\mathbf{Grph}}(G) = G$ by definition, so we need $\eta_G\colon G \to \mathrm{Paths}(G)$. Recall that $\mathrm{Paths}(G)$ has the same vertices as G, and every arrow in G counts as a path (of length 1). So there is an obvious graph homomorphism from G to $\mathrm{Paths}(G)$. It is not hard to see that the necessary naturality squares commute.

Example 5.3.1.16. For any graph G we can associate the paths-graph $\mathrm{Paths}(G)$, and can do that twice to yield a new graph $\mathrm{Paths}(\mathrm{Paths}(G))$. Let's think through what a path of paths in G is. It is a head-to-tail sequence of arrows in $\mathrm{Paths}(G)$, meaning a head-to-tail sequence of paths in G. These composable sequences of paths (or "paths of paths") are the individual arrows in $\mathrm{Paths}(\mathrm{Paths}(G))$. The vertices in $\mathrm{Paths}(G)$ and $\mathrm{Paths}(\mathrm{Paths}(G))$ are the same as those in G, and all source and target functions are as expected.

Clearly, given such a sequence of paths in G, we could compose them to one big path in G with the same endpoints. In other words, for every $G \in \mathrm{Ob}(\mathbf{Grph})$, there is graph homomorphism $\mu_G\colon \mathrm{Paths}(\mathrm{Paths}(G)) \to \mathrm{Paths}(G)$ that is called *concatenation*. In fact, this concatenation extends to a natural transformation

$$\mu\colon \mathrm{Paths} \circ \mathrm{Paths} \to \mathrm{Paths}$$

between functors $\mathbf{Grph} \to \mathbf{Grph}$. Example 5.3.1.15 compared a graph to its paths-graph using a natural transformation $\mathrm{id}_{\mathbf{Grph}} \to \mathrm{Paths}$; here we are making a similar kind of comparison.

Remark 5.3.1.17. Example 5.3.1.15 showed that there is a natural transformation comparing each graph to its paths-graph. There is a formal sense in which a category is nothing more than a kind of reverse mapping. That is, to specify a category is the same thing as to specify a graph G together with a graph homomorphism $\text{Paths}(G) \to G$. The formalities involve monads (see Section 7.3).

Exercise 5.3.1.18.

Let X and Y be sets, and let $h \colon X \to Y$. There is a functor $C_X \colon \mathbf{Grph} \to \mathbf{Set}$ that sends every graph to the set X and sends every morphism of graphs to the identity morphism $\text{id}_X \colon X \to X$. This functor is called *the constant functor at X*. Similarly, there is a constant functor $C_Y \colon \mathbf{Grph} \to \mathbf{Set}$.

a. Use h to construct the components of a questionably natural transformation $\alpha \colon C_X \to C_Y$.

b. Is α natural?

◊

Solution 5.3.1.18.

a. For each graph $G \in \text{Ob}(\mathbf{Grph})$, we need a component $\alpha_G \colon C_X(G) \to C_Y(G)$, but $C_X(G) = X$ and $C_Y(G) = Y$, so we need a morphism $\alpha_G \colon X \to Y$. Let's set every component to be $\alpha_G := h$. We now have a questionably natural transformation.

b. For any graph morphism $f \colon G \to G'$, we have $C_X(f) = \text{id}_X$ and $C_Y(f) = \text{id}_Y$. The naturality square for f obviously commutes (because $\text{id}_Y \circ f = f \circ \text{id}_X$), so α is indeed natural.

♦

Exercise 5.3.1.19.

For any graph (V, A, src, tgt) we can extract the set of arrows or the set of vertices. Since each morphism of graphs includes a function between their arrow sets and a function between their vertex sets, we actually have functors $Ar \colon \mathbf{Grph} \to \mathbf{Set}$ and $Ve \colon \mathbf{Grph} \to \mathbf{Set}$.

a. If someone said, "Taking source vertices gives a natural transformation from Ar to Ve," what questionably natural transformation might she be referring to?

b. Is she correct, i.e., is it natural?

c. If a different person, say, from a totally different city and in a totally different frame of mind, were to hear this and say, "Taking target vertices also gives a natural transformation from Ar to Ve," would they also be correct?

◊

Solution 5.3.1.19.

a. To give a questionably natural transformation, we need to provide, for each graph $G \in \mathrm{Ob}(\mathbf{Grph})$, a morphism $\alpha_G \colon Ar(G) \to Ve(G)$ in $\mathbf{Set}$. In other words, if $G = (V_G, A_G, src_G, tgt_G)$, we need to provide a function $\alpha_G \colon A \to V$. This person seems to be suggesting we use G's source function, $\alpha_G := src_G \colon A \to V$.

b. We need to check that for each graph morphism $f \colon G \to H$, the square

$$
\begin{array}{ccc}
Ar(G) & \xrightarrow{\ src_G\ } & Ve(G) \\
{\scriptstyle Ar(f)}\big\downarrow & & \big\downarrow{\scriptstyle Ve} \\
Ar(H) & \xrightarrow[\ src_H\ ]{} & Ve(H)
\end{array}
$$

commutes. Since $Ar(f) = f_1$ and $Ve(f) = f_0$ are the "on arrows" and "on-vertices" parts of f, the square does commute by definition of graph morphism, Definition 4.3.3.1.

c. Yes, as astonishing as that may be. With category theory, we are truly one human race.

♦

Example 5.3.1.20 (Graph homomorphisms are natural transformations). As discussed (see diagram (5.8)), there is a category $\mathbf{GrIn}$ for which a functor $G \colon \mathbf{GrIn} \to \mathbf{Set}$ is the same thing as a graph. Namely, we have

$$
\mathbf{GrIn} := \boxed{ \begin{array}{ccc} Ar & \xrightarrow[\ tgt\ ]{\ src\ } & Ve \\ \bullet & & \bullet \end{array} }
$$

A natural transformation of two such functors $\alpha \colon G \to G'$ involves two components, $\alpha_{Ar} \colon G(Ar) \to G'(Ar)$ and $\alpha_{Ve} \colon G(Ve) \to G'(Ve)$, and two naturality squares, one for src and one for tgt. This is precisely the same thing as a graph homomorphism, as defined in Definition 4.3.3.1.

5.3.2   Vertical and horizontal composition

This section discusses two types of compositions for natural transformations. The terms *vertical* and *horizontal* are used to describe them; these terms come from the following pictures:

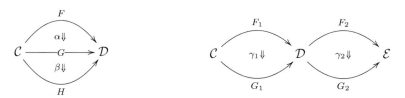

We use the symbol $\circ$ to denote vertical composition, so we have $\beta \circ \alpha \colon F \to H$ in the left-hand diagram. We use the symbol $\diamond$ for horizontal composition, so we have $\gamma_2 \diamond \gamma_1 \colon F_2 \circ F_1 \longrightarrow G_2 \circ G_1$ in the right-hand diagram. Of course, the actual arrangement of things on a page of text does not correlate with verticality or horizontality—these are just names. We define them more carefully in the following.

5.3.2.1   Vertical composition of natural transformations

The following proposition proves that functors and natural transformations (using vertical composition) form a category.

Proposition 5.3.2.2. *Let $\mathcal{C}$ and $\mathcal{D}$ be categories. There exists a category, called the category of functors from $\mathcal{C}$ to $\mathcal{D}$ and denoted $\mathrm{Fun}(\mathcal{C}, \mathcal{D})$, whose objects are the functors $\mathcal{C} \to \mathcal{D}$ and whose morphisms are the natural transformations,*

$$\mathrm{Hom}_{\mathrm{Fun}(\mathcal{C}, \mathcal{D})}(F, G) = \{\alpha \colon F \to G \mid \alpha \text{ is a natural transformation}\}.$$

Under this setup, there are indeed identity natural transformations and a composition formula for natural transformations, so we have defined a questionable category $\mathrm{Fun}(\mathcal{C}, \mathcal{D})$. The category laws hold, so it is indeed a category.

Proof. Exercise 5.3.1.13 showed that for any functor $F \colon \mathcal{C} \to \mathcal{D}$, there is an identity natural transformation $\mathrm{id}_F \colon F \to F$ (its component at $c \in \mathrm{Ob}(\mathcal{C})$ is $\mathrm{id}_{F(c)} \colon F(c) \to F(c)$).

Given a natural transformation $\alpha \colon F \to G$ and a natural transformation $\beta \colon G \to H$, we need a composite $\beta \circ \alpha$. We propose the transformation $\gamma \colon F \to H$ having components $\beta_c \circ \alpha_c$ for every $c \in \mathrm{Ob}(\mathcal{C})$. To see that γ is indeed a natural transformation, one simply puts together naturality squares for α and β to get naturality squares for $\beta \circ \alpha$.

One proves the associativity and identity laws in $\mathrm{Fun}(\mathcal{C}, \mathcal{D})$ using the fact that they hold in $\mathcal{D}$. $\square$

Notation 5.3.2.3. We sometimes denote the category $\mathrm{Fun}(\mathcal{C}, \mathcal{D})$ by $\mathcal{D}^{\mathcal{C}}$.

Example 5.3.2.4. Recall from Exercise 5.1.2.41 that there is a functor $\mathrm{Ob}\colon \mathbf{Cat} \to \mathbf{Set}$ sending a category to its set of objects. And recall from Example 5.1.2.38 that there is a functor $\mathbf{Set} \xrightarrow{Disc} \mathbf{Cat}$ sending a set to the discrete category with that set of objects (all morphisms in $Disc(S)$ are identity morphisms). Let $P\colon \mathbf{Cat} \to \mathbf{Cat}$ be the composition $P = Disc \circ \mathrm{Ob}$. Then P takes a category and makes a new category with the same objects but no morphisms. It is like crystal meth for categories.

Let $\mathrm{id}_{\mathbf{Cat}}\colon \mathbf{Cat} \to \mathbf{Cat}$ be the identity functor. There is a natural transformation $i\colon P \to \mathrm{id}_{\mathbf{Cat}}$. For any category $\mathcal{C}$, the component $i_{\mathcal{C}}\colon P(\mathcal{C}) \to \mathcal{C}$ is pretty easily understood. It is a morphism of categories, i.e., a functor. The two categories $P(\mathcal{C})$ and $\mathcal{C}$ have the same set of objects, namely, $\mathrm{Ob}(\mathcal{C})$, so the functor is identity on objects; and $P(\mathcal{C})$ has no nonidentity morphisms, so nothing else needs be specified.

Exercise 5.3.2.5.

Let $\mathcal{D} = \begin{array}{c} A \\ \bullet \end{array}$ be the category with $\mathrm{Ob}(\mathcal{D}) = \{A\}$, and $\mathrm{Hom}_{\mathcal{D}}(A, A) = \{\mathrm{id}_A\}$. What is $\mathrm{Fun}(\mathcal{D}, \mathbf{Set})$? In particular, characterize the objects and the morphisms. ◊

Solution 5.3.2.5.

A functor $\mathcal{D} \to \mathbf{Set}$ requires only knowing to which object A is sent. In other words, there is a natural bijection $\mathrm{Ob}(\mathrm{Fun}(\mathcal{D}, \mathbf{Set})) \xrightarrow{\cong} \mathrm{Ob}(\mathbf{Set})$. Given two functors $X, Y\colon \mathcal{D} \to \mathbf{Set}$, identified with sets $X, Y \in \mathrm{Ob}(\mathbf{Set})$, a questionably natural transformation $\alpha\colon X \to Y$ consists of only one component, a function $\alpha_A\colon X \to Y$. By Proposition 5.3.1.5 we find that α is natural, so there is a bijection

$$\mathrm{Hom}_{\mathrm{Fun}(\mathcal{D}, \mathbf{Set})}(X, Y) = \mathrm{Nat}(X, Y) = \mathrm{Hom}_{\mathbf{Set}}(X, Y).$$

In fact, if $\underline{1}$ is the terminal category, and $\mathcal{C}$ is any category, then there is an isomorphism in $\mathbf{Cat}$:

$$\mathrm{Fun}(\underline{1}, \mathcal{C}) \xrightarrow{\cong} \mathcal{C}. \tag{5.14}$$

♦

Notation 5.3.2.6. Recall from Notation 2.1.2.9 that if X is a set, we can represent an element $x \in X$ as a function $\{\odot\} \xrightarrow{x} X$. Similarly, suppose that $\mathcal{C}$ is a category and $c \in \mathrm{Ob}(\mathcal{C})$ is an object. There is a functor $\underline{1} \to \mathcal{C}$ that sends $1 \mapsto c$. We say that this functor *represents* $c \in \mathrm{Ob}(\mathcal{C})$. We may denote it $c\colon \underline{1} \to \mathcal{C}$.

Exercise 5.3.2.7.

Let $n \in \mathbb{N}$, and let $\underline{n}$ be the set with n elements, considered as a discrete category.[13] In other words, we write $\underline{n}$ to mean what should really be called $Disc(n)$. Describe the category $\mathrm{Fun}(\underline{3}, \underline{2})$.   ◇

Solution 5.3.2.7.

To describe the category $\mathrm{Fun}(\underline{3}, \underline{2})$, we describe its objects and then its morphisms. An object in $\mathrm{Fun}(\underline{3}, \underline{2})$ is a functor $F \colon \underline{3} \to \underline{2}$. To describe a functor, we need to say what it does on objects, $\mathrm{Ob}(\underline{3}) = \{1, 2, 3\}$, and what it does on morphisms; however the only morphisms in $\underline{3}$ are identity morphisms, and we know how a functor behaves on identities. So to give a functor $F \colon \underline{3} \to \underline{2}$ is the same thing as giving a function $F \colon \underline{3} \to \underline{2}$, and we know there are $2^3 = 8$ of these.

Suppose we have two functors $F, G \colon \underline{3} \to \underline{2}$; what is a natural transformation between them? For each object $x \in \mathrm{Ob}(\underline{3})$, we need to give a morphism $F(x) \to G(x)$ in $\underline{2}$. However, the only morphisms in $\underline{2}$ are the identities, so to have a natural transformation $F \to G$, we need that $F(x) = G(x)$ for all $x \in \mathrm{Ob}(\underline{3})$. It follows that $\mathrm{Fun}(\underline{3}, \underline{2}) \cong \underline{8}$ is the discrete category on eight objects.   ◆

Example 5.3.2.8. Let $\underline{1}$ denote the discrete category with one object (also known as the trivial monoid). For any category $\mathcal{C}$, we investigate the category $\mathcal{D} := \mathrm{Fun}(\mathcal{C}, \underline{1})$. Its objects are functors $\mathcal{C} \to \underline{1}$. Such a functor F assigns to each object in $\mathcal{C}$ an object in $\underline{1}$, of which there is one; so there is no choice in what F does on objects. And there is only one morphism in $\underline{1}$, so there is no choice in what F does on morphisms. The upshot is that there is only one object in $\mathcal{D}$, let's call it F, so $\mathcal{D}$ is a monoid. What are its morphisms?

A morphism $\alpha \colon F \to F$ in $\mathcal{D}$ is a natural transformation of functors. For every $c \in \mathrm{Ob}(\mathcal{C})$, we need a component $\alpha_c \colon F(c) \to F(c)$, which is a morphism $1 \to 1$ in $\underline{1}$. But there is only one morphism in $\underline{1}$, namely, id_1, so there is no choice about what these components should be: they are all id_1. The necessary naturality squares commute, so α is indeed a natural transformation. Thus the monoid $\mathcal{D}$ is the trivial monoid; that is, $\mathrm{Fun}(\mathcal{C}, \underline{1}) \cong \underline{1}$ for any category $\mathcal{C}$.

---

[13]When we have a functor, such as $Disc \colon \mathbf{Set} \to \mathbf{Cat}$, we sometimes say, "Let S be a set, considered as a category." This means that we want to take ideas and methods available in $\mathbf{Cat}$ and use them on the set S. Having the functor $Disc$, we use it to move S into $\mathbf{Cat}$, as $Disc(S) \in \mathrm{Ob}(\mathbf{Cat})$, upon which we can use the intended methods. However, $Disc(S)$ is bulky, e.g., $\mathrm{Fun}(Disc(\underline{3}), Disc(\underline{2}))$ is harder to read than $\mathrm{Fun}(\underline{3}, \underline{2})$. So we abuse notation and write S instead of $Disc(S)$, and talk about S as though it were still a set, e.g., discussing its elements rather than its objects. This kind of conceptual abbreviation is standard practice in mathematical discussion because it eases the mental burden, but when one says "Let S be an X considered as a Y," the other may always ask, "How are you considering X's to be Y's?" and expect a functor .

Exercise 5.3.2.9.

Let $\underline{0}$ represent the discrete category on 0 objects; it has no objects and no morphisms. Let $\mathcal{C}$ be any category.

a. What is $\mathrm{Fun}(\underline{0}, \mathcal{C})$?

b. What is $\mathrm{Fun}(\mathcal{C}, \underline{0})$?

$\Diamond$

Solution 5.3.2.9.

a. It is isomorphic to the terminal category $\underline{1}$. There is only one functor $\underline{0} \to \mathcal{C}$ because the definition of functor is vacuously satisfied in only one way. Let's call the unique object in $\mathrm{Ob}(\mathrm{Fun}(\underline{0}, \mathcal{C}))$ the "empty functor to $\mathcal{C}$." A natural transformation between the empty functor and itself requires no components, so there is again vacuously one such natural transformation, and it must be the identity.

b. If $\mathcal{C}$ is not empty, $\mathrm{Ob}(\mathcal{C}) \neq \varnothing$, then since $\mathrm{Ob}(\underline{0}) = \varnothing$ and there is no way to give a function $\mathrm{Ob}(\mathcal{C}) \to \varnothing$, we have $\mathrm{Fun}(\mathcal{C}, \underline{0}) = \underline{0}$. If $\mathcal{C} = \underline{0}$, then by the result in part (a) we know that $\mathrm{Fun}(\underline{0}, \underline{0}) = \underline{1}$.

$\blacklozenge$

Exercise 5.3.2.10.

Let [1] denote the free arrow category as in Exercise 5.1.2.34, and let **GrIn** be the graph-indexing category (see (5.8)). Draw the underlying graph of the category $\mathrm{Fun}([1], \mathbf{GrIn})$. $\Diamond$

Solution 5.3.2.10.

$\blacklozenge$

5.3.2.11   Natural isomorphisms

Let $\mathcal{C}$ and $\mathcal{D}$ be categories. We have defined a category $\text{Fun}(\mathcal{C}, \mathcal{D})$ whose objects are functors $\mathcal{C} \to \mathcal{D}$ and whose morphisms are natural transformations. What are the isomorphisms in this category?

Proposition 5.3.2.12 (Natural isomorphism). *Let $\mathcal{C}$ and $\mathcal{D}$ be categories, and let $F, G \colon \mathcal{C} \to \mathcal{D}$ be functors. A natural transformation $\alpha \colon F \to G$ is an isomorphism in $\text{Fun}(\mathcal{C}, \mathcal{D})$ if and only if the component $\alpha_c \colon F(c) \to G(c)$ is an isomorphism for each object $c \in \text{Ob}(\mathcal{C})$. In this case α is called a* natural isomorphism.

Proof. First, suppose that α is an isomorphism with inverse $\beta \colon G \to F$, and let $\beta_c \colon G(c) \to F(c)$ denote its c component. We know that $\alpha \circ \beta = \text{id}_G$ and $\beta \circ \alpha = \text{id}_F$. Using the definitions of composition and identity given in Proposition 5.3.2.2, this means that for every $c \in \text{Ob}(\mathcal{C})$, we have $\alpha_c \circ \beta_c = \text{id}_{G(c)}$ and $\beta_c \circ \alpha_c = \text{id}_{F(c)}$; in other words, α_c is an isomorphism.

Second, suppose that each α_c is an isomorphism with inverse $\beta_c \colon G(c) \to F(c)$. We need to see that these components assemble into a natural transformation, i.e., for every morphism $h \colon c \to c'$ in $\mathcal{C}$, the right-hand square

$$
\begin{array}{ccc}
F(c) & \xrightarrow{\ \alpha_c\ } & G(c) \\
{\scriptstyle F(h)}\big\downarrow & \checkmark & \big\downarrow{\scriptstyle G(h)} \\
F(c') & \xrightarrow[\ \alpha_{c'}\ ]{} & G(c')
\end{array}
\qquad\qquad
\begin{array}{ccc}
G(c) & \xrightarrow{\ \beta_c\ } & F(c) \\
{\scriptstyle G(h)}\big\downarrow & ? & \big\downarrow{\scriptstyle F(h)} \\
G(c') & \xrightarrow[\ \beta_{c'}\ ]{} & F(c')
\end{array}
$$

commutes. We know that the left-hand square commutes because α is a natural transformation; each square is labeled with a ? or a ✓ accordingly. In the following diagram we want to show that the left-hand square commutes. We know that the middle square commutes.

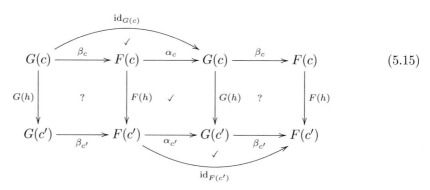

$$\tag{5.15}$$

To complete the proof we need only show that $F(h) \circ \beta_c = \beta_{c'} \circ G(h)$. This can be shown by a "diagram chase." We go through it symbolically, for demonstration. The following three equalities come from the three check marks in the (5.15).

$$F(h) \circ \beta_c = \beta_{c'} \circ \alpha_{c'} \circ F(h) \circ \beta_c = \beta_{c'} \circ G(h) \circ \alpha_c \circ \beta_c = \beta_{c'} \circ G(h).$$

□

Exercise 5.3.2.13.

Recall from Application 5.3.1.12 that a finite state machine on alphabet Σ can be understood as a functor $\mathcal{M} \to \textbf{Set}$, where $\mathcal{M} = \text{List}(\Sigma)$ is the free monoid generated by Σ. That example also discussed how natural transformations provide a language for changing state machines. Describe what kinds of changes are made by natural isomorphisms. ◊

Solution 5.3.2.13.

Let $F, G \colon \mathcal{M} \to \textbf{Set}$ be two state machines, with $X = F(\blacktriangle)$ and $Y = G(\blacktriangle)$. A natural isomorphism $\alpha \colon F \to G$ consists of a single component function $\alpha_{\blacktriangle} \colon X \to Y$. It must be an isomorphism in $\textbf{Set}$, which one can think of as a simple renaming of the states in the machine. The fact that α is natural means that for any $\sigma \in \Sigma$, we have $\alpha_{\blacktriangle} \circ F(\sigma) = G(\sigma) \circ \alpha_{\blacktriangle}$. In other words, the renaming is consistent with the $\mathcal{M}$ action.

For example, consider the following state machines $X, Y, Z \colon \text{List}(a, b) \to \textbf{Set}$:

Original model X		
ID	**a**	**b**
State 0	State 1	State 2
State 1	State 2	State 1
State 2	State 0	State 0

Isomorphic model Y		
ID	**a**	**b**
Charles	Ursula	Garfield
Ursula	Garfield	Ursula
Garfield	Charles	Charles

Nonisomorphic model Z_1		
ID	**a**	**b**
Land	Land	Air
Sea	Air	Sea
Air	Air	Air

Nonisomorphic model Z_2		
ID	**a**	**b**
Charles	Ursula	Garfield
Ursula	Garfield	Ursula
Garfield	Charles	Charles
Mary	Charles	Garfield

The first two are isomorphic, $X \cong Y$, but neither Z_1 nor Z_2 is isomorphic to any of the others. ♦

5.3.2.14   Horizontal composition of natural transformations

Example 5.3.2.15 (Whiskering). Suppose that $\mathcal{M} = \text{List}(a, b)$ and $\mathcal{M}' = \text{List}(m, n, p)$ are free monoids, and let $F \colon \mathcal{M}' \to \mathcal{M}$ be given by sending $[m] \mapsto [a], [n] \mapsto [b]$, and $[p] \mapsto [b, a, a]$. An application of this might be if the sequence $[b, a, a]$ were commonly used in practice and one wanted to add a new button just for that sequence.

Recall Application 5.3.1.12 and Figure 5.1, which is reproduced here. Let $X \colon \mathcal{M} \to$ **Set** and $Y \colon \mathcal{M} \to$ **Set** be the functors, and let $\alpha \colon Y \to X$ be the natural transformation.

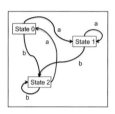
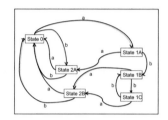

Original model $X \colon \mathcal{M} \to$ **Set**		
ID	**a**	**b**
State 0	State 1	State 2
State 1	State 2	State 1
State 2	State 0	State 0

Proposed model $Y \colon \mathcal{M} \to$ **Set**		
ID	**a**	**b**
State 0	State 1A	State 2A
State 1A	State 2A	State 1B
State 1B	State 2B	State 1C
State 1C	State 2B	State 1B
State 2A	State 0	State 0
State 2B	State 0	State 0

We can compose X and Y with F as in the diagram below

$$\mathcal{M}' \xrightarrow{\ F\ } \mathcal{M} \quad \overset{Y}{\underset{X}{\rightrightarrows}} \quad \alpha\Downarrow \quad \textbf{Set}$$

to get functors $Y \circ F$ and $X \circ F$, both of type $\mathcal{M}' \to$ **Set**. These would be as follows:[14]

	$X \circ F$		
ID	**m**	**n**	**p**
State 0	State 1	State 2	State 1
State 1	State 2	State 1	State 0
State 2	State 0	State 0	State 2

	$Y \circ F$		
ID	**m**	**n**	**p**
State 0	State 1A	State 2A	State 1A
State 1A	State 2A	State 1B	State 0
State 1B	State 2B	State 1C	State 0
State 1C	State 2B	State 1B	State 0
State 2A	State 0	State 0	State 2A
State 2B	State 0	State 0	State 2A

---

[14]The p column comes from applying b, then a, then a, as specified by F.

The map α is what sent both State 1A and State 1B in Y to State 1 in X, and so on. We can see that the same α works now: the p columns of the tables respect that mapping; that is, they act like $[b, a, a]$ or equivalently $[n, m, m]$. This is called *whiskering*. We used $\alpha \colon Y \to X$ to get a natural transformation $Y \circ F \to X \circ F$. It is a kind of horizontal composition of natural transformation.

Definition 5.3.2.16 (Whiskering). Let $\mathcal{B}, \mathcal{C}, \mathcal{D}$, and $\mathcal{E}$ be categories, let $G_1, G_2 \colon \mathcal{C} \to \mathcal{D}$ be functors, and let $\alpha \colon G_1 \to G_2$ be a natural transformation. Suppose that $F \colon \mathcal{B} \to \mathcal{C}$ (resp. $H \colon \mathcal{D} \to \mathcal{E}$) is a functor as depicted here:

$$
\mathcal{B} \xrightarrow{\ F\ } \mathcal{C} \underset{G_2}{\overset{G_1}{\Rrightarrow}} \alpha\Downarrow \mathcal{D}
\qquad
\left(\text{resp.} \quad \mathcal{C} \underset{G_2}{\overset{G_1}{\Rrightarrow}} \alpha\Downarrow \mathcal{D} \xrightarrow{\ H\ } \mathcal{E} \right),
$$

Then the *prewhiskering of α by F*, denoted $\alpha \diamond F \colon G_1 \circ F \to G_2 \circ F$ (resp. the *postwhiskering of α by H*, denoted $H \diamond \alpha \colon H \circ G_1 \to H \circ G_2$),

$$
\mathcal{B} \underset{G_2 \circ F}{\overset{G_1 \circ F}{\Rrightarrow}} \alpha \diamond F \Downarrow \mathcal{D}
\qquad
\left(\text{resp.} \quad \mathcal{C} \underset{H \circ G_2}{\overset{H \circ G_1}{\Rrightarrow}} H \diamond \alpha \Downarrow \mathcal{E} \right),
$$

is defined as follows.

For each $b \in \mathrm{Ob}(\mathcal{B})$ the component $(\alpha \diamond F)_b \colon G_1 \circ F(b) \to G_2 \circ F(b)$ is defined to be $\alpha_{F(b)}$ (resp. for each $c \in \mathrm{Ob}(\mathcal{C})$, the component $(H \diamond \alpha)_c \colon H \circ G_1(c) \to H \circ G_2(c)$ is defined to be $H(\alpha_c)$). Checking that the naturality squares commute (in each case) is straightforward.

Exercise 5.3.2.17.

Suppose given functors $\mathcal{B} \xrightarrow{F} \mathcal{C} \xrightarrow{G} \mathcal{D}$, and let $\mathrm{id}_G \colon G \to G$ be the identity natural isomorphism. Show that $\mathrm{id}_G \diamond F = \mathrm{id}_{G \circ F}$.   ◊

Solution 5.3.2.17.

By Definition 5.3.2.16, for each object $b \in \mathrm{Ob}(\mathcal{B})$, the component $(\mathrm{id}_G \diamond F)_b$ is the identity morphism $(\mathrm{id}_G)_{F(b)} \colon G(F(b)) \to G(F(b))$. But there can be only one identity morphism, so $(\mathrm{id}_G)_{F(b)} = \mathrm{id}_{G \circ F(b)} = \mathrm{id}_{G \circ F}(b)$.   ♦

Definition 5.3.2.18 (Horizontal composition of natural transformations). Let $\mathcal{B}, \mathcal{C}$, and $\mathcal{D}$ be categories, let $F_1, F_2 \colon \mathcal{B} \to \mathcal{C}$ and $G_1, G_2 \colon \mathcal{C} \to \mathcal{D}$ be functors, and let $\alpha \colon F_1 \to F_2$ and $\beta \colon G_1 \to G_2$ be natural transformations, as depicted here:

$$\mathcal{B} \;\; \alpha \Downarrow \;\; \mathcal{C} \;\; \beta \Downarrow \;\; \mathcal{D}$$

with F_1, G_1 above and F_2, G_2 below.

By pre- and postwhiskering in one order or the other we get the following diagram:

$$
\begin{array}{ccc}
G_1 \circ F_1 & \xrightarrow{G_1 \diamond \alpha} & G_1 \circ F_2 \\
\beta \diamond F_1 \downarrow & & \downarrow \beta \diamond F_2 \\
G_2 \circ F_1 & \xrightarrow[G_2 \diamond \alpha]{} & G_2 \circ F_2
\end{array}
$$

It is straightforward to show that this diagram commutes, so we can take the composition to be the definition of the horizontal composition:

$$\beta \diamond \alpha \colon G_1 \circ F_1 \to G_2 \circ F_2.$$

Remark 5.3.2.19. Whiskering a natural transformation α with a functor F is the same thing as horizontally composing α with the identity natural transformation id_F. This is true for both pre- and postwhiskering. For example, in the notation of Definition 5.3.2.16, we have

$$\alpha \diamond F = \alpha \diamond \mathrm{id}_F \qquad \text{and} \qquad H \diamond \alpha = \mathrm{id}_H \diamond \alpha.$$

Theorem 5.3.2.20 (Interchange).

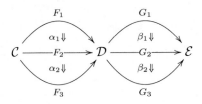

Given a setup of categories, functors, and natural transformations as shown, we have

$$(\beta_2 \circ \beta_1) \diamond (\alpha_2 \circ \alpha_1) \; = \; (\beta_2 \diamond \alpha_2) \circ (\beta_1 \diamond \alpha_1).$$

Proof. One need only observe that each square commutes in the following diagram, so taking either outer path to get $(\beta_2 \circ \beta_1) \diamond (\alpha_2 \circ \alpha_1)$ yields the same morphism as taking the diagonal path, $(\beta_2 \diamond \alpha_2) \circ (\beta_1 \diamond \alpha_1)$:

$$
\begin{array}{ccccc}
G_1 F_1 & \xrightarrow{G_1 \diamond \alpha_1} & G_1 F_2 & \xrightarrow{G_1 \diamond \alpha_2} & G_1 F_3 \\
{\scriptstyle \beta_1 \diamond F_1}\downarrow & & {\scriptstyle \beta_1 \diamond F_2}\downarrow & & \downarrow{\scriptstyle \beta_1 \diamond F_3} \\
G_2 F_1 & \xrightarrow{G_2 \diamond \alpha_1} & G_2 F_2 & \xrightarrow{G_2 \diamond \alpha_2} & G_2 F_3 \\
{\scriptstyle \beta_2 \diamond F_1}\downarrow & & {\scriptstyle \beta_2 \diamond F_2}\downarrow & & \downarrow{\scriptstyle \beta_2 \diamond F_3} \\
G_3 F_1 & \xrightarrow[G_3 \diamond \alpha_1]{} & G_3 F_2 & \xrightarrow[G_3 \diamond \alpha_2]{} & G_3 F_3
\end{array}
$$

□

Exercise 5.3.2.21.

Suppose given categories, functors, and natural transformations as shown:

$$
\mathcal{C} \; \overset{F}{\underset{F'}{\rightrightarrows}} \;{\scriptstyle \alpha\Downarrow}\; \mathcal{D} \; \overset{G}{\underset{G'}{\rightrightarrows}} \;{\scriptstyle \beta\Downarrow}\; \mathcal{E}
$$

such that $\alpha \colon F \to F'$ and $\beta \colon G \to G'$ are natural isomorphisms. Show that $\beta \diamond \alpha \colon G \circ F \to G' \circ F'$ is a natural isomorphism. ◊

Solution 5.3.2.21.

Let $\alpha' \colon F' \to \alpha$ and $\beta' \colon G' \to G$ be the inverses of α and β respectively. To check that $\beta \diamond \alpha$ is an isomorphism, we use Theorem 5.3.2.20 (and Exercise 5.3.2.17) to see that

$$(\beta \diamond \alpha) \circ (\beta' \diamond \alpha') = (\beta \circ \beta') \diamond (\alpha \circ \alpha') = \mathrm{id}_{G'} \diamond \mathrm{id}_{F'} = \mathrm{id}_{G' \circ F'}$$

and similarly for the other order, $(\beta' \diamond \alpha') \circ (\beta \diamond \alpha) = \mathrm{id}_{G \circ f}$.   ♦

5.3.3   The category of instances on a database schema

Section 5.2.2 showed that schemas are presentations of categories, and Section 5.4 shows that in fact the category of schemas is equivalent to the category of categories. This section therefore takes license to blur the distinction between schemas and categories.

If $\mathcal{C}$ is a schema, i.e., a category, then as discussed in Section 5.2.2.6, an instance on $\mathcal{C}$ is a functor $I \colon \mathcal{C} \to \mathbf{Set}$. But now we have a notion beyond categories and functors, namely, that of natural transformations. So we make the following definition.

Definition 5.3.3.1. Let $\mathcal{C}$ be a schema (or category). The *category of instances on $\mathcal{C}$*, denoted $\mathcal{C}$–$\mathbf{Set}$, is $\mathrm{Fun}(\mathcal{C}, \mathbf{Set})$. Its objects are $\mathcal{C}$-instances (i.e., functors $\mathcal{C} \to \mathbf{Set}$), and its morphisms are natural transformations.

Remark 5.3.3.2. One might object to Definition 5.3.3.1 on the grounds that database instances should not be infinite. This is a reasonable perspective, and the definition can be modified easily to accommodate it. The subcategory $\mathbf{Fin}$ (see Example 5.1.1.4) of finite sets can be substituted for $\mathbf{Set}$ in Definition 5.3.3.1. One could define the *category of finite instances on $\mathcal{C}$* as $\mathcal{C}$–$\mathbf{Fin} = \mathrm{Fun}(\mathcal{C}, \mathbf{Fin})$. Almost all of the ideas in this book will make perfect sense in $\mathcal{C}$–$\mathbf{Fin}$.

Natural transformations should serve as some kind of morphism between instances on the same schema. How are we to interpret a natural transformation $\alpha \colon I \to J$ between database instances $I, J \colon \mathcal{C} \to \mathbf{Set}$?

A first clue comes from Application 5.3.1.12. There we considered the case of a monoid $\mathcal{M}$, and we thought about a natural transformation between two functors $X, Y \colon \mathcal{M} \to \mathbf{Set}$, considered as different finite state machines. The notion of natural transformation captured the idea of one model being a refinement of another. This same kind of idea works for databases with more than one table (categories with more than one object). Let's work it through slowly.

Example 5.3.3.3. Consider the terminal schema, $\underline{1} \cong \boxed{\bullet^{\mathrm{Grapes}}}$. An instance is a functor $\underline{1} \to \mathbf{Set}$, which represents a set (see Notation 5.3.2.6). A natural transformation $\alpha \colon I \to J$ is a function from set I to set J. In the standard table view, we might have I and J as shown here:

Grapes (I)
ID
Grape 1
Grape 3
Grape 4

Grapes (J)
ID
Jan1-01
Jan1-02
Jan1-03
Jan1-04
Jan3-01
Jan4-01
Jan4-02

There are 343 natural transformations $I \to J$. Perhaps some of them make more sense than others, e.g., we could hope that the numbers in I corresponded to the numbers after the hyphen in J or perhaps to what seems to be the date in January. Knowing something like this would reduce this to only a few options out of 343 possible mappings. But it could be that the rows in J correspond to batches, and all three grapes in I are part of the first batch on Jan-01.

The point is that the notion of natural transformation is a mathematical one; it has nothing to do with the kinds of associations we might find natural, unless we have found a categorical encoding for this intuition.

Exercise 5.3.3.4.

Recall the notion of set-indexed sets from Definition 3.4.6.11. Let A be a set, and devise a schema $\mathcal{A}$ such that instances on $\mathcal{A}$ are A-indexed sets. Is our current notion of morphism between instances (i.e., natural transformations) well aligned with this definition of mapping of A-indexed sets? ◊

Solution 5.3.3.4.

Definition 3.4.6.11 actually gives us the objects and morphisms of a category, say, the *category of A-indexed sets*, in that it tells us that the objects and morphisms are merely the A-indexed sets and the A-indexed functions. Let us denote the category of A-indexed sets A–**Set**; this exercise is asking for a category $\mathcal{A}$ for which there is an isomorphism

$$A\text{–}\mathbf{Set} \xrightarrow{\cong} \mathrm{Fun}(\mathcal{A}, \mathbf{Set}).$$

And indeed there is. Let $\mathcal{A} = Disc(A)$ be the discrete category on A objects. Then a functor $S \colon \mathcal{A} \to \mathbf{Set}$ is just a set $S(a)$ for every $a \in A$, and a morphism $S \to S'$ is just a component $f_a \colon S(a) \to S'(a)$ for each $a \in A$. These coincide exactly with the notions of A-indexed set and of mappings between them. ♦

For a general schema (or category) $\mathcal{C}$, let us think through what a morphism $\alpha\colon I \to J$ between instances $I, J\colon \mathcal{C} \to \mathbf{Set}$ is. For each object $c \in \mathrm{Ob}(\mathcal{C})$, there is a component $\alpha_c\colon I(c) \to J(c)$. This means that just as in Example 5.3.3.3, there is for each table c a function from the rows in I's manifestation of c to the rows in J's manifestation of c. So to make a natural transformation, such a function has to be specified table by table. But then we have to contend with naturality squares, one for every arrow in $\mathcal{C}$. Arrows in $\mathcal{C}$ correspond to foreign key columns in the database. The naturality requirement was already covered in Application 5.3.1.12 (see especially how (5.11) is checked in (5.12) and (5.13)).

Example 5.3.3.5. We saw in Section 5.2.1.21 that graphs can be regarded as functors $\mathcal{G} \to \mathbf{Set}$, where $\mathcal{G} \cong \mathbf{GrIn}$ is the schema for graphs shown here:

A database instance $I\colon \mathcal{G} \to \mathbf{Set}$ on $\mathcal{G}$ consists of two tables. Here is an example instance:

Arrow (I)		
ID	**src**	**tgt**
f	v	w
g	w	x
h	w	x

Vertex (I)
ID
v
w
x

To discuss natural transformations, we need two instances. Here is another, $J\colon \mathcal{G} \to \mathbf{Set}$:

Arrow (J)		
ID	**src**	**tgt**
i	q	r
j	r	s
k	s	r
ℓ	s	t

Vertex (J)
ID
q
r
s
t
u

To give a natural transformation $\alpha\colon I \to J$, we give two components: one for `Arrow` and one for `Vertex`. We need to say where each vertex in I goes in J, and we need to say where each arrow in I goes in J. The naturality squares insist that if we specify that $g \mapsto j$, for example, then we had better specify that $w \mapsto r$ and that $x \mapsto s$. What a computer is very good at, but a human is fairly slow at, is checking that a given pair of components (arrows and vertices) really is natural.

There are 8000 ways to devise component functions $\alpha_{\texttt{Arrow}}$ and $\alpha_{\texttt{Vertex}}$, but precisely six natural transformations, i.e., six graph homomorphisms, $I \to J$; the other 7,994 are haphazard flingings of arrows to arrows and vertices to vertices without any regard to sources and targets. The six are briefly described now. The reader should look at the graph diagrams of I and J while following along.

Every vertex in I has to be sent to some vertex in J, so we think about where to send v and proceed from there.

- If we try to send $v \mapsto^? u$, we fail because u touches no arrows, so there is nowhere for f to go. (0)

- If we send $v \mapsto q$, then f must map to i, and w must map to r, and both g and h must map to j, and x must map to s. (1)

- If we send $v \mapsto r$, then there are two choices for g times two choices for h. (4)

- If we send $v \mapsto s$, then there is one way to obtain a graph morphism. (1)

- If we try to send $v \mapsto^? t$, we fail as before. (0)

Humans may follow the diagrams better than the tables, whereas computers probably understand the tables better.

Exercise 5.3.3.6.

If $I, J\colon \mathcal{G} \to \mathbf{Set}$, as in Example 5.3.3.5, how many natural transformations are there $J \to I$? ◊

Solution 5.3.3.6.

A computer can find all the natural transformations between two database instances (if the schemas are finite) relatively quickly, but it can be significantly more challenging for humans. However, a natural transformation $\alpha\colon J \to I$ is a graph homomorphism

$J \to I$. Luckily the human visual system is equipped to look for graph homomorphisms.

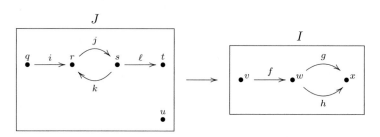

We start by asking where we can send q.

Work through the following informal reasoning, and then work the rest out for yourself:

> If q is sent to v, then i must be sent to f, so r must be sent to w. And then
> s would have to go to x. But then k does not make sense because there is no
> arrow $x \to w$ in I. So it turns out that q cannot be sent to v after all.

From here one can see that there are no natural transformations, $\mathrm{Nat}(J, I) = \varnothing$, basically
because of k.   ♦

Exercise 5.3.3.7.

Let **GrIn** be the graph-indexing category, and let $Y_A \colon \mathbf{GrIn} \to \mathbf{Set}$ denote the following instance:

Arrow (Y_A)		
ID	**src**	**tgt**
a	v_0	v_1

Vertex (Y_A)
ID
v_0
v_1

Let $I \colon \mathbf{GrIn} \to \mathbf{Set}$ be as in Example 5.3.3.5.

a. How many natural transformations are there $Y_A \to I$?

b. With J as previously, how many natural transformations are there $Y_A \to J$?

c. Do you have any conjecture about the way natural transformations $Y_A \to X$ behave
   for arbitrary graphs $X \colon \mathcal{G} \to \mathbf{Set}$?

◇

Solution 5.3.3.7.

It is useful to see Y_A as a graph so we can visualize the graph morphisms $Y_A \to I$ or $Y_A \to J$.

$$Y_A$$

v_0	a	v_1
$\bullet$	$\longrightarrow$	$\bullet$

a. A graph morphism $Y_A \to I$ amounts to an arrow in graph I. In other words, there is a natural isomorphism

$$\mathrm{Nat}(Y_A, I) \cong \{f, g, h\}.$$

How does this works? What might g mean as a natural transformation $Y_A \to I$?

To give a questionably natural transformation $\alpha\colon Y_A \to I$, we need to give a component $\alpha_{Ar}\colon \{a\} \to \{f, g, h\}$ and a component $\alpha_{Ve}\colon \{v_0, v_1\} \to \{v, w, x\}$. Since we have g in mind, let's put $\alpha_{Ar}(a) := g$. There are 3^2 choices for α_{Ve}, but only one is natural because the two morphisms $src, tgt\colon Ar \to Ve$ demand two naturality equations,

$$\alpha_{Ve}(v_0) = \alpha_{Ve} \circ src(a) = src \circ \alpha_{Ar}(a) = src(g) = w;$$
$$\alpha_{Ve}(v_1) = \alpha_{Ve} \circ tgt(a) = tgt \circ \alpha_{Ar}(a) = tgt(g) = x.$$

In other words, once we choose $\alpha_{Ar}(a)$ to be g, the rest is forced on us. In the same way, we could have chosen $\alpha_{Ar}(a)$ to be any of f, g, h, which is why we said $\mathrm{Nat}(Y_A, I) \cong \{f, g, h\}$.

b. There are four, $\mathrm{Nat}(Y_A, J) \cong \{i, j, k, \ell\}$.

♦

In terms of databases, this notion of instance morphism $\alpha\colon I \to J$ on a schema $\mathcal{C}$ is sometimes called a *database homomorphism*. It is related to what is known as *provenance*, in that it tells us how every row in I relates to a counterpart row in J. More precisely, for every table in $\mathcal{C}$, the morphism α gives a mapping from the set of rows in I's version of the table to J's version of the table, such that all the foreign keys are respected. This notion of morphism has excellent formal properties, so projections, unions, and joins of tables (the typical database operations) would be predicted to be interesting by a category theorist who has no idea what a database is.[15]

---

[15]More precisely, given a functor between schemas $F\colon \mathcal{C} \to \mathcal{D}$, the pullback $\Delta_F\colon \mathcal{D}\text{–}\mathbf{Set} \to \mathcal{C}\text{–}\mathbf{Set}$, its left Σ_F and its right adjoint Π_F constitute these important queries. See Section 7.1.4.

5.3.4   Equivalence of categories

We have a category **Cat** of categories, and in every category there is a notion of isomorphism between objects: one morphism each way, such that each round-trip composition is the identity. An isomorphism in **Cat**, therefore, takes place between two categories, say, $\mathcal{C}$ and $\mathcal{D}$: it is a functor $F\colon \mathcal{C} \to \mathcal{D}$ and a functor $G\colon \mathcal{D} \to \mathcal{C}$ such that $G \circ F = \mathrm{id}_{\mathcal{C}}$ and $F \circ G = \mathrm{id}_{\mathcal{D}}$.

It turns out that categories are often similar enough to be considered equivalent without being isomorphic. For this reason, the notion of isomorphism is considered too strong to be useful for categories, akin to saying that two material samples are the same if there is an atom by atom matching, or that two words are the same if they are written in the same font and size, by the same person, in the same state of mind.

As reasonable as isomorphism is as a notion *in* most categories, it fails to be the right notion *about* categories. The reason is that *in* categories there are objects and morphisms, whereas when we talk *about* categories, we have categories and functors plus natural transformations. Natural transformations serve as mappings between mappings, and this is not part of the structure of an ordinary category. In cases where a category $\mathcal{C}$ does have such mappings between mappings, it is often best to take that extra structure into account, as we do for $\mathcal{C} = $ **Cat**. This whole subject leads to the study of 2-categories (or n-categories, or ∞-categories), not discussed in this book. See, for example, Leinster [25] for an introduction.

The purpose now is to explain this "good notion" of sameness for categories, namely, *equivalence of categories*, which appropriately takes natural transformations into account. Instead of functors going both ways with round-trips equal to identity, which is required in order to be an isomorphism of categories, equivalence of categories demands functors going both ways with roundtrips *naturally isomorphic* to identity.

Definition 5.3.4.1 (Equivalence of categories). Let $\mathcal{C}$ and $\mathcal{C}'$ be categories. A functor $F\colon \mathcal{C} \to \mathcal{C}'$ is called *an equivalence of categories* and denoted $F\colon \mathcal{C} \xrightarrow{\simeq} \mathcal{C}'$[16] if there exists a functor $F'\colon \mathcal{C}' \to \mathcal{C}$ and natural isomorphisms $\alpha\colon \mathrm{id}_{\mathcal{C}} \xrightarrow{\cong} F' \circ F$ and $\alpha'\colon \mathrm{id}_{\mathcal{C}'} \xrightarrow{\cong} F \circ F'$. In this case we say that F and F' are *mutually inverse equivalences*.

Suppose we are given functors $F\colon \mathcal{C} \to \mathcal{C}'$ and $F'\colon \mathcal{C}' \to \mathcal{C}$. We want to know something about the round-trips on $\mathcal{C}$ and on $\mathcal{C}'$; we want to know the same kind of information about each round-trip, so let's concentrate on the $\mathcal{C}$ side. We want to know something about $F' \circ F\colon \mathcal{C} \to \mathcal{C}$, so let's name it $i\colon \mathcal{C} \to \mathcal{C}$; we want to know that i is a natural isomorphism. That is, for every $c \in \mathrm{Ob}(\mathcal{C})$, we want an isomorphism $\alpha_c\colon c \xrightarrow{\cong} i(c)$, and we want to know that these isomorphisms are picked carefully enough that given $g\colon c \to c'$

---

[16]The notation $\simeq$ has already been used for equivalences of paths in a schema. I do not mean to equate these ideas; I am just reusing the symbol. Hopefully, no confusion will arise.

in $\mathcal{C}$, the choice of isomorphisms for c and c' are compatible:

$$
\begin{array}{ccc}
c & \xrightarrow{\;\alpha_c\;} & i(c) \\
{\scriptstyle g}\big\downarrow & & \big\downarrow{\scriptstyle i(g)} \\
c' & \xrightarrow[\;\alpha_{c'}\;]{} & i(c').
\end{array}
$$

To be an equivalence, the same has to hold for the other round-trip, $i' = F \circ F' \colon \mathcal{C}' \to \mathcal{C}'$.

Exercise 5.3.4.2.

Let $\mathcal{C}$ and $\mathcal{C}'$ be categories. Suppose that $F \colon \mathcal{C} \to \mathcal{C}'$ is an isomorphism of categories.

a. Is it an equivalence of categories?

b. If not, why? If so, what are the components of α and α' (with notation as in Definition 5.3.4.1)?

◊

Solution 5.3.4.2.

a. Yes.

b. If a functor $F \colon \mathcal{C} \to \mathcal{C}'$ is an isomorphism of categories, then there exists a functor $F' \colon \mathcal{C}' \to \mathcal{C}$ such that $F' \circ F = \mathrm{id}_{\mathcal{C}}$ and $F \circ F' = \mathrm{id}_{\mathcal{C}'}$. We might hope that F and F' are mutually inverse equivalences of categories as well. We need natural transformations $\alpha \colon \mathrm{id}_{\mathcal{C}} \to F' \circ F$ and $\alpha' \colon \mathrm{id}_{\mathcal{C}'} \to F \circ F'$. But since $F' \circ F = \mathrm{id}_{\mathcal{C}}$ and $F \circ F' = \mathrm{id}_{\mathcal{C}'}$, we can take α and α' to be the identity transformations. Thus F and F' are indeed mutually inverse equivalences of categories.

◆

Example 5.3.4.3. Let S be a set, and let $S \times S \subseteq S \times S$ be the complete relation on S, which is a preorder K_S. Recall from Proposition 5.2.1.13 that there is a functor $i \colon \mathbf{PrO} \to \mathbf{Cat}$, and the resulting category $i(K_S)$ is called the *indiscrete category on S*; it has objects S and a single morphism between every pair of objects. Here is a diagram of $K_{\{1,2,3\}}$:

$$(5.16)$$

It is easy check that $K_{\underline{1}}$, the indiscrete category on one element, is isomorphic to $\underline{1}$, the discrete category on one object, also known as the terminal category (see Exercise 5.1.2.40). The category $\underline{1}$ consists of one object, its identity morphism, and nothing else. Let's think about the difference between isomorphism and equivalence using $K_S \in$ Ob(**Cat**).

The only way that K_S can be isomorphic to $\underline{1}$ is if S has one element.[17] On the other hand, there is an equivalence of categories

$$K_S \simeq \underline{1}$$

for every set $S \neq \emptyset$. So for example, $K_{\{1,2,3\}}$ from (5.16) is equivalent to the terminal category, $\underline{1}$.

In fact, there are many such equivalences, one for each element of S. To see this, let S be a nonempty set, and choose an element $s_0 \in S$. For every $s \in S$, there is a unique isomorphism $k_s \colon s \xrightarrow{\cong} s_0$ in K_S. Let $F \colon K_S \to \underline{1}$ be the only possible functor (see Exercise 5.1.2.40), and let $F' \colon \underline{1} \to K_S$ represent the object s_0. Note that $F' \circ F = \mathrm{id}_{\underline{1}} \colon \underline{1} \to \underline{1}$ is the identity, but that $F \circ F' \colon K_S \to K_S$ sends everything to s_0. So F is not an isomorphism. We need to show that it is an equivalence.

Let $\alpha = \mathrm{id}_{\underline{1}}$, and define $\alpha' \colon \mathrm{id}_{K_S} \to F \circ F'$ by $\alpha'_s = k_s$. Note that α'_s is an isomorphism for each $s \in \mathrm{Ob}(K_S)$ and that α' is a natural transformation (hence, a natural isomorphism) because every possible square commutes in K_S. This completes the proof, initiated in the preceding paragraph, that the category K_S is equivalent to $\underline{1}$ for every nonempty set S and that this fact can be witnessed by any element $s_0 \in S$.

Example 5.3.4.4. Consider the category **FLin**, described in Example 5.1.1.13, of finite nonempty linear orders. For every natural number $n \in \mathbb{N}$, let $[n] \in \mathrm{Ob}(\mathbf{FLin})$ denote the linear order shown in Example 4.4.1.7. Define a category **Δ** whose objects are given by

---

[17]One way to see this is that by Exercise 5.1.2.41, we have a functor Ob: **Cat** → **Set**, and we know by Exercise 5.1.2.27 that functors preserve isomorphisms, so an isomorphism between categories must restrict to an isomorphism between their sets of objects. The only sets that are isomorphic to $\underline{1}$ have one element.

$Ob(\mathbf{\Delta}) = \{[n] \mid n \in \mathbb{N}\}$ and with $Hom_{\mathbf{\Delta}}([m], [n]) = Hom_{\mathbf{FLin}}([m], [n])$. The difference between $\mathbf{FLin}$ and $\mathbf{\Delta}$ is only that objects in $\mathbf{FLin}$ may have odd labels, e.g.,

$$\overset{5}{\bullet} \longrightarrow \overset{x}{\bullet} \longrightarrow \overset{\text{``Sam''}}{\bullet}$$

whereas objects in $\mathbf{\Delta}$ all have standard labels, e.g.,

$$\overset{0}{\bullet} \longrightarrow \overset{1}{\bullet} \longrightarrow \overset{2}{\bullet}$$

Clearly, $\mathbf{FLin}$ is a much larger category, and yet it feels as if it is pretty much the same as $\mathbf{\Delta}$. Actually, they are equivalent, $\mathbf{FLin} \simeq \mathbf{\Delta}$. We will find functors F and F' which witness this equivalence.

Let $F' : \mathbf{\Delta} \to \mathbf{FLin}$ be the inclusion; and let $F : \mathbf{FLin} \to \mathbf{\Delta}$ send every finite nonempty linear order $X \in Ob(\mathbf{FLin})$ to the object $F(X) := [n] \in \mathbf{\Delta}$, where $Ob(X) \cong \{0, 1, \ldots, n\}$. For each such X, there is a unique isomorphism $\alpha_X : X \xrightarrow{\cong} [n]$, and these fit together into[18] the required natural isomorphism $\mathrm{id}_{\mathbf{FLin}} \to F' \circ F$. The other natural isomorphism $\alpha' : \mathrm{id}_{\mathbf{\Delta}} \to F \circ F'$ is the identity.

Exercise 5.3.4.5.

Recall from Definition 2.1.2.23 that a set X is called finite if there exists a natural number $n \in \mathbb{N}$ and an isomorphism of sets $X \to \underline{n}$. Let $\mathbf{Fin}$ denote the category whose objects are the finite sets and whose morphisms are the functions. Let $\mathcal{S}$ denote the category whose objects are the sets $\underline{n}$ and whose morphisms are again the functions. The difference between $\mathbf{Fin}$ and $\mathcal{S}$ is that every object in $\mathcal{S}$ is one of these $\underline{n}$'s, whereas every object in $\mathbf{Fin}$ is just isomorphic to one of these $\underline{n}$'s.

For every object $X \in Ob(\mathbf{Fin})$, there exists an isomorphism $p_X : X \xrightarrow{\cong} \underline{m}$ for some unique object $\underline{m} \in Ob(\mathcal{S})$. Find an equivalence of categories $\mathbf{Fin} \xrightarrow{\simeq} \mathcal{S}$. ◊

Solution 5.3.4.5.

There is an obvious inclusion functor $i : \mathcal{S} \to \mathbf{Fin}$, i.e., $\underline{n} \mapsto \underline{n} \in Ob(\mathbf{Fin})$. Define a functor $Q : \mathbf{Fin} \to \mathcal{S}$. On a finite set X of cardinality $|X| = m$, we take $Q(X) = \underline{m}$. But given a function $f : X \to Y$, where $|Y| = n$, how do we get a function $Q(f) : \underline{m} \xrightarrow{?} \underline{n}$, especially after what was learned in Exercise 2.1.2.19?

Luckily, we are provided with isomorphisms $p_X : X \xrightarrow{\cong} \underline{m}$ and $p_Y : Y \xrightarrow{\cong} \underline{n}$. Let $q_X : \underline{m} \to X$ be the inverse of p_X, so $q_X \circ p_X = p_X \circ q_X = \mathrm{id}_X$. Then to define $Q(f)$ we

---

[18]The phrase "these fit together into" is shorthand for, and can be replaced by, "the naturality squares commute for these components, so together they constitute."

just follow the other arrows in this diagram:

$$X \xrightarrow{\ f\ } Y$$

with q_X, p_Y, $m \dashrightarrow n$ labeled $Q(f)$

In other words, for $f \colon X \to Y$, we define $Q(f) = p_Y \circ f \circ q_X$. We have now given a questionable functor Q. In fact, the construction of Q is sufficiently odd that it pays to check carefully that it follows both the identity law and the composition law for functors. On identity morphisms id_X, we have $Q(\mathrm{id}_X) = p_X \circ \mathrm{id}_X \circ q_X = \mathrm{id}_X$, so the identity law holds. And for $g \colon Y \to Z$, we have

$$Q(g) \circ Q(f) = p_Z \circ g \circ q_Y \circ p_Y \circ f \circ q_X = p_Z \circ g \circ f \circ q_X = Q(g \circ f),$$

so the composition law holds too; i.e., Q is indeed a functor.

We still need to show that Q and i are mutually inverse equivalences of categories. We give a natural transformation $\alpha \colon \mathrm{id}_{\mathbf{Fin}} \to i \circ Q$, and a natural transformation $\alpha' \colon \mathrm{id}_{\mathcal{S}} \to Q \circ i$. Roughly, we take $\alpha = p$ and $\alpha' = q$. We go through the details for α.

For each $X \in \mathrm{Ob}(\mathbf{Fin})$, we set α_X to be the component $p_X \colon X \to i \circ Q(X) = \underline{m}$. We simply need to check that the naturality square commutes for each $f \colon X \to Y$, namely,

$$X \xrightarrow{\ p_X\ } \underline{m}$$

with f, $p_Y \circ f \circ q_X$, $Y \xrightarrow{p_Y} \underline{n}$

But $p_Y \circ f \circ q_X \circ p_X = p_Y \circ f$ holds because $q_X \circ p_X = \mathrm{id}_X$.   ♦

Exercise 5.3.4.6.

We say that two categories $\mathcal{C}$ and $\mathcal{D}$ are equivalent if there exists an equivalence of categories between them. Show that the relation of being equivalent is an equivalence relation on $\mathrm{Ob}(\mathbf{Cat})$.   ◊

Solution 5.3.4.6.

For categories $\mathcal{C}$ and $\mathcal{D}$, we write $\mathcal{C} \simeq \mathcal{D}$ if there exists an equivalence between them. Clearly, this relation is reflexive because the identity $\mathcal{C} \xrightarrow{\ \mathrm{id}_{\mathcal{C}}\ } \mathcal{C}$ is an isomorphism by Exercise 5.1.1.21, and isomorphisms of categories are equivalences by Exercise 5.3.4.2.

To see that it is symmetric, suppose $\mathcal{C} \simeq \mathcal{D}$. Then there exists an equivalence $F\colon \mathcal{C} \to \mathcal{D}$, which by Definition 5.3.4.1 has an inverse $F'\colon \mathcal{D} \to \mathcal{C}$. Then F' is also an equivalence, so $\mathcal{D} \simeq \mathcal{C}$.

To see that it is transitive, suppose $\mathcal{C} \simeq \mathcal{D}$ and $\mathcal{D} \simeq \mathcal{E}$. Then we have inverse equivalences (F, F') and (G, G') as shown:

$$\mathcal{C} \underset{F'}{\overset{F}{\rightleftarrows}} \mathcal{D} \underset{G'}{\overset{G}{\rightleftarrows}} \mathcal{E}$$

We need to show that $G \circ F\colon \mathcal{C} \to \mathcal{E}$ is an equivalence. Before we can do that, we should name the natural isomorphisms that establish the equivalences:

$$\alpha\colon F' \circ F \overset{\cong}{\rightarrow} \mathrm{id}_{\mathcal{C}}, \qquad \alpha'\colon F \circ F' \overset{\cong}{\rightarrow} \mathrm{id}_{\mathcal{D}}, \qquad \beta\colon G' \circ G \overset{\cong}{\rightarrow} \mathrm{id}_{\mathcal{D}}, \qquad \beta'\colon G \circ G' \overset{\cong}{\rightarrow} \mathrm{id}_{\mathcal{E}}.$$

Using whiskering (Definition 5.3.2.16), we have natural transformations

$$G \circ F \circ F' \circ G' \xrightarrow{G \diamond \alpha' \diamond G} G \circ G' \xrightarrow{\beta'} \mathrm{id}_{\mathcal{E}} \qquad (5.17)$$

and similarly we obtain a natural transformation $F' \circ G' \circ G \circ F \to \mathrm{id}_{\mathcal{C}}$. In fact, each morphism in (5.17) is a natural isomorphism by Exercise 5.3.2.21, so the composites $G \circ F \circ F' \circ G' \overset{\cong}{\rightarrow} \mathrm{id}_{\mathcal{E}}$ and $F' \circ G' \circ G \circ F \overset{\cong}{\rightarrow} \mathrm{id}_{\mathcal{C}}$ are natural isomorphisms too. This proves that $G \circ F$ is an equivalence of categories, so $\mathcal{C} \simeq \mathcal{E}$ as desired. ♦

Example 5.3.4.7. Consider the group $\mathbb{Z}_2 := (\{0, 1\}, 0, +)$, where $1 + 1 = 0$. As a category, $\mathbb{Z}_2$ has one object ▲ and two morphisms, namely, $0, 1$, such that 0 is the identity. Since $\mathbb{Z}_2$ is a group, every morphism is an isomorphism.

Let $\mathcal{C} = \underline{1}$ be the terminal category, as in Exercise 5.1.2.40. One might accidentally believe that $\mathcal{C}$ is equivalent to $\mathbb{Z}_2$, but this is not the case. The argument in favor of the accidental belief is that we have unique functors $F\colon \mathbb{Z}_2 \to \mathcal{C}$ and $F'\colon \mathcal{C} \to \mathbb{Z}_2$ (and this is true); the round-trip $F \circ F'\colon \mathcal{C} \to \mathcal{C}$ is the identity (and this is true); and for the round-trip $F' \circ F\colon \mathbb{Z}_2 \to \mathbb{Z}_2$ both morphisms in $\mathbb{Z}_2$ are isomorphisms, so any choice of morphism $\alpha_{\blacktriangle}\colon \blacktriangle \to F' \circ F(\blacktriangle)$ will be an isomorphism (and this is true). The problem is that whatever one does with $\alpha_{\blacktriangle}$, one gets a questionably natural isomorphism, but it will never be natural.

When we round-trip $F' \circ F\colon \mathbb{Z}_2 \to \mathbb{Z}_2$, the image of $1\colon \blacktriangle \to \blacktriangle$ is $F' \circ F(1) = 0 = \mathrm{id}_{\blacktriangle}$. So the naturality square for the morphism 1 looks like this:

$$
\begin{array}{ccc}
\blacktriangle & \xrightarrow{\alpha_{\blacktriangle}} & \blacktriangle \\
{\scriptstyle 1}\big\downarrow & & \big\downarrow {\scriptstyle 0 = F' \circ F(1)} \\
\blacktriangle & \xrightarrow[\alpha_{\blacktriangle}]{} & \blacktriangle
\end{array}
$$

where it is undecided whether $\alpha_{\blacktriangle}$ is to be 0 or 1. Unfortunately, neither choice works (i.e., for neither choice will the diagram commute) because $x + 1 \neq x + 0$ in $\mathbb{Z}_2$.

Definition 5.3.4.8 (Full and faithful functors). Let $\mathcal{C}$ and $\mathcal{D}$ be categories, and let $F\colon \mathcal{C} \to \mathcal{D}$ be a functor. For any two objects $c, c' \in \mathrm{Ob}(\mathcal{C})$, there is a function

$$\mathrm{Hom}_F(c, c')\colon \mathrm{Hom}_{\mathcal{C}}(c, c') \to \mathrm{Hom}_{\mathcal{D}}(F(c), F(c'))$$

guaranteed by the definition of functor. We say that F is *a full functor* if $\mathrm{Hom}_F(c, c')$ is surjective for every $c, c' \in \mathrm{Ob}(\mathcal{C})$. We say that F is *a faithful functor* if $\mathrm{Hom}_F(c, c')$ is injective for every c, c'. We say that F is *a fully faithful functor* if $\mathrm{Hom}_F(c, c')$ is bijective for every c, c'.

Exercise 5.3.4.9.

Let $\underline{1}$ and $\underline{2}$ be the discrete categories on one and two objects respectively. There is only one functor $F\colon \underline{2} \to \underline{1}$.

a. Is it full?

b. Is it faithful?

$\diamond$

Solution 5.3.4.9.

a. No. Take the objects $c = 1$ and $c' = 2$ in $\mathrm{Ob}(\underline{2})$. We have $\mathrm{Hom}_{\underline{2}}(1, 2) = \varnothing$, whereas $\mathrm{Hom}_{\underline{1}}(F(1), F(2)) = \mathrm{Hom}_{\underline{1}}(1, 1) = \{\mathrm{id}_1\}$ has one element. The function $\mathrm{Hom}_F(1, 2)$ cannot be surjective.

b. Yes. Every hom-set in $\underline{2}$ has at most one element, and every function out of a 0- or 1-element set is injective.

$\blacklozenge$

Exercise 5.3.4.10.

Let $\underline{0}$ denote the empty category, and let $\mathcal{C}$ be any category. There is a unique functor $F\colon \underline{0} \to \mathcal{C}$.

a. For general $\mathcal{C}$, will F be full?

b. For general $\mathcal{C}$, will F be faithful?

c. For general $\mathcal{C}$, will F be an equivalence of categories?

$\diamond$

Solution 5.3.4.10.

a. Yes, vacuously.

b. Yes, vacuously.

c. No, generally there will be no functor $\mathcal{C} \xrightarrow{?} \underline{0}$ (unless $\mathcal{C} = 0$), so F could not be an equivalence.

♦

Proposition 5.3.4.11. *Let $\mathcal{C}$ and $\mathcal{C}'$ be categories, and let $F \colon \mathcal{C} \to \mathcal{C}'$ be an equivalence of categories. Then F is fully faithful.*

Sketch of proof. Suppose F is an equivalence, so we can find a functor $F' \colon \mathcal{C}' \to \mathcal{C}$ and natural isomorphisms $\alpha \colon \mathrm{id}_{\mathcal{C}} \xrightarrow{\cong} F' \circ F$ and $\alpha' \colon \mathrm{id}_{\mathcal{C}'} \xrightarrow{\cong} F \circ F'$. We need to know that for any objects $c, d \in \mathrm{Ob}(\mathcal{C})$, the map

$$\mathrm{Hom}_F(c, d) \colon \mathrm{Hom}_{\mathcal{C}}(c, d) \to \mathrm{Hom}_{\mathcal{C}'}(Fc, Fd)$$

is bijective. Consider the following diagram

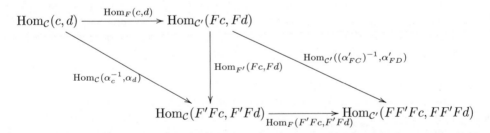

One can check that $\mathrm{Hom}_{\mathcal{C}}(\alpha_c^{-1}, \alpha_d)$ is bijective, so the vertical function is surjective by Exercise 3.4.5.3. The fact that $\mathrm{Hom}_{\mathcal{C}'}((\alpha'_{FC})^{-1}, \alpha'_{FD})$ is bijective implies that the vertical function is injective. Thus we know that $\mathrm{Hom}_{F'}(Fc, Fd)$ is bijective. This implies that $\mathrm{Hom}_F(c, d)$ is bijective as well.

□

Exercise 5.3.4.12.

Let $\mathbb{Z}_2$ be the group (as category) from Example 5.3.4.7. Are there any fully faithful functors $\mathbb{Z}_2 \to \underline{1}$?

◊

Solution 5.3.4.12.

No. As a category, $\mathbb{Z}_2$ has one object $\mathrm{Ob}(\mathbb{Z}_2) = \{\blacktriangle\}$, and $\mathrm{Hom}_{\mathbb{Z}_2}(\blacktriangle, \blacktriangle) = \{0, 1\}$ has two elements. Since $\mathrm{Hom}_{\underline{1}}(1, 1) = \{\mathrm{id}_1\}$ has only one element, there is no way a functor $\mathbb{Z}_2 \to \underline{1}$ could be faithful.   ♦

5.4   Categories and schemas are equivalent, Cat $\simeq$ Sch

Perhaps it is intuitively clear that schemas are somehow equivalent to categories. In fact, this is a reason that so much attention has been given to databases (and ologs). This section makes the equivalence between schemas and categories precise; it is proved in Section 5.4.2. The basic idea was laid out in Section 5.2.2.

5.4.1   The category Sch of schemas

Recall from Definition 4.5.2.7 that a schema consists of a pair $\mathcal{C} := (G, \simeq)$, where $G = (V, A, src, tgt)$ is a graph and $\simeq$ is a congruence, meaning a kind of equivalence relation on the paths in G (see Definition 4.5.2.3). If we think of a schema as being analogous to a category, what in schema-land should fulfill the role of functors? That is, what are to be the morphisms in **Sch**?

Unfortunately, one's first guess may give the wrong idea if we want an equivalence **Sch** $\simeq$ **Cat**. Since an object in **Sch** is a graph with a congruence, one might imagine that a morphism $\mathcal{C} \to \mathcal{C}'$ in **Sch** should be a graph homomorphism (as in Definition 4.3.3.1) that preserves the congruence. But graph homomorphisms require that arrows be sent to arrows, whereas we are more interested in paths than in individual arrows—the arrows are merely useful for presentation.

If instead we define morphisms between schemas to be maps that send paths in $\mathcal{C}$ to paths in $\mathcal{C}'$, subject to the requirements that path endpoints, path concatenations, and path equivalences are preserved, this will turn out to give the correct notion. In fact, since a path is a concatenation of its arrows, it is more concise to give a function F from the arrows of $\mathcal{C}$ to the paths of $\mathcal{C}'$. This is how we proceed.

Recall from Examples 5.1.2.25 and 5.3.1.16 the paths-graph functor Paths: **Grph** $\to$ **Grph**, the paths of paths functor Paths $\circ$ Paths: **Grph** $\to$ **Grph**, and the natural transformations for any graph G,

$$\eta_G \colon G \to \mathrm{Paths}(G) \qquad \text{and} \qquad \mu_G \colon \mathrm{Paths}(\mathrm{Paths}(G)) \to \mathrm{Paths}(G). \tag{5.18}$$

The function η_G spells out the fact that every arrow in G counts as a path in G, and the function μ_G spells out the fact that a head-to-tail sequence of paths (a path of paths) in G can be concatenated to a single path in G.

Exercise 5.4.1.1.

Let [2] denote the graph $\overset{0}{\bullet} \overset{e_1}{\longrightarrow} \overset{1}{\bullet} \overset{e_2}{\longrightarrow} \overset{2}{\bullet}$, and let $\mathcal{L}oop$ denote the unique graph having one vertex and one arrow

$$\mathcal{L}oop := \boxed{\begin{array}{c} f \\ \circlearrowright_s \\ \bullet \end{array}}.$$

a. Find a graph homomorphism $f\colon [2] \to \mathrm{Paths}(\mathcal{L}oop)$ that is injective on arrows (i.e., such that no two arrows in the graph [2] are sent by f to the same arrow in $\mathrm{Paths}(\mathcal{L}oop)$).

b. The graph [2] has six paths, so $\mathrm{Paths}([2])$ has six arrows. What are the images of these arrows under the graph homomorphism $\mathrm{Paths}(f)\colon \mathrm{Paths}([2]) \to \mathrm{Paths}(\mathrm{Paths}(\mathcal{L}oop))$, where f is the morphism you chose in part (a)?

c. Finally, using $\mu_{\mathcal{L}oop}\colon \mathrm{Paths}(\mathrm{Paths}(\mathcal{L}oop)) \to \mathrm{Paths}(\mathcal{L}oop)$, a path of paths in $\mathcal{L}oop$ can be concatenated to a path. Write what the composite graph homomorphism

$$\mathrm{Paths}([2]) \xrightarrow{\mathrm{Paths}(f)} \mathrm{Paths}(\mathrm{Paths}(\mathcal{L}oop)) \xrightarrow{\mu_{\mathcal{L}oop}} \mathrm{Paths}(\mathcal{L}oop)$$

does to the six arrows in $\mathrm{Paths}([2])$.

$\diamond$

Solution 5.4.1.1.

a. The graph $\mathrm{Paths}(\mathcal{L}oop)$ has one vertex, s, and its set of arrows can be identified with the set of natural numbers $\mathbb{N}$. To give a graph morphism $f\colon [2] \to \mathrm{Paths}(\mathcal{L}oop)$, we have no choice on vertices, and we can send each of e_1 and e_2 to any path we choose. In other words, we have a bijection,

$$\mathrm{Hom}_{\mathbf{Grph}}([2], \mathrm{Paths}(\mathcal{L}oop)) \xrightarrow{\cong} \mathbb{N} \times \mathbb{N}.$$

Let's arbitrarily choose f to be $(4, 3)$:

$f\colon [2] \to \mathrm{Paths}(\mathcal{L}oop)$	
[2]	$\mathrm{Paths}(\mathcal{L}oop)$
e_1	$_s[f, f, f, f]$
e_2	$_s[f, f, f]$

b.

Paths(f)	
Paths([2])	Paths(Paths($\mathcal{L}oop$))
$_0[\,]$	$_s[\,]$
$_0[e_1]$	$_s[_s[f,f,f,f]]$
$_0[e_1,e_2]$	$_s[_s[f,f,f,f],_s[f,f,f]]$
$_1[\,]$	$_s[\,]$
$_1[e_2]$	$_s[_s[f,f,f]]$
$_2[\,]$	$_s[\,]$

c.

$\mu_{\mathcal{L}oop} \circ$ Paths(f)	
Paths([2])	Paths(Paths($\mathcal{L}oop$))
$_0[\,]$	$_s[\,]$
$_0[e_1]$	$_s[f,f,f,f]$
$_0[e_1,e_2]$	$_s[f,f,f,f,f,f,f]$
$_1[\,]$	$_s[\,]$
$_1[e_2]$	$_s[f,f,f]$
$_2[\,]$	$_s[\,]$

♦

Before we look at the definition of schema morphism, let's return to the original question. Given graphs G, G' (underlying schemas $\mathcal{C}, \mathcal{C}'$) we wanted a function from the paths in G to the paths in G', but it was more concise to speak of a function from arrows in G to paths in G'. How do we get what we originally wanted from the concise version?

Given a graph homomorphism $f\colon G \to \mathrm{Paths}(G')$, we use (5.18) to form the following composition, denoted simply $\mathrm{Paths}_f\colon \mathrm{Paths}(G) \to \mathrm{Paths}(G')$:

$$\mathrm{Paths}(G) \xrightarrow{\ \mathrm{Paths}(f)\ } \mathrm{Paths}(\mathrm{Paths}(G')) \xrightarrow{\ \mu_{G'}\ } \mathrm{Paths}(G') \qquad (5.19)$$

This says that given a function from arrows in G to paths in G', a path in G becomes a path of paths in G', which can be concatenated to a path in G'.

Definition 5.4.1.2 (Schema morphism). Let $G = (V, A, src, tgt)$ and $G' = (V', A', src', tgt')$ be graphs, and let $\mathcal{C} = (G, \simeq_G)$ and $\mathcal{C}' = (G', \simeq_{G'})$ be schemas. A *schema morphism F from $\mathcal{C}$ to $\mathcal{D}$*, denoted $F\colon \mathcal{C} \to \mathcal{D}$, is a graph homomorphism [19]

$$F\colon G \to \mathrm{Paths}(G')$$

---

[19] By Definition 4.3.3.1, a graph homomorphism $F\colon G \to \mathrm{Paths}(G')$ will consist of a vertex part $F_0\colon V \to V'$ and an arrows part $F_1\colon E \to \mathrm{Path}(G')$. See also Definition 4.3.2.1.

that satisfies the following condition for any paths p and q in G:

$$\text{if} \quad p \simeq_G q \quad \text{then} \quad \text{Paths}_F(p) \simeq_{G'} \text{Paths}_F(q). \tag{5.20}$$

Two schema morphisms $E, F \colon \mathcal{C} \to \mathcal{C}'$ are considered identical if they agree on vertices (i.e., $E_0 = F_0$) and if, for every arrow f in G, there is a path equivalence in G'

$$E_1(f) \simeq_{G'} F_1(f).$$

We now define the *category of schemas*, denoted **Sch**, to be the category whose objects are schemas as in Definition 4.5.2.7 and whose morphisms are schema morphisms, as in Definition 5.4.1.2. The identity morphism on schema $\mathcal{C} = (G, \simeq_G)$ is the schema morphism $\text{id}_{\mathcal{C}} := \eta_G \colon G \to \text{Paths}(G)$, as defined in Equation (5.18). We need only understand how to compose schema morphisms $F \colon \mathcal{C} \to \mathcal{C}'$ and $F' \colon \mathcal{C}' \to \mathcal{C}''$. On objects their composition is clear. Given an arrow in $\mathcal{C}$, it is sent to a path in $\mathcal{C}'$; each arrow in that path is sent to a path in $\mathcal{C}''$. We then have a path of paths, which we can concatenate (via $\mu_{G''} \colon \text{Paths}(\text{Paths}(G'')) \to \text{Paths}(G'')$, as in (5.18)) to get a path in $\mathcal{C}''$ as desired.

Slogan 5.4.1.3.

   A schema morphism sends vertices to vertices, arrows to paths, and path equivalences to path equivalences.

Example 5.4.1.4. Let $[2]$ be the linear order graph of length 2, at the left, and let $\mathcal{C}$ denote the diagram at the right:

$$[2] := \boxed{\begin{array}{ccccc} 0 & \xrightarrow{\ f_1\ } & 1 & \xrightarrow{\ f_2\ } & 2 \\ \bullet & & \bullet & & \bullet \end{array}} \qquad\qquad \mathcal{C} := \boxed{\begin{array}{c} \overset{a}{\bullet} \xrightarrow{\ g\ } \overset{b}{\bullet} \\ \text{\small$i$} \searrow \quad \downarrow h \\ \underset{c}{\bullet} \end{array}} \tag{5.21}$$

We impose on $\mathcal{C}$ the path equivalence declaration $_a[g, h] \simeq {_a[i]}$ and show that in this case $\mathcal{C}$ and $[2]$ are isomorphic in **Sch**. There is a unique schema morphism $F \colon [2] \to \mathcal{C}$ such that $0 \mapsto a, 1 \mapsto b, 2 \mapsto c$; it sends each arrow in $[2]$ to a path of length 1 in $\mathcal{C}$. And we have a schema morphism $F' \colon \mathcal{C} \to [2]$, which reverses this mapping on vertices; note that F' must send the arrow i in $\mathcal{C}$ to the path $_0[f_1, f_2]$ in $[2]$, which is okay. The round-trip $F' \circ F \colon [2] \to [2]$ is identity. The round-trip $F \circ F' \colon \mathcal{C} \to \mathcal{C}$ may look like it is not the identity; indeed it sends vertices to themselves and sends i to the path $_a[g, h]$. But according to Definition 5.4.1.2, this schema morphism is considered identical to $\text{id}_{\mathcal{C}}$ because there is a path equivalence $\text{id}_{\mathcal{C}}(i) = {_a[i]} \simeq {_a[g, h]} = F \circ F'(i)$.

Exercise 5.4.1.5.

Consider the schema [2] and the schema C pictured in (5.21); this time we *do not* impose any path equivalence declarations on C, so $_a[g, h] \neq {}_a[i]$ in the current version of C.

a. How many schema morphisms are there $[2] \to C$ that send 0 to a?

b. How many schema morphisms are there $C \to [2]$ that send a to 0?

$\Diamond$

Solution 5.4.1.5.

a. There are eight, shown as columns:

The eight schema morphisms $[2] \to C$ that send $0 \to a$								
f_1	$_a[\,]$	$_a[\,]$	$_a[\,]$	$_a[\,]$	$_a[g]$	$_a[g]$	$_a[g,h]$	$_a[i]$
f_2	$_a[\,]$	$_a[g]$	$_a[g,h]$	$_a[i]$	$_b[\,]$	$_b[h]$	$_c[\,]$	$_c[\,]$

b. There are six, shown as columns:

The six schema morphisms $C \to [2]$ that send $a \to 0$						
g	$_0[\,]$	$_0[\,]$	$_0[\,]$	$_0[f_1]$	$_0[f_1]$	$_0[f_1,f_2]$
h	$_0[\,]$	$_0[f_1]$	$_0[f_1,f_2]$	$_1[\,]$	$_1[f_2]$	$_2[\,]$
i	$_0[\,]$	$_0[f_1]$	$_0[f_1,f_2]$	$_1[\,]$	$_1[f_2]$	$_2[\,]$

$\blacklozenge$

Exercise 5.4.1.6.

Consider the graph $\mathcal{L}oop$ as follows:

$$\mathcal{L}oop := \boxed{\overset{f}{\curvearrowright}\,\underset{s}{\bullet}}$$

and for any natural number $n \in \mathbb{N}$, let $\mathcal{L}_n$ denote the schema $(\mathcal{L}oop, \simeq_n)$, where $\simeq_n$ is the PED $f^{n+1} \simeq f^n$. Then $\mathcal{L}_n$ is the "finite hierarchy of height n" schema of Example 4.5.2.12. Let $\underline{1}$ denote the graph with one vertex and no arrows; consider it a schema.

a. Is $\underline{1}$ isomorphic to $\mathcal{L}_1$ in **Sch**?

b. Is $\underline{1}$ isomorphic to any (other) $\mathcal{L}_n$?

◇

Solution 5.4.1.6.

a. No. The schema $\mathcal{L}_1$ is the graph $\mathcal{L}oop$ with the PED $f^2 = f$, so there is still one nontrivial arrow in $\mathcal{L}_1$, namely, $f^1 \neq f^0$, whereas $\underline{1}$ has only the identity arrow.

b. Yes, there is an isomorphism of schemas $\underline{1} \cong \mathcal{L}_0$, because $f \simeq f^0 = \mathrm{id}_s$ in $\mathcal{L}_0$.

◆

Exercise 5.4.1.7.

Let $\mathcal{L}oop$ and $\mathcal{L}_n$ be schemas as defined in Exercise 5.4.1.6.

a. What is the cardinality of the set $\mathrm{Hom}_{\mathbf{Sch}}(\mathcal{L}_3, \mathcal{L}_5)$?

b. What is the cardinality of the set $\mathrm{Hom}_{\mathbf{Sch}}(\mathcal{L}_5, \mathcal{L}_3)$? Hint: The cardinality of the set $\mathrm{Hom}_{\mathbf{Sch}}(\mathcal{L}_4, \mathcal{L}_9)$ is 8.

◇

Solution 5.4.1.7.

In general, a schema morphism $G \colon \mathcal{L}_m \to \mathcal{L}_n$ is determined by where it sends the generating morphism $f \colon s \to s$. At first glance, G can send f to f^k for any $k \in \mathbb{N}$, and we note that these are all the same for $k \geq n$. But the functor laws impose a rule they must follow, because $f^{m+1} = f^m$, namely,

$$f^{km+k} = (f^k)^{m+1} = G(f^{m+1}) = G(f^m) = (f^k)^m = f^{km}.$$

The only way we could have $f^{km+k} = f^{km}$ is if $k = 0$ or if $km \geq n$. Thus we have a bijection

$$\mathrm{Hom}_{\mathbf{Sch}}(\mathcal{L}_m, \mathcal{L}_n) \xrightarrow{\cong} \{k \leq n \mid k = 0 \text{ or } km \geq n\}.$$

We use this criterion to arrive at each of the following answers.

a. Here $m = 3$ and $n = 5$, so the cardinality is $|\mathrm{Hom}_{\mathbf{Sch}}(\mathcal{L}_3, \mathcal{L}_5)| = |\{0, 2, 3, 4, 5\}| = 5$.

b. Here $m = 5$ and $n = 3$, so the cardinality is $|\mathrm{Hom}_{\mathbf{Sch}}(\mathcal{L}_5, \mathcal{L}_3)| = |\{0, 1, 2, 3\}| = 4$.

◆

5.4.2   Proving the equivalence

This section proves the equivalence of categories, **Sch** $\simeq$ **Cat**. We construct the two functors **Sch** $\to$ **Cat** and **Cat** $\to$ **Sch** and then prove that these are mutually inverse equivalences (see Theorem 5.4.2.3).

Construction 5.4.2.1 (From schema to category). We first define a functor $L\colon$ **Sch** $\to$ **Cat**. Let $\mathcal{C} = (G, \simeq)$ be a schema, where $G = (V, A, src, tgt)$. Define $L(\mathcal{C})$ to be the category with $\mathrm{Ob}(L(\mathcal{C})) = V$, and with $\mathrm{Hom}_{L(\mathcal{C})}(v_1, v_2) := \mathrm{Path}_G(v, w)/\simeq$, i.e., the set of paths in G modulo the path equivalence relation for $\mathcal{C}$. The composition of morphisms is defined by concatenation of paths, and part (4) of Definition 4.5.2.3 implies that such composition is well defined. We have thus defined L on objects of **Sch**.

Given a schema morphism $F\colon \mathcal{C} \to \mathcal{C}'$, where $\mathcal{C}' = (G', \simeq')$, we need to produce a functor $L(F)\colon L(\mathcal{C}) \to L(\mathcal{C}')$. The objects of $L(\mathcal{C})$ and $L(\mathcal{C}')$ are the vertices of G and G' respectively, and F provides the necessary function on objects. Diagram (5.19) provides a function $\mathrm{Paths}_F\colon \mathrm{Paths}(G) \to \mathrm{Paths}(G')$ provides the requisite function for morphisms.

A morphism in $L(\mathcal{C})$ is an equivalence class of paths in $\mathcal{C}$. For any representative path $p \in \mathrm{Paths}(G)$, we have $\mathrm{Paths}_F(p) \in \mathrm{Paths}(G')$, and if $p \simeq q$, then $\mathrm{Paths}_F(p) \simeq' \mathrm{Paths}_F(q)$ by condition (5.20). Thus Paths_F indeed provides us with a function $\mathrm{Hom}_{L(\mathcal{C})} \to \mathrm{Hom}_{L(\mathcal{C}')}$. This defines L on morphisms in **Sch**. It is clear that L preserves composition and identities, so it is a functor.

Construction 5.4.2.2 (From category to schema). We first define a functor $R\colon$ **Cat** $\to$ **Sch**. Let $\mathcal{C} = (\mathrm{Ob}(\mathcal{C}), \mathrm{Hom}_{\mathcal{C}}, dom, cod, ids, comp)$ be a category (see Exercise 5.1.1.27). Let $R(\mathcal{C}) = (G, \simeq)$, where G is the graph

$$G = (\mathrm{Ob}(\mathcal{C}), \mathrm{Hom}_{\mathcal{C}}, dom, cod),$$

and with $\simeq$ defined as the congruence generated by the following path equivalence declarations: for any composable sequence of morphisms $f_1, f_2, \ldots, f_n$ (with $dom(f_{i+1}) = cod(f_i)$ for each $1 \leqslant i \leqslant n - 1$), we put

$$_{dom(f_1)}[f_1, f_2, \ldots, f_n] \simeq {}_{dom(f_1)}[f_n \circ \cdots \circ f_2 \circ f_1], \tag{5.22}$$

equating a path of length n with a path of length 1. This defines R on objects of **Cat**.

A functor $F\colon \mathcal{C} \to \mathcal{D}$ induces a schema morphism $R(F)\colon R(\mathcal{C}) \to R(\mathcal{D})$, because vertices are sent to vertices, arrows are sent to arrows (as paths of length 1), and path equivalence is preserved by (7.17) and the fact that F preserves the composition formula. This defines R on morphisms in **Cat**. It is clear that R preserves compositions, so it is a functor.

Theorem 5.4.2.3. *The functors*

$$L\colon \mathbf{Sch} \rightleftarrows \mathbf{Cat}\colon R$$

are mutually inverse equivalences of categories.

Sketch of proof. It is clear that there is a natural isomorphism $\alpha\colon \mathrm{id}_{\mathbf{Cat}} \overset{\cong}{\longrightarrow} L \circ R$; i.e., for any category $\mathcal{C}$, there is an isomorphism $\mathcal{C} \cong L(R(\mathcal{C}))$.

Before giving an isomorphism $\beta\colon \mathrm{id}_{\mathbf{Sch}} \overset{\cong}{\longrightarrow} R \circ L$, we look at $R(L(\mathcal{G})) =: (G', \simeq')$ for a schema $\mathcal{G} = (G, \simeq)$. Write $G = (V, A, src, tgt)$ and $G' = (V', A', src', tgt')$. On vertices we have $V = V'$. On arrows we have $A' = \mathrm{Path}_G/\simeq$. The congruence $\simeq'$ for $R(L(\mathcal{G}))$ is imposed in (5.22). Under $\simeq'$, every path of paths in G is made equivalent to its concatenation, considered as a single path of length 1 in G'.

There is a natural transformation $\beta\colon \mathrm{id}_{\mathbf{Sch}} \to R \circ L$ whose $\mathcal{G}$ component sends each arrow in G to a certain path of length 1 in G'. We need to see that $\beta_{\mathcal{G}}$ has an inverse. But this is straightforward: every arrow f in $R \circ L(\mathcal{G})$ is an equivalence class of paths in $\mathcal{G}$; choose any one, and have β^{-1} send f there; by Definition 5.4.1.2, any other choice will give the identical morphism of schemas. It is easy to show that each round-trip is equal to the identity (again up to the notion of equality of schema morphism given in Definition 5.4.1.2).

$\square$

$X:=$

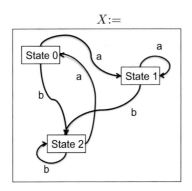

$Y:=$

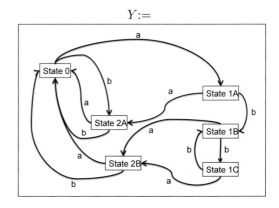

Original model X		
ID	**a**	**b**
State 0	State 1	State 2
State 1	State 2	State 1
State 2	State 0	State 0

Proposed model Y		
ID	**a**	**b**
State 0	State 1A	State 2A
State 1A	State 2A	State 1B
State 1B	State 2B	State 1C
State 1C	State 2B	State 1B
State 2A	State 0	State 0
State 2B	State 0	State 0

Figure 5.1 Finite state machines X and Y with alphabet $\Sigma = \{a, b\}$ and three states (left) or six states (right), and their associated action tables.

Chapter 6

Fundamental Considerations of Categories

This chapter focuses mainly on limits and colimits in a given category $\mathcal{C}$. It also discusses other important and interesting categorical constructions, such as the simple notion of opposite categories and the Grothendieck construction, which gives something like the histogram of a set-valued functor. As usual, the work relies as often as possible on a grounding in databases.

This chapter is in some sense parallel to Chapter 3, Fundamental Considerations in **Set**. When attention is restricted to $\mathcal{C} = $ **Set**, the discussion of limits and colimits in this chapter subsumes the earlier work (which focused on certain finite limits and colimits). Also, this chapter ends with a section called Arithmetic of Categories, Section 6.2.5, which is tightly parallel with Section 3.4.3. This shows that in terms of grade school arithmetic expressions like

$$\mathcal{A} \times (\mathcal{B} + \mathcal{C}) =^? (\mathcal{C} \times \mathcal{A}) + (\mathcal{B} \times \mathcal{A}),$$

the behavior of categories is predictable: the rules for categories are well aligned with those of sets, which are well aligned with those of natural numbers.

6.1 Limits and colimits

Limits and colimits are universal constructions, meaning they represent certain ideals of behavior in a category. When it comes to sets that map to A and B, the $A \times B$ grid is ideal—it projects on to both A and B as straightforwardly as possible. When it comes to sets that can interpret the elements of both A and B, the disjoint union $A \sqcup B$ is ideal—it includes both A and B without confusion or superfluity. These are limits and colimits in **Set**. Limits and colimits exist in other categories as well.

Limits in a preorder are meets; colimits in a preorder are joins. Limits and colimits also exist for database instances and monoid actions, allowing us to discuss, for example, the product or union of different finite state machines. Limits and colimits exist for topological spaces, giving rise to products and unions as well as to quotients.

Limits and colimits do not exist in every category. However, when $\mathcal{C}$ is complete with respect to limits (or colimits), these limits always seem to mean something valuable to human intuition. For example, when a subject had already been studied for a long time before category theory came to promenance, it often turned out that classically interesting constructions in the subject corresponded with limits and colimits in its categorification $\mathcal{C}$. For example, products, unions, and quotients by equivalence relations are classical ideas in set theory that are naturally captured by limits and colimits in **Set**.

6.1.1   Products and coproducts in a category

Section 3.1 discussed products and coproducts in the category **Set** of sets. Now we discuss the same notions in an arbitrary category. For both products and coproducts, we begin with examples and then write the general concept.

6.1.1.1   Products

The product of two sets is a grid, which projects down onto each of the two sets. This is a good intuition for products in general.

Example 6.1.1.2. Given two preorders, $\mathcal{X}_1 := (X_1, \leqslant_1)$ and $\mathcal{X}_2 := (X_2, \leqslant_2)$, we can take their product and get a new preorder $\mathcal{X}_1 \times \mathcal{X}_2$. Both $\mathcal{X}_1$ and $\mathcal{X}_2$ have underlying sets (namely, X_1 and X_2), so we might hope that the underlying set of $\mathcal{X}_1 \times \mathcal{X}_2$ is the set $X_1 \times X_2$ of ordered pairs, and this turns out to be true. We have a notion of less-than on $\mathcal{X}_1$, and we have a notion of less-than on $\mathcal{X}_2$; we need to construct a notion of less-than on $\mathcal{X}_1 \times \mathcal{X}_2$. So, given two ordered pairs (x_1, x_2) and (x_1', x_2'), when should we say that $(x_1, x_2) \leqslant_{1,2} (x_1', x_2')$ holds? A guess is that it holds iff both $x_1 \leqslant_1 x_1'$ and $x_2 \leqslant_2 x_2'$ hold, and this works:[1]

$$\mathcal{X}_1 \times \mathcal{X}_2 := (X_1 \times X_2, \leqslant_{1,2}).$$

Note that the projection functions $X_1 \times X_2 \to X_1$ and $X_1 \times X_2 \to X_2$ induce morphisms of preorders. That is, if $(x_1, x_2) \leqslant_{1,2} (x_1', x_2')$, then in particular, $x_1 \leqslant_1 x_1'$ and

---

[1] Given $R_1 \subseteq X_1 \times X_1, R_2 \subseteq X_2 \times X_2$, take $R_1 \times R_2 \subseteq (X_1 \times X_2) \times (X_1 \times X_2)$.

$x_2 \leqslant_2 x_2'$. So we have preorder morphisms

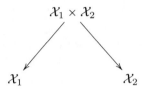

Exercise 6.1.1.3.

Suppose you have a partial order $\mathcal{S} = (S, \leqslant_S)$ on songs (you prefer some songs over others, but sometimes you cannot compare). And suppose you have a partial order $\mathcal{A} = (A, \leqslant_A)$ on pieces of art. You are about to be given two pairs (s, a) and (s', a'), each including a song and an art piece. Does the product partial order $\mathcal{S} \times \mathcal{A}$ provide a reasonable guess for your preferences on these pairs? ◇

Solution 6.1.1.3.

According to the product partial order $\mathcal{S} \times \mathcal{A}$, I would like a pair (s, a) more than a pair (s', a') if and only if I liked s more than s', and a more than a'. It is agnostic (i.e., puts no preference) on cases where $s \geqslant s'$ and $a \leqslant a'$ or where $s \leqslant s'$ and $a \geqslant a'$. This is perfectly reasonable, at least as a first guess. What better could an automatic procedure do? ♦

Exercise 6.1.1.4.

Consider the partial order $\leqslant$ on $\mathbb{N}$ given by standard less-than-or-equal-to, so $5 \leqslant 9$, and let `divides` be the partial order from Example 4.4.3.2, where 6 `divides` 12. If we call the product order $(X, \leq) := (\mathbb{N}, \leqslant) \times (\mathbb{N}, \texttt{divides})$, which of the following are true?

$$(2, 4) \leq (3, 4) \qquad (2, 4) \leq (3, 5) \qquad (2, 4) \leq (8, 0) \qquad (2, 4) \leq (0, 0)$$

◇

Solution 6.1.1.4.

Among these, the true ones are

$$(2, 4) \leq (3, 4) \qquad \text{and} \qquad (2, 4) \leq (8, 0).$$

♦

Example 6.1.1.5. Given two graphs $G_1 = (V_1, A_1, src_1, tgt_1)$ and $G_2 = (V_2, A_2, src_2, tgt_2)$, we can take their product and get a new graph $G_1 \times G_2$. The vertices are the grid of vertices $V_1 \times V_2$, so each vertex in $G_1 \times G_2$ is labeled by a pair of vertices, one from G_1 and one from G_2. When should an arrow connect (v_1, v_2) to (v_1', v_2')? Whenever we can find an arrow in G_1 connecting v_1 to v_1' and we can find an arrow in G_2 connecting v_2 to v_2'. It turns out there is a simple formula for the set of arrows in $G_1 \times G_2$, namely, $A_1 \times A_2$.

Let's write $G := G_1 \times G_2$ and say, $G = (V, A, src, tgt)$. We said that $V = V_1 \times V_2$ and $A = A_1 \times A_2$. What should the source and target functions $A \to V$ be? Given a function $src_1 \colon A_1 \to V_1$ and a function $src_2 \colon A_2 \to V_2$, the universal property for products in **Set** (Proposition 3.1.1.10 or, better, Example 3.1.1.15) provides a unique function

$$src := src_1 \times src_2 \colon A_1 \times A_2 \to V_1 \times V_2.$$

Namely, the source of arrow (a_1, a_2) will be the vertex $(src_1(a_1), src_2(a_2))$. Similarly, we have a ready-made choice of target function $tgt = tgt_1 \times tgt_2$. We have now defined the product graph,

$$G = G_1 \times G_2 = (V_1 \times V_2, A_1 \times A_2, src_1 \times src_2, tgt_1 \times tgt_2).$$

Here is a concrete example. Let I and J be drawn as follows:

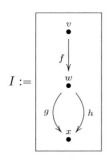

Arrow (I)		
ID	**src**	**tgt**
f	v	w
g	w	x
h	w	x

Vertex (I)
ID
v
w
x

Arrow (J)		
ID	**src**	**tgt**
i	q	r
j	r	s
k	s	r
ℓ	s	t

Vertex (J)
ID
q
r
s
t

The product $I \times J$ has, as expected, $3 * 4 = 12$ vertices and $3 * 4 = 12$ arrows:

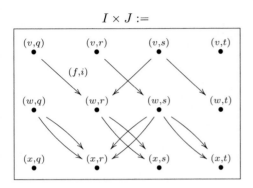

$I \times J :=$

Arrow $(I \times J)$		
ID	**src**	**tgt**
(f, i)	(v, q)	(w, r)
(f, j)	(v, r)	(w, s)
(f, k)	(v, s)	(w, r)
(f, ℓ)	(v, s)	(w, t)
(g, i)	(w, q)	(x, r)
(g, j)	(w, r)	(x, s)
(g, k)	(w, s)	(x, r)
(g, ℓ)	(w, s)	(x, t)
(h, i)	(w, q)	(x, r)
(h, j)	(w, r)	(x, s)
(h, k)	(w, s)	(x, r)
(h, ℓ)	(w, s)	(x, t)

Vertex $(I \times J)$
ID
(v, q)
(v, r)
(v, s)
(v, t)
(w, q)
(w, r)
(w, s)
(w, t)
(x, q)
(x, r)
(x, s)
(x, t)

Here is the most important thing to notice. Look at the `Arrow` table for $I \times J$, and for each ordered pair, look only at the first entry in all three columns; you will see something that matches with the `Arrow` table for I. For example, in the $I \times J$ table, the first row's first entries are f, v, w. Then do the same for the second entry in each column, and again you will see a match with the `Arrow` table for J. These matches are readily visible graph homomorphisms $I \times J \to I$ and $I \times J \to J$ in **Grph**.

Exercise 6.1.1.6.

Let $[1]$ denote the linear order graph of length 1,

$$[1] := \boxed{\overset{0}{\bullet} \overset{f}{\longrightarrow} \overset{1}{\bullet}}$$

and let $P = \mathrm{Paths}([1])$ be its paths-graph, as in Example 5.1.2.25 (so P should have three arrows and two vertices). Draw the graph $P \times P$. ◊

Solution 6.1.1.6.

As a check, $P \times P$ should have four vertices and nine arrows. Here is the solution:

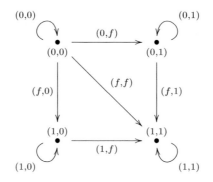

Exercise 6.1.1.7.

Recall from Example 4.5.2.10 that a discrete dynamical system (DDS) is a set s together with a function $f\colon s \to s$. It is clear that if

$$\mathcal{Loop} := \boxed{\begin{array}{c} f \\ \circlearrowright \, s \\ \bullet \end{array}}$$

is the loop schema, then a DDS is simply an instance (a functor) $I\colon \mathcal{Loop} \to \mathbf{Set}$. We have not yet discussed DDS products, but perhaps you can guess how they should work. For example, consider these instances $I, J\colon \mathcal{Loop} \to \mathbf{Set}$:

s	(I)
ID	**f**
A	C
B	C
C	C

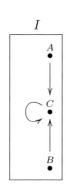

s	(J)
ID	**f**
x	y
y	x
z	z

a. Make a guess and tabulate $I \times J$. Then draw it.[2]

b. Recall the notion of natural transformations between functors (see Example 5.3.3.5), which in the case of functors $\mathcal{L}oop \to \mathbf{Set}$ are the morphisms of instances. Do you see clearly that there is a morphism of instances $I \times J \to I$ and $I \times J \to J$? Check that if you look only at the left-hand coordinates in your $I \times J$, you see something compatible with I.

◊

Solution 6.1.1.7.

a.

s	$(I \times J)$
ID	**f**
(A,x)	(C,y)
(A,y)	(C,x)
(A,z)	(C,z)
(B,x)	(C,y)
(B,y)	(C,x)
(B,z)	(C,z)
(C,x)	(C,y)
(C,y)	(C,x)
(C,z)	(C,z)

(Note that the order of (C,x) and (C,y) is switched in the picture of $I \times J$, for readability reasons.)

b. We can see the projection $I \times J \to I$ (resp. $I \times J \to J$) given by looking at the first entry (resp. the second entry) in each pair.

♦

In each case what is most important to recognize is that there are projection maps $I \times J \to I$ and $I \times J \to J$, and that the construction of $I \times J$ seems as straightforward as possible, subject to having these projections.

Definition 6.1.1.8. Let $\mathcal{C}$ be a category, and let $X, Y \in \mathrm{Ob}(\mathcal{C})$ be objects. A *span on X and Y* consists of three constituents (Z, p, q), where $Z \in \mathrm{Ob}(\mathcal{C})$ is an object, and where

---

[2]The result is not necessarily inspiring, but at least computing it is straightforward.

$p\colon Z \to X$ and $q\colon Z \to Y$ are morphisms in $\mathcal{C}$.

A *product of* X *and* Y is a span $X \xleftarrow{\pi_1} X \times Y \xrightarrow{\pi_2} Y$, such that for any other span $X \xleftarrow{p} Z \xrightarrow{q} Y$ there *exists a unique* morphism $t_{p,q}\colon Z \to X \times Y$ such that the following diagram commutes:[3]

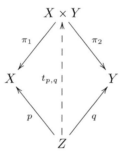

We often denote the morphism $t_{p,q}$ by $\langle p, q \rangle\colon Z \to X \times Y$.

Remark 6.1.1.9. Definition 6.1.1.8 endows the product of two objects with a *universal property*. It says that a product of two objects X and Y maps to those two objects and serves as a gateway for all that do the same. "None shall map to X and Y except through me!" This grandiose property is held by products in all the various categories discussed so far. It is what is meant by "$X \times Y$ maps to both X and Y and does so as straightforwardly as possible." The grid of dots obtained as the product of two sets has such a property (see Example 3.1.1.11).

Example 6.1.1.10. Example 6.1.1.2 discussed products of preorders. This example discusses products in an individual preorder. That is, by Proposition 5.2.1.13, there is a functor **PrO** $\to$ **Cat** that realizes each individual preorder as a category. If $\mathcal{P} = (P, \leqslant)$ is a preorder, what are products in $\mathcal{P}$? Given two objects $a, b \in \mathrm{Ob}(\mathcal{P})$, we first consider $\{a, b\}$ spans, i.e., $a \leftarrow z \to b$. That is some z such that $z \leqslant a$ and $z \leqslant b$. The product is a span $a \geqslant a \times b \leqslant b$, but such that every other spanning object z is less than or equal to $a \times b$. In other words, $a \times b$ is as big as possible subject to the condition of being less than a and less than b. This is precisely their meet, $a \wedge b$ (see Definition 4.4.2.1).

---

[3]The names $X \times Y$ and π_1, π_2 are not mathematically important; they are pedagogically useful.

Example 6.1.1.11. Note that the product of two objects in a category $\mathcal{C}$ may not exist. Let's return to preorders to see this phenomenon.

Consider the set $\mathbb{R}^2$, and say that $(x_1, y_1) \leqslant (x_2, y_2)$ if there exists $\ell \geqslant 1$ such that $x_1\ell = x_2$ and $y_1\ell = y_2$; in other words, point p is less than point q if, in order to travel from q to the origin along a straight line, one must pass through p along the way. [4] We have given a perfectly good partial order, but $p := (1, 0)$ and $q := (0, 1)$ do not have a product. Indeed, it would have to be a nonzero point that was on the same line through the origin as p and the same line through the origin as q, of which there are none.

Example 6.1.1.12. Note that there can be more than one product of two objects in a category $\mathcal{C}$ but that any two choices will be canonically isomorphic. Let's return once more to preorders to see this phenomenon.

Consider the set $\mathbb{R}^2$, and say that $(x_1, y_1) \leqslant (x_2, y_2)$ if $x_1^2 + y_1^2 \leqslant x_2^2 + y_2^2$, in other words, if the former is closer to the origin. For any point $p = (x_0, y_0)$, let $C_p = \{(x, y) \in \mathbb{R}^2 \mid x^2 + y^2 = x_0^2 + y_0^2)\}$, and call it the orbit circle of p.

For any two points p, q, there will be lots of points that serve as products $p \times q$: any point a on the smaller of their two orbit circles will suffice. Given any two points a, a' on this smaller circle, we have a unique isomorphism $a \cong a'$ because $a \leqslant a'$ and $a' \leqslant a$.

Exercise 6.1.1.13.

Consider the preorder $\mathcal{P}$ of cards in a deck, shown in Example 4.4.1.3; it is not the whole story of cards in a deck, but take it to be so. Consider this preorder $\mathcal{P}$ as a category (by way of the functor **PrO** $\rightarrow$ **Cat**).

a. For each of the following pairs, what is their product in $\mathcal{P}$ (if it exists)?

⌜a diamond⌝ × ⌜a heart⌝	⌜a queen⌝ × ⌜a black card⌝
⌜a card⌝ × ⌜a red card⌝	⌜a face card⌝ × ⌜a black card⌝

b. How would these answers differ if $\mathcal{P}$ were completed to the "whole story" partial order classifying cards in a deck?

◊

Solution 6.1.1.13.

a. The product of two elements in any preorder, such as $\mathcal{P}$, is their meet (if it exists). For example, the meet of ⌜a queen⌝ and ⌜a black card⌝ should be a card that is both

---

[4]Note that $(0, 0)$ is not related to anything else.

a queen and a black card, if this notion exists in $\mathcal{P}$. Then we have

$$\ulcorner \text{a diamond} \urcorner \times \ulcorner \text{a heart} \urcorner = \text{Does Not Exist}$$
$$\ulcorner \text{a queen} \urcorner \times \ulcorner \text{a black card} \urcorner = \ulcorner \text{a black queen} \urcorner$$
$$\ulcorner \text{a card} \urcorner \times \ulcorner \text{a red card} \urcorner = \ulcorner \text{a red card} \urcorner$$
$$\ulcorner \text{a face card} \urcorner \times \ulcorner \text{a black card} \urcorner = \ulcorner \text{a black queen} \urcorner$$

Some of these are strange because $\mathcal{P}$ is not the "whole story."

b. If every subset of cards were put into the order, i.e., if we took $\mathcal{P}$ to be the power-set of the set of cards, it would have all meets. Let's write the empty set of cards as $\ulcorner$a nonexistent card$\urcorner$ because (by Rules 2.3.1.2) the label on a box is what one should call *each example* of that class.

$$\ulcorner \text{a diamond} \urcorner \times \ulcorner \text{a heart} \urcorner = \ulcorner \text{a nonexistent card} \urcorner$$
$$\ulcorner \text{a queen} \urcorner \times \ulcorner \text{a black card} \urcorner = \ulcorner \text{a black queen} \urcorner$$
$$\ulcorner \text{a card} \urcorner \times \ulcorner \text{a red card} \urcorner = \ulcorner \text{a red card} \urcorner$$
$$\ulcorner \text{a face card} \urcorner \times \ulcorner \text{a black card} \urcorner = \ulcorner \text{a black face card} \urcorner$$

◆

Exercise 6.1.1.14.

Let X be a set, and consider it as a discrete category. Given two objects $x, y \in \text{Ob}(X)$, under what conditions will there exist a product $x \times y$ in X?   ◇

Solution 6.1.1.14.

The only morphisms in a discrete category are identities. Since a product needs to project to both factors, we must have $x = x \times y = y$. So the condition is $x = y$. Indeed, in that case, any other object z mapping to x and y maps uniquely to $x \times y$ by $z \xrightarrow{\text{id}} x \times y$.

◆

Exercise 6.1.1.15.

Let $f : \mathbb{R} \to \mathbb{R}$ be a function like one that you would see in grade school (e.g., $f(x) = x+7$). A typical thing to do is to graph f as a curve running through the plane $\mathbb{R}^2 := \mathbb{R} \times \mathbb{R}$. For example, f is graphed as a straight line with slope 1 and y-intercept 7. In general, the graph of f is a curve that be understood as a function $F : \mathbb{R} \to \mathbb{R}^2$.

a. For an arbitrary function $f : \mathbb{R} \to \mathbb{R}$ with graph $F : \mathbb{R} \to \mathbb{R}^2$ and an arbitrary $r \in \mathbb{R}$, what are the (x, y) coordinates of $F(r) \in \mathbb{R}^2$?

b. Obtain $F\colon \mathbb{R} \to \mathbb{R}^2$ using the universal property given in Definition 6.1.1.8.

◊

Solution 6.1.1.15.

a. The coordinates are $(r, f(r))$. For example, if $f(x) = x + 7$ and $r = -12$, we have
   $F(r) = (-12, -5)$.

b. To obtain a function $\mathbb{R} \to \mathbb{R} \times \mathbb{R}$, we need two functions $\mathbb{R} \to \mathbb{R}$. We take them to be
   $\mathrm{id}_\mathbb{R}\colon \mathbb{R} \to \mathbb{R}$ and $f\colon \mathbb{R} \to \mathbb{R}$. The universal property gives the desired

$$\mathbb{R} \xrightarrow{\ \langle \mathrm{id}_\mathbb{R}, f \rangle\ } \mathbb{R} \times \mathbb{R}.$$

Note that $\langle \mathrm{id}_\mathbb{R}, f \rangle(r) = (r, f(r))$.

♦

Exercise 6.1.1.16.

Consider the preorder $(\mathbb{N}, \mathtt{divides})$, discussed in Example 4.4.3.2, where, e.g., $5 \leqslant 15$,
but $5 \nleqslant 6$. Consider it as a category, using the functor $\mathbf{PrO} \to \mathbf{Cat}$.

a. What is the product of 9 and 12 in this category?

b. Is there a standard name for products in this category?

◊

Solution 6.1.1.16.

a. $9 \times 12 = 3$.

b. Greatest common divisor. But note that 0 is the biggest element, so, for example,
   $0 \times 17 = 0$, even though 17 is a common divisor of 0 and 17, and even though one
   might say, "17 is greater than 0" (in the usual ordering).

♦

Example 6.1.1.17. Products do not have to exist in an arbitrary category, but they do
exist in $\mathbf{Cat}$, the category of categories. That is, given two categories $\mathcal{C}$ and $\mathcal{D}$, there is a
product category $\mathcal{C} \times \mathcal{D}$. We have $\mathrm{Ob}(\mathcal{C} \times \mathcal{D}) = \mathrm{Ob}(\mathcal{C}) \times \mathrm{Ob}(\mathcal{D})$, and for any two objects
(c, d) and (c', d'), we have

$$\mathrm{Hom}_{\mathcal{C} \times \mathcal{D}}((c, d), (c', d')) = \mathrm{Hom}_{\mathcal{C}}(c, c') \times \mathrm{Hom}_{\mathcal{C}}(d, d').$$

The composition formula is clear.

Let $[1] \in \mathrm{Ob}(\mathbf{Cat})$ denote the linear order category of length 1:

$$[1] := \boxed{\begin{array}{ccc} 0 & f & 1 \\ \bullet & \longrightarrow & \bullet \end{array}}$$

As a schema it has one arrow, but as a category it has three morphisms. So we expect $[1] \times [1]$ to have nine morphisms, and that is true. In fact, $[1] \times [1]$ looks like a commutative square:

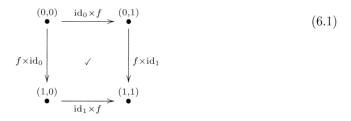

$$(6.1)$$

We see only four morphisms here, but there are also four identities and one morphism $(0,0) \to (1,1)$ given by composition of either direction. It is a minor miracle that the categorical product somehow "knows" that this square should commute; however, this is not a mere preference but follows rigorously from the definitions we already gave of $\mathbf{Cat}$ and products.

6.1.1.18   Coproducts

The coproduct of two sets is their disjoint union, which includes nonoverlapping copies of each of the two sets. This is a good intuition for coproducts in general.

Example 6.1.1.19. Given two preorders, $\mathcal{X}_1 := (X_1, \leqslant_1)$ and $\mathcal{X}_2 := (X_2, \leqslant_2)$, we can take their coproduct and get a new preorder $\mathcal{X}_1 \sqcup \mathcal{X}_2$. Both $\mathcal{X}_1$ and $\mathcal{X}_2$ have underlying sets (namely, X_1 and X_2), so we might hope that the underlying set of $\mathcal{X}_1 \times \mathcal{X}_2$ is the disjoint union $X_1 \sqcup X_2$, and that turns out to be true. We have a notion of less-than on $\mathcal{X}_1$ and a notion of less-than on $\mathcal{X}_2$.

Given an element $x \in X_1 \sqcup X_2$ and an element $x' \in X_1 \sqcup X_2$, how can we use $\leqslant_1$ and $\leqslant_2$ to compare x_1 and x_2? The relation $\leqslant_1$ only knows how to compare elements of X_1, and the relation $\leqslant_2$ only knows how to compare elements of X_2. But x and x' may come from different homes, e.g., $x \in X_1$ and $x' \in X_2$, in which case neither $\leqslant_1$ nor $\leqslant_2$ gives any clue about which should be bigger.

So when should we say that $x \leqslant_{1 \sqcup 2} x'$ holds? The obvious guess is to say that x is less than x' iff both x and x' are from the same home and the local ordering has $x \leqslant x'$. To be precise, we say $x \leqslant_{1 \sqcup 2} x'$ if and only if either one of the following conditions hold:

- $x \in X_1$ and $x' \in X_1$ and $x \leqslant_1 x'$, or

- $x \in X_2$ and $x' \in X_2$ and $x \leqslant_2 x'$.

With $\leqslant_{1 \sqcup 2}$ so defined, one checks that it is not only a preorder but that it serves as a coproduct of $\mathcal{X}_1$ and $\mathcal{X}_2$,[5]

$$\mathcal{X}_1 \sqcup \mathcal{X}_2 := (X_1 \sqcup X_2, \leqslant_{1 \sqcup 2}).$$

Note that the inclusion functions $X_1 \to X_1 \sqcup X_2$ and $X_2 \to X_1 \sqcup X_2$ induce morphisms of preorders. That is, if $x, x' \in X_1$ are elements such that $x \leqslant_1 x'$ in $\mathcal{X}_1$, then the same will hold in $\mathcal{X}_1 \sqcup \mathcal{X}_2$, and similarly for $\mathcal{X}_2$. So we have preorder morphisms

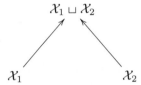

Exercise 6.1.1.20.

Suppose you have a partial order $\mathcal{A} := (A, \leqslant_A)$ on apples (you prefer some apples to others, but sometimes you cannot compare). And suppose you have a partial order $\mathcal{O} := (O, \leqslant_O)$ on oranges. You are about to be given two pieces of fruit from a basket of apples and oranges. Is the coproduct partial order $\mathcal{A} \sqcup \mathcal{O}$ a reasonable guess for your preferences, or does it seem biased?   ◊

Solution 6.1.1.20.

You cannot compare apples and oranges. If the two fruits are both apples, the coproduct ordering will match your preference; if the two fruits are both oranges, the coproduct ordering will match your preference; but if one is an apple and the other is an orange, the coproduct ordering will be agnostic about which you prefer.   ♦

Example 6.1.1.21. Given two graphs $G_1 = (V_1, A_1, src_1, tgt_1)$ and $G_2 = (V_2, A_2, src_2, tgt_2)$, we can take their coproduct and get a new graph $G_1 \sqcup G_2$. The vertices will be the disjoint

---

[5]Given $R_1 \subseteq X_1 \times X_1, R_2 \subseteq X_2 \times X_2$, take

$$\begin{aligned}
R_1 \sqcup R_2 &\subseteq (X_1 \times X_1) \sqcup (X_2 \times X_2) \\
&\subseteq (X_1 \times X_1) \sqcup (X_1 \times X_2) \sqcup (X_2 \times X_1) \sqcup (X_2 \times X_2) \\
&\cong (X_1 \sqcup X_2) \times (X_1 \sqcup X_2).
\end{aligned}$$

union of vertices $V_1 \sqcup V_2$, so each vertex in $G_1 \sqcup G_2$ is labeled either by a vertex in G_1 or by one in G_2 (if any labels are shared, then something must be done to differentiate them). When should an arrow connect v to v'? Whenever both are from the same component (i.e., either $v, v' \in V_1$ or $v, v' \in V_2$) and we can find an arrow connecting them in that component. It turns out there is a simple formula for the set of arrows in $G_1 \sqcup G_2$, namely, $A_1 \sqcup A_2$.

Let's write $G := G_1 \sqcup G_2$ and say, $G = (V, A, src, tgt)$. We now know that $V = V_1 \sqcup V_2$ and $A = A_1 \sqcup A_2$. What should the source and target functions $A \to V$ be? Given a function $src_1 \colon A_1 \to V_1$ and a function $src_2 \colon A_2 \to V_2$, the universal property for coproducts in **Set** can be used to specify a unique function

$$src := src_1 \sqcup src_2 \colon A_1 \sqcup A_2 \to V_1 \sqcup V_2.$$

Namely, for any arrow $a \in A$, we know either $a \in A_1$ or $a \in A_2$ (and not both), so the source of a will be the vertex $src_1(a)$ if $a \in A_1$ and $src_2(a)$ if $a \in A_2$. Similarly, we have a ready-made choice of target function $tgt = tgt_1 \sqcup tgt_2$. We have now defined the coproduct graph.

Here is an example. Let I and J be as in Example 5.3.3.5:

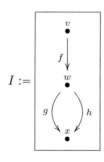

Arrow (I)		
ID	**src**	**tgt**
f	v	w
g	w	x
h	w	x

Vertex (I)
ID
v
w
x

Arrow (J)		
ID	**src**	**tgt**
i	q	r
j	r	s
k	s	r
ℓ	s	t

Vertex (J)
ID
q
r
s
t
u

The coproduct $I \sqcup J$ has, as expected, $3 + 5 = 8$ vertices and $3 + 4 = 7$ arrows:

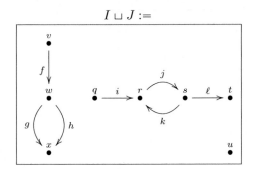

$I \sqcup J :=$

Arrow $(I \sqcup J)$		
ID	**src**	**tgt**
f	v	w
g	w	x
h	w	x
i	q	r
j	r	s
k	s	r
ℓ	s	t

Vertex $(I \sqcup J)$
ID
v
w
x
q
r
s
t
u

Here is the most important thing to notice. Look at the `Arrow` tables and notice that there is a way to send each row in I to a row in $I \sqcup J$ such that all the foreign keys match, and similarly for J. This also works for the vertex tables. These matches are readily visible graph homomorphisms $I \to I \sqcup J$ and $J \to I \sqcup J$ in **Grph**.

Exercise 6.1.1.22.

Recall from Example 4.5.2.10 that a discrete dynamical system (DDS) is a set s together with a function $f \colon s \to s$; if

$$\mathcal{L}oop := \boxed{\begin{array}{c} f \\ \circlearrowright \\ s \\ \bullet \end{array}}$$

is the loop schema, then a DDS is simply an instance (a functor) $I \colon \mathcal{L}oop \to \mathbf{Set}$. We have not yet discussed DDS coproducts but perhaps you can guess how they should work. For example, consider these instances $I, J \colon \mathcal{L}oop \to \mathbf{Set}$:

s	(I)
ID	**f**
A	C
B	C
C	C

s	(J)
ID	**f**
x	y
y	x
z	z

Make a guess and tabulate $I \sqcup J$. Then draw it. ◊

Solution 6.1.1.22.

s	$(I \times J)$
ID	**f**
A	C
B	C
C	C
x	y
y	x
z	z

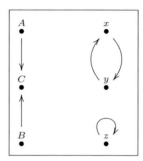

♦

In each case (preorders, graphs, DDSs), what is most important to recognize is that there are inclusion maps $I \to I \sqcup J$ and $J \to I \sqcup J$, and that the construction of $I \sqcup J$ seems as straightforward as possible, subject to having these inclusions.

Definition 6.1.1.23. Let $\mathcal{C}$ be a category, and let $X, Y \in \mathrm{Ob}(\mathcal{C})$ be objects. A *cospan on X and Y* consists of three constituents (Z, i, j), where $Z \in \mathrm{Ob}(\mathcal{C})$ is an object, and where $i \colon X \to Z$ and $j \colon Y \to Z$ are morphisms in $\mathcal{C}$.

A *coproduct of X and Y* is a cospan $X \xrightarrow{\iota_1} X \sqcup Y \xleftarrow{\iota_2} Y$, such that for any other cospan $X \xrightarrow{i} Z \xleftarrow{j} Y$ there *exists a unique* morphism $s_{i,j} \colon X \sqcup Y \to Z$ such that the following diagram commutes:[6]

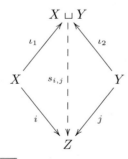

---

[6]The names $X \sqcup Y$ and ι_1, ι_2 are not mathematically important; they are pedagogically useful.

The morphism $s_{i,j}$ is often denoted $\begin{cases} i \\ j \end{cases} : X \sqcup Y \to Z$.

Remark 6.1.1.24. Definition 6.1.1.8 endows the coproduct of two objects with a *universal property*. It says that a coproduct of two objects X and Y receives maps from those two objects, and serves as a gateway for all that do the same. "None shall receive maps from X and Y except through me!" This grandiose property is held by all the coproducts discussed so far. It is what is meant by "$X \sqcup Y$ receives maps from both X and Y and does so as straightforwardly as possible." The disjoint union of dots obtained as the coproduct of two sets has such a property (see Example 3.1.2.5).

Example 6.1.1.25. By Proposition 5.2.1.13, there is a functor $\mathbf{PrO} \to \mathbf{Cat}$ that realizes every preorder as a category. If $\mathcal{P} = (P, \leqslant)$ is a preorder, what are coproducts in $\mathcal{P}$? Given two objects $a, b \in \mathrm{Ob}(\mathcal{P})$, we first consider $\{a, b\}$ cospans, i.e., $a \to z \leftarrow b$. A cospan of a and b is any z such that $a \leqslant z$ and $b \leqslant z$. The coproduct will be such a cospan $a \leqslant a \sqcup b \geqslant b$, but such that every other cospanning object z is greater than or equal to $a \sqcup b$. In other words, $a \sqcup b$ is as small as possible subject to the condition of being bigger than a and bigger than b. This is precisely their join, $a \vee b$ (see Definition 4.4.2.1).

Just as for products, the coproduct of two objects in a category $\mathcal{C}$ may not exist, or it may not be unique. The nonuniqueness is much less "bad" because given two candidate coproducts, they will be canonically isomorphic. They may not be equal, but they are isomorphic. But coproducts might not exist at all in certain categories.

Example 6.1.1.26. Consider the set $\mathbb{R}^2$ and partial order from Example 6.1.1.11, where $(x_1, y_1) \leqslant (x_2, y_2)$ if there exists $\ell \geqslant 1$ such that $x_1 \ell = x_2$ and $y_1 \ell = y_2$. Again the points $p := (1, 0)$ and $q := (0, 1)$ do not have a coproduct. Indeed, it would have to be a nonzero point that was on the same line through the origin as p and the same line through the origin as q, of which there are none.

Exercise 6.1.1.27.

Consider the preorder $\mathcal{P}$ of cards in a deck, shown in Example 4.4.1.3; it is not the whole story of cards in a deck, but take it to be so. Consider this preorder $\mathcal{P}$ as a category (by way of the functor $\mathbf{PrO} \to \mathbf{Cat}$).

a. For each of the following pairs, what is their coproduct in $\mathcal{P}$ (if it exists)?

⌜a diamond⌝ ⊔ ⌜a heart⌝   ⌜a queen⌝ ⊔ ⌜a black card⌝

⌜a card⌝ ⊔ ⌜a red card⌝   ⌜a face card⌝ ⊔ ⌜a black card⌝

b. How would these answers differ if $\mathcal{P}$ were completed to the "whole story" partial order classifying cards in a deck?

◇

Solution 6.1.1.27.

a. The product of two elements in any preorder, such as $\mathcal{P}$, is their join (if it exists). For example, the join of ⌜a queen⌝ and ⌜a black card⌝ should be a card that is either a queen or a black card, if this notion exists in $\mathcal{P}$. Then we have

$$\text{⌜a diamond⌝} \sqcup \text{⌜a heart⌝} = \text{⌜a red card⌝}$$
$$\text{⌜a queen⌝} \sqcup \text{⌜a black card⌝} = \text{⌜a card⌝}$$
$$\text{⌜a card⌝} \sqcup \text{⌜a red card⌝} = \text{⌜a card⌝}$$
$$\text{⌜a face card⌝} \sqcup \text{⌜a black card⌝} = \text{⌜a card⌝}$$

b. If every subset of cards were put into the order, i.e., if we took $\mathcal{P}$ to be the power-set of the set of cards, it would have more intuitive joins:

$$\text{⌜a diamond⌝} \sqcup \text{⌜a heart⌝} = \text{⌜a red card⌝}$$
$$\text{⌜a queen⌝} \sqcup \text{⌜a black card⌝} = \text{⌜a queen or a black card⌝}$$
$$\text{⌜a card⌝} \sqcup \text{⌜a red card⌝} = \text{⌜a card⌝}$$
$$\text{⌜a face card⌝} \sqcup \text{⌜a black card⌝} = \text{⌜a face card or a black card⌝}$$

◆

Exercise 6.1.1.28.

Let X be a set, and consider it as a discrete category. Given two objects $x, y \in \text{Ob}(X)$, under what conditions will there exist a coproduct $x \sqcup y$? ◇

Solution 6.1.1.28.

The only morphisms in a discrete category are identities. Since a coproduct needs morphisms including both its summands, we must have $x = x \sqcup y = y$. So the condition is $x = y$. Indeed, in that case, any other object z receiving a map from x and y receives a unique map from $x \sqcup y$ by $x \sqcup y \xrightarrow{\text{id}} z$. ◆

Exercise 6.1.1.29.

Consider the preorder $(\mathbb{N}, \texttt{divides})$, discussed in Example 4.4.3.2, where, e.g., $5 \leqslant 15$, but $5 \nleqslant 6$.

a. What is the coproduct of 9 and 12 in that category?

b. Is there a standard name for coproducts in that category?

◊

Solution 6.1.1.29.

a. $9 \sqcup 12 = 36$.

b. Least common multiple.

♦

6.1.2 Diagrams in a category

Diagrams have illustrated the text throughout the book. What is the mathematical foundation of these illustrations? The answer is functors.

Definition 6.1.2.1 (Diagrams). Let C and I be categories. An *I-shaped diagram in* C is simply a functor $d: I \to C$. In this case I is called the *indexing category* for the diagram.[7]

Here are some rules for drawing diagrams as in Definition 6.1.2.1.

Rules of good practice 6.1.2.2. Suppose given an indexing category I and an I-shaped diagram $X: I \to C$. One draws this as follows:

(i) For each object in $q \in I$, draw a dot labeled by $X(q)$; if several objects in I point to the same object in C, then several dots are labeled the same way.

(ii) For each morphism $f: q \to q'$ in I, draw an arrow between dots $X(q)$ and $X(q')$, and label it $X(f)$ in C. Again, if several morphisms in I are sent to the same morphism in C, then several arrows are labeled the same way.

(iii) One can abridge this process by not drawing *every* morphism in I, as long as every morphism in I is represented by a unique path in C, i.e., as long as the drawing is sufficiently unambiguous as a depiction of $X: I \to C$.

(iv) One may choose to draw a dash box around the finished diagram X to indicate that it is referencing an ambient category C.

---

[7]The indexing category I is usually assumed to be small in the sense of Remark 5.1.1.2, meaning that its collection of objects is a set.

Example 6.1.2.3. Consider the commutative diagram in **Set**:

$$\begin{array}{ccc} \mathbb{N} & \xrightarrow{\ +1\ } & \mathbb{N} \\ {\scriptstyle *2}\big\downarrow & & \big\downarrow{\scriptstyle *2} \\ \mathbb{N} & \xrightarrow[\ +2\ ]{} & \mathbb{Z} \end{array} \qquad\qquad (6.2)$$

This is the drawing of a functor $d\colon [1]\times[1]\to$ **Set** (see Example 6.1.1.17). With notation for the objects and morphisms of $[1]\times[1]$, as shown in diagram (6.1), we have $d(0,0) = d(0,1) = d(1,0) = \mathbb{N}$ and $d(1,1) = \mathbb{Z}$ (for some reason) and $d(\mathrm{id}_0, f)\colon \mathbb{N}\to\mathbb{N}$ given by $n \mapsto n+1$, and so on. The fact that d is a functor means it must respect composition formulas, which implies that diagram (6.2) commutes. We call $[1]\times[1]$ the *commutative square indexing category*. [8]

Example 6.1.2.4. Recall from Section 2.2 that not all diagrams commute; one must specify that a given diagram commutes if one wishes to communicate this fact. But then, how is a *noncommuting diagram* to be understood as a functor?

Let $G \in \mathrm{Ob}(\mathbf{Grph})$ denote the following graph:

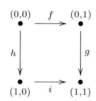

Recall the free category functor $F\colon \mathbf{Grph}\to\mathbf{Cat}$ (see Example 5.1.2.33). The free category $F(G) \in \mathrm{Ob}(\mathbf{Cat})$ on G looks almost like $[1]\times[1]$ in Example 6.1.2.3 except that since $_{(0,0)}[f,g]$ is a different path in G than is $_{(0,0)}[h,i]$, they become different morphisms in $F(G)$. A functor $F(G)\to$ **Set** might be drawn the same way that (6.2) is, but it would be a diagram that would *not* be said to commute.

Exercise 6.1.2.5.

Consider $[2]$, the linear order category of length 2.

a. Is $[2]$ the appropriate indexing category for commutative triangles?

b. If not, what is? If so, what might lead someone to be skeptical, and why would the skeptic be wrong?

$\diamond$

---

[8]What is here denoted $F(G)$ might be called the *noncommutative square indexing category*.

Solution 6.1.2.5.

a. Yes.

b. One might picture [2] and the commutative triangle as shown here at the left and the right (respectively),

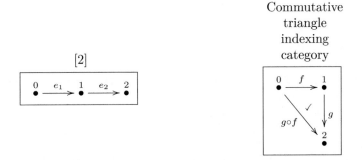

and think of these as different. But that is not correct because the drawing of [2] does not show the composite $e_2 \circ e_1$, which is implicitly there.

◆

Example 6.1.2.6. Recall that an equalizer in **Set** is a diagram of sets that looks like this:

$$ E \xrightarrow{\ f\ } A \underset{g_2}{\overset{g_1}{\rightrightarrows}} B \tag{6.3} $$

where $g_1 \circ f = g_2 \circ f$. What is the indexing category for such a diagram? It is the schema (6.3) with the PED $_E[f, g_1] \simeq \, _E[f, g_2]$. That is, in some sense one sees the indexing category, but the PED needs to be declared.

Exercise 6.1.2.7.

Let $\mathcal{C}$ be a category, $A \in \mathrm{Ob}(\mathcal{C})$ an object, and $f \colon A \to A$ a morphism in $\mathcal{C}$. Consider the following two diagrams in $\mathcal{C}$:

$$ d_1 := \left[A \xrightarrow{\ f\ } A \xrightarrow{\ f\ } A \xrightarrow{\ f\ } \cdots \right] \qquad\qquad d_2 := \left[f \, \circlearrowleft \, A \right] $$

a. Should these two diagrams have the same indexing category?

b. Write the indexing category for both.

c. If they have the same indexing category, what is causing or allowing the pictures to appear different?

d. If they do not have the same indexing category, what coincidence makes the two pictures have so much in common?

◇

Solution 6.1.2.7.

a. No.

b. The left-hand diagram is indexed by the linear order $[\mathbb{N}]$ as a category (under the functor $\mathbf{PrO} \to \mathbf{Cat}$), i.e., having objects $\mathrm{Ob}([\mathbb{N}]) = \mathbb{N}$ and a morphism $i \to j$ if $i \leqslant j$. The right-hand diagram is indexed by the category $\mathcal{L}oop$ as in (4.17).

c. This question does not apply; they do not have the same indexing category.

d. The issue arises from the fact that a diagram in $\mathcal{C}$ is not just an indexing category but also the functor to $\mathcal{C}$. The coincidence is in the functor $[\mathbb{N}] \xrightarrow{d_1} \mathbf{Set}$, which happens to send each object $i \in \mathrm{Ob}([\mathbb{N}])$ to the object $A \in \mathrm{Ob}(\mathcal{C})$, and which happens to send each morphism in $[\mathbb{N}]$ to the morphism f in $\mathcal{C}$. In other words, the coincidence is that there is a functor $[\mathbb{N}] \xrightarrow{F} \mathcal{L}oop$, under which the left-hand diagram $[\mathbb{N}] \xrightarrow{d_1} \mathcal{C}$ is the composite $[\mathbb{N}] \xrightarrow{F} \mathcal{L}oop \xrightarrow{d_2} \mathbf{Set}$ with the right-hand diagram, $d_1 = d_2 \circ F$.

♦

Definition 6.1.2.8. Let $I \in \mathrm{Ob}(\mathbf{Cat})$ be a category. The *left cone on I*, denoted $I^{\triangleleft}$, is the category defined as follows. On objects we put $\mathrm{Ob}(I^{\triangleleft}) = \{LC_I\} \sqcup \mathrm{Ob}(I)$, and we call the new object LC_I the *cone point of $I^{\triangleleft}$*. On morphisms we add a single new morphism $s_b \colon LC_I \to b$ for every object $b \in \mathrm{Ob}(I)$; more precisely,

$$\mathrm{Hom}_{I^{\triangleleft}}(a, b) = \begin{cases} \mathrm{Hom}_I(a, b) & \text{if } a, b \in \mathrm{Ob}(I), \\ \{s_b\} & \text{if } a = LC_I, b \in \mathrm{Ob}(I), \\ \{\mathrm{id}_{LC_I}\} & \text{if } a = b = LC_I, \\ \varnothing & \text{if } a \in \mathrm{Ob}(I), b = LC_I. \end{cases}$$

The composition formula is in some sense obvious. To compose two morphisms both in I, compose as dictated by I; if one has LC_I as source, then there will be a unique choice of composite.

There is an obvious inclusion of categories,

$$I \to I^{\triangleleft}. \tag{6.4}$$

Remark 6.1.2.9. Note that the specification of $I^{\triangleleft}$ given in Definition 6.1.2.8 works just as well if I is considered a schema and we are constructing a schema $I^{\triangleleft}$: add the new object LC_I and the new arrows $s_b \colon LC_I \to b$ for each $b \in \mathrm{Ob}(I)$, and for every morphism $f \colon b \to b'$ in I, add a PED $_{LC_I}[s_{b'}] \simeq {}_{LC_I}[s_b, f]$. We generally do not distinguish between categories and schemas, since they are equivalent, by Theorem 5.4.2.3.

Example 6.1.2.10. For a natural number $n \in \mathbb{N}$, define the *n-leaf star schema*, denoted $\mathbf{Star}_n$, to be the category (or schema; see Remark 6.1.2.9) $\underline{n}^{\triangleleft}$, where $\underline{n}$ is the discrete category on n objects. The following illustrate the categories $\mathbf{Star}_0, \mathbf{Star}_1, \mathbf{Star}_2$, and $\mathbf{Star}_3$:

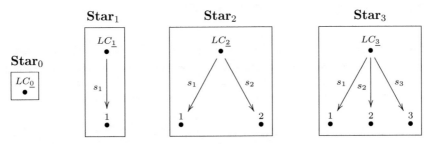

Exercise 6.1.2.11.

Let $\mathcal{C}_0 := \underline{0}$ denote the empty category, and for any natural number $n \in \mathbb{N}$, let $\mathcal{C}_{n+1} = (\mathcal{C}_n)^{\triangleleft}$. Draw $\mathcal{C}_4$. ◊

Solution 6.1.2.11.

We draw $\mathcal{C}_n$ for all $0 \leqslant n \leqslant 4$:

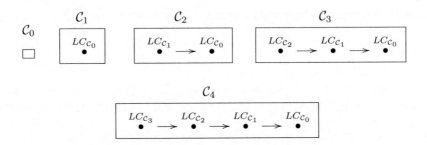

Exercise 6.1.2.12.

Let $\mathcal{C}$ be the graph-indexing schema as in (5.8). What is $\mathcal{C}^{\triangleleft}$, and how does it compare to the indexing category for equalizers, (6.3)?   ◊

Solution 6.1.2.12.

They are the same,

$$\mathcal{C}^{\triangleleft} \cong \boxed{\begin{array}{ccccc} E & \xrightarrow{\;f\;} & A & \underset{g_2}{\overset{g_1}{\rightrightarrows}} & B \\ \bullet & & \bullet & & \bullet \end{array}}$$

where the latter is understood to include the PED $_E[f, g_1] = {}_E[f, g_2]$.   ♦

Definition 6.1.2.13. Let $I \in \mathrm{Ob}(\mathbf{Cat})$ be a category. The *right cone on I*, denoted $I^{\triangleright}$, is the category defined as follows. On objects we put $\mathrm{Ob}(I^{\triangleright}) = \mathrm{Ob}(I) \sqcup \{RC_I\}$, and we call the new object RC_I the *cone point of $I^{\triangleright}$*. On morphisms we add a single new morphism $t_b \colon b \to RC_I$ for every object $b \in \mathrm{Ob}(I)$; more precisely,

$$\mathrm{Hom}_{I^{\triangleright}}(a, b) = \begin{cases} \mathrm{Hom}_I(a, b) & \text{if } a, b \in \mathrm{Ob}(I), \\ \{t_b\} & \text{if } a \in \mathrm{Ob}(I), b = RC_I, \\ \{\mathrm{id}_{RC_I}\} & \text{if } a = b = RC_I, \\ \varnothing & \text{if } a = RC_I, b \in \mathrm{Ob}(I). \end{cases}$$

The composition formula is in some sense obvious. To compose two morphisms both in I, compose as dictated by I; if one has RC_I as target, then there will be a unique choice of composite.

There is an obvious inclusion of categories $I \to I^{\triangleright}$.

Exercise 6.1.2.14.

Let $\mathcal{C}$ be the category $(\underline{2}^{\triangleleft})^{\triangleright}$, where $\underline{2}$ is the discrete category on two objects. Then $\mathcal{C}$ is somehow square-shaped, but what category is it exactly? Is $\mathcal{C}$ the commutative square indexing category $[1] \times [1]$ (see Example 6.1.2.3), is it the noncommutative square indexing category $F(G)$ (see Example 6.1.2.4), or is it something else?   ◊

Solution 6.1.2.14.

It is the commutative square indexing category,

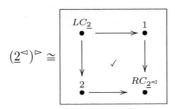

$$(\underline{2}^{\lhd})^{\rhd} \cong$$

Exercise 6.1.2.15.

Let $I = \underline{2}$, let $\mathcal{C}$ be an arbitrary category, and let $D = \mathrm{Fun}(I^{\lhd}, \mathcal{C})$.

a. Using Rules 6.1.2.2, draw an object $d \in \mathrm{Ob}(D)$.

b. How might you draw a morphism $f \colon d \to d'$ in D?

Solution 6.1.2.15.

a. We have $I^{\lhd} = \mathbf{Star}_2$, as in Example 6.1.2.10. We can draw an object $d \colon I^{\lhd} \to \mathcal{C}$ as a span,

$$d_1 \xleftarrow{\ i\ } d_0 \xrightarrow{\ j\ } d_2.$$

b. We could draw $f \colon d \to d'$ as

$$
\begin{array}{ccccc}
d_1 & \xleftarrow{\ i\ } & d_0 & \xrightarrow{\ j\ } & d_2 \\
\downarrow{\scriptstyle f_1} & & \downarrow{\scriptstyle f_0} & & \downarrow{\scriptstyle f_2} \\
d'_1 & \xleftarrow{\ i'\ } & d'_0 & \xrightarrow{\ j'\ } & d'_2
\end{array}
$$

6.1.3 Limits and colimits in a category

Let $\mathcal{C}$ be a category, let I be an indexing category (which means that I is a category that we use as the indexing category for a diagram), and let $D \colon I \to \mathcal{C}$ be an I-shaped diagram (which means a functor). It is in relation to this setup that we can discuss the limit or colimit. In general, the limit of a diagram $D\colon I \to C$ is a $I^\triangleleft$ shaped diagram, $\lim D\colon I^\triangleleft \to C$. In the case of products we have $I = \underline{2}$, and the limit looks like a span, the shape of $I^\triangleleft$ (see Exercise 6.1.2.15). For general I, D we may have many $I^\triangleleft$-shaped diagrams; which of them is the limit of D? Answer: The one with the universal gateway property; see Remark 6.1.1.9.

6.1.3.1   Universal objects

Definition 6.1.3.2. Let C be a category. An object $a \in \mathrm{Ob}(C)$ is called *initial* if, for all objects $c \in \mathrm{Ob}(C)$, there exists a unique morphism $a \to c$, i.e., $|\mathrm{Hom}_C(a, c)| = 1$. An object $z \in \mathrm{Ob}(C)$ is called *terminal* if, for all objects $c \in \mathrm{Ob}(C)$, there is exists a unique morphism $c \to z$, i.e., $|\mathrm{Hom}_C(c, z)| = 1$.

Example 6.1.3.3. For any category I, the left cone $I^\triangleleft$ has a unique initial object, and the right cone $I^\triangleright$ has a unique terminal object; in both cases it is the cone point. See Definitions 6.1.2.8 and 6.1.2.13.

Example 6.1.3.4. The initial object in **Set** is the set a for which there is always one way to map from a to anything else. Given $c \in \mathrm{Ob}(\mathbf{Set})$, there is exactly one function $\varnothing \to c$, because there are no choices to be made, so the empty set $\varnothing$ is the initial object in **Set**.

    The terminal object in **Set** is the set z for which there is always one way to map to z from anything else. Given $c \in \mathrm{Ob}(\mathbf{Set})$, there is exactly one function $c \to \{\odot\}$, where $\{\odot\}$ is any set with one element, because there are no choices to be made: everything in c must be sent to the single element in $\{\odot\}$. There are lots of terminal objects in **Set**, and they are all isomorphic to $\underline{1}$.

Example 6.1.3.5. The initial object in **Grph** is the graph a for which there is always one way to map from a to anything else. Given $c \in \mathrm{Ob}(\mathbf{Grph})$, there is exactly one graph homomorphism $\varnothing \to c$, where $\varnothing \in \mathrm{Ob}(\mathbf{Grph})$ is the empty graph; so $\varnothing$ is the initial object.

    The terminal object in **Grph** is more interesting. It is

the graph with one vertex and one arrow. In fact, there are infinitely many terminal objects in **Grph**, but all of them are isomorphic to $\mathcal{L}oop$, meaning one can change the names of the vertex (s) and the arrow (f) and get another terminal object.

Exercise 6.1.3.6.

Let X be a set, let $\mathbb{P}(X)$ be the set of subsets of X (see Definition 3.4.4.9). We can regard $\mathbb{P}(X)$ as a preorder under inclusion of subsets (see, for example, Section 4.4.2). And we can regard preorders as categories using a functor $\mathbf{PrO} \to \mathbf{Cat}$ (see Proposition 5.2.1.13).

a. What is the initial object in $\mathbb{P}(X)$?

b. What is the terminal object in $\mathbb{P}(X)$?

◊

Solution 6.1.3.6.

a. The object $\varnothing$ is initial.

b. The object X is terminal.

◆

Example 6.1.3.7. The initial object in the category $\mathbf{Mon}$ of monoids is the trivial monoid, $\underline{1}$. Indeed, for any monoid M, a morphism of monoids $\underline{1} \to M$ is a functor between one-object categories and these are determined by where they send morphisms. Since $\underline{1}$ has only the identity morphism and functors must preserve identities, there is no choice involved in finding a monoid morphism $\underline{1} \to M$.

Similarly, the terminal object in $\mathbf{Mon}$ is also the trivial monoid, $\underline{1}$. For any monoid M, a morphism of monoids $M \to \underline{1}$ sends everything to the identity; there is no choice.

Exercise 6.1.3.8.

a. What is the initial object in $\mathbf{Grp}$, the category of groups?

b. What is the terminal object in $\mathbf{Grp}$?

◊

Solution 6.1.3.8.

a. The initial object in $\mathbf{Grp}$ is the trivial group. It could be denoted $(\underline{1}, 1, !)$.

b. The terminal object is again the trivial group.

◆

Example 6.1.3.9. Recall the preorder **Prop** of logical propositions from Section 5.2.4.1. The initial object is a proposition that implies all others. It turns out that "FALSE" is such a proposition. The proposition "FALSE" is like "$1 \neq 1$"; in logical formalism it can be shown that if "FALSE" is true, then everything is true.

The terminal object in **Prop** is a proposition that is implied by all others. It turns out that "TRUE" is such a proposition. In logical formalism, everything implies that "TRUE" is true.

Example 6.1.3.10. The discrete category $\underline{2}$ has no initial object and no terminal object. The reason is that it has two objects $1, 2$, but no maps from one to the other, so $\mathrm{Hom}_{\underline{2}}(1, 2) = \mathrm{Hom}_{\underline{2}}(2, 1) = \emptyset$.

Exercise 6.1.3.11.

Recall the `divides` preorder (see Example 4.4.3.2), where 5 `divides` 15.

a. Considering this preorder as a category, does it have an initial object?

b. Does it have a terminal object?

$\diamond$

Solution 6.1.3.11.

a. Yes, 1 divides everything.

b. Yes, everything divides 0.

$\blacklozenge$

Exercise 6.1.3.12.

Let $\mathcal{M} = (\mathrm{List}(\{a, b\}), [\ ], +\!\!+\ )$ denote the free monoid on the set $\{a, b\}$ (see Definition 4.1.1.15) considered as a category via the functor **Mon** $\to$ **Cat** (see Theorem 5.2.1.3).

a. Does $\mathcal{M}$ have an initial object?

b. Does $\mathcal{M}$ have a terminal object?

c. Which monoids $\mathcal{M}$, considered as one-object categories, have initial (resp. terminal) objects?

$\diamond$

Solution 6.1.3.12.

As a category, a monoid $\mathcal{M}$ has only one object, $\blacktriangle$. This object would be initial or terminal if there is only one morphism, $|\text{Hom}_{\mathcal{M}}(\blacktriangle, \blacktriangle)| = 1$. But this occurs only if $\mathcal{M} = (\underline{1}, 1, !)$ is the trivial monoid.

a. No.

b. No.

c. Only $\mathcal{M} = (\underline{1}, 1, !)$.

$\blacklozenge$

Exercise 6.1.3.13.

Let S be a set, and consider the indiscrete category $K_S \in \text{Ob}(\mathbf{Cat})$ on objects S (see Example 5.3.4.3).

a. For what S does K_S have an initial object?

b. For what S does K_S have a terminal object?

$\lozenge$

Solution 6.1.3.13.

For every two objects $s, s' \in S$, we have $|\text{Hom}_{K_S}(s, s')| = 1$, so every object is both initial and terminal.

a. The indiscrete category K_S has an initial object if and only if $S \neq \varnothing$. In this case every element of S is initial.

b. The indiscrete category K_S has an terminal object if and only if $S \neq \varnothing$. In this case every element of S is terminal.

$\blacklozenge$

An object in a category is sometimes called *universal* if it is either initial or terminal, but we rarely use that term in practice, preferring to be specific about whether the object is initial or terminal. The word *final* is synonymous with the word *terminal*, but we will use the latter.

Universal properties refer to either initial or terminal objects in a specially-designed category. Colimits end up having an initial sort of universal property, and limits end up having a terminal sort of universal property. See Section 6.1.3.16.

Warning 6.1.3.14. A category $\mathcal{C}$ may have more than one initial object; similarly a category $\mathcal{C}$ may have more than one terminal object. As shown in Example 6.1.3.4, any set with one element, e.g., $\{*\}$ or $\{\odot\}$ or $\{43\}$, is a terminal object in **Set**. Each of these terminal sets has the same number of elements, i.e., there exists an isomorphism between them, but they are not exactly the same set.

In fact, Proposition 6.1.3.15 shows that in any category $\mathcal{C}$, any two terminal objects in $\mathcal{C}$ are isomorphic (similarly, any two initial objects in $\mathcal{C}$ are isomorphic). While there are many isomorphisms in **Set** between $\{1, 2, 3\}$ and $\{a, b, c\}$, there is only one isomorphism between $\{*\}$ and $\{\odot\}$. This is always the case for universal objects: there is a unique isomorphism between any two terminal (resp. initial) objects in any category.

As a result, we often speak of *the* initial object in $\mathcal{C}$ or *the* terminal object in $\mathcal{C}$, as though there were only one. "It is unique up to unique isomorphism" is put forward as the justification for using *the* rather than *a*. This is not too misleading, because just as a person today does not contain exactly the same atoms as that person yesterday, the difference is unimportant.

This book uses either the definite or the indefinite article, as is convenient, when speaking about initial or terminal objects. For example, Example 6.1.3.4 discussed *the* initial object in **Set** and *the* terminal object in **Set**. This usage is common throughout mathematical literature.

Proposition 6.1.3.15. *Let $\mathcal{C}$ be a category, and let $a_1, a_2 \in \mathrm{Ob}(\mathcal{C})$ both be initial objects. Then there is a unique isomorphism $f \colon a_1 \xrightarrow{\cong} a_2$. (Similarly, for any two terminal objects in $\mathcal{C}$, there is a unique isomorphism between them.)*

Proof. Suppose a_1 and a_2 are initial. Since a_1 is initial, there is a unique morphism $f \colon a_1 \to a_2$; there is also a unique morphism $a_1 \to a_1$, which must be id_{a_1}. Since a_2 is initial, there is a unique morphism $g \colon a_2 \to a_1$; there is also a unique morphism $a_2 \to a_2$, which must be id_{a_2}. So $g \circ f = \mathrm{id}_{a_1}$ and $f \circ g = \mathrm{id}_{a_2}$, which means that f is the desired (unique) isomorphism.

The proof for terminal objects is appropriately dual.

$\square$

6.1.3.16   Examples of limits

We are moving toward defining limits and colimits in full generality. We have assembled most of the pieces we will need: indexing categories, their left and right cones, and the notion of initial and terminal objects. Relying on the now familiar notion of products, we put these pieces in place and motivate one more construction, the slice category over a diagram.

Let $\mathcal{C}$ be a category, and let $X, Y \in \mathrm{Ob}(\mathcal{C})$ be objects. Definition 6.1.1.8 defines a product of X and Y to be a span $X \xleftarrow{\pi_1} X \times Y \xrightarrow{\pi_2} Y$ such that for every other span $X \xleftarrow{p} Z \xrightarrow{q} Y$, there exists a unique morphism $Z \to X \times Y$ making the triangles commute. It turns out that we can enunciate this in the language of universal objects by saying that the span $X \xleftarrow{\pi_1} X \times Y \xrightarrow{\pi_2} Y$ is itself a terminal object in the category of $\{X, Y\}$ spans. Phrasing the definition of products in this way is generalizable to defining arbitrary limits.

Construction 6.1.3.17 (Products). Let $\mathcal{C}$ be a category, and let X_1, X_2 be objects. We can consider this setup as a diagram $X\colon \underline{2} \to \mathcal{C}$, where $X(1) = X_1$ and $X(2) = X_2$. Consider the category $\underline{2}^{\triangleleft} = \mathbf{Star}_2$ (see Example 6.1.2.10), the inclusion $i\colon \underline{2} \to \underline{2}^{\triangleleft}$ (see (6.4)), and the category of functors $\mathrm{Fun}(\underline{2}^{\triangleleft}, \mathcal{C})$. The objects in $\mathrm{Fun}(\underline{2}^{\triangleleft}, \mathcal{C})$ are spans in $\mathcal{C}$, and the morphisms are natural transformations between them (see Exercise 6.1.2.15).

Given a functor $S\colon \underline{2}^{\triangleleft} \to \mathcal{C}$, we can compose with $i\colon \underline{2} \to \underline{2}^{\triangleleft}$ to get a functor $\underline{2} \to \mathcal{C}$. We want that to be X. That is, to get the product of X_1 and X_2, we are looking among those $S\colon \underline{2}^{\triangleleft} \to \mathcal{C}$ for which the following diagram commutes:

We are ready to define the category of $\{X_1, X_2\}$ spans.

Define the *category of X spans in $\mathcal{C}$*, denoted $\mathcal{C}_{/X}$, to be the category whose objects and morphisms are as follows:

$$\mathrm{Ob}(\mathcal{C}_{/X}) = \{S\colon \underline{2}^{\triangleleft} \to \mathcal{C} \mid S \circ i = X\} \qquad (6.5)$$
$$\mathrm{Hom}_{\mathcal{C}_{/X}}(S, S') = \{\alpha\colon S \to S' \mid \alpha \diamond i = \mathrm{id}_X\}.$$

The product of X_1 and X_2 was defined in Definition 6.1.1.8; we can now recast $X_1 \times X_2$ as the terminal object in $\mathcal{C}_{/X}$.

An object in $\mathcal{C}_{/X}$ can be pictured as a diagram in $\mathcal{C}$ of the following form:

$$X_1 \xleftarrow{\;p\;} Z \xrightarrow{\;q\;} X_2.$$

In other words, the objects of $\mathcal{C}_{/X}$ are spans. A morphism in $\mathcal{C}_{/X}$ from object $X_1 \xleftarrow{p} Z \xrightarrow{q} X_2$ to object $X_1 \xleftarrow{p'} Z' \xrightarrow{q'} X_2$ consists of a morphism $\ell\colon Z \to Z'$, such that $p' \circ \ell = p$ and $q' \circ \ell = q$. So the set of such morphisms in $\mathcal{C}_{/X}$ are all the ℓ's that make both squares commute in the right-hand diagram:

$$\text{Hom}_{\mathcal{C}_{/X}}\left(X_1 \xleftarrow{p} Z \xrightarrow{q} X_2 \quad , \quad X_1 \xleftarrow{p'} Z' \xrightarrow{q'} X_2 \right) = \left\{ \begin{array}{c} X_1 \xleftarrow{p} Z \xrightarrow{q} X_2 \\ \parallel \quad \checkmark \; \downarrow \ell \; \checkmark \quad \parallel \\ X_1 \xleftarrow{p'} Z' \xrightarrow{q'} X_2 \end{array} \right\}$$

(6.6)

Each object in $\mathcal{C}_{/X}$ is a span on X_1 and X_2, and each morphism in $\mathcal{C}_{/X}$ is a morphism of cone points in $\mathcal{C}$ making everything commute. The terminal object in $\mathcal{C}_{/X}$ is the product of X_1 and X_2 (see Definition 6.1.1.8).

It may be strange to have a category in which the objects are spans in another category. But once one admits this possibility, the notion of morphism between spans becomes totally sensible.

Example 6.1.3.18. Consider the following arbitrary six-object category $\mathcal{C}$, in which the three diagrams that can commute do so:

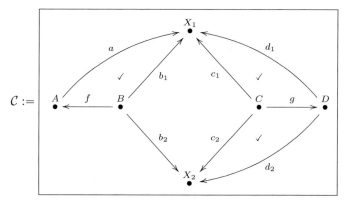

Let $X \colon \underline{2} \to \mathcal{C}$ be given by $X(1) = X_1$ and $X(2) = X_2$. Then the category of X spans might be drawn

$$\mathcal{C}_{/X} \cong \boxed{\quad \underset{\bullet}{(B,b_1,b_2)} \qquad\qquad \underset{\bullet}{(C,c_1,c_2)} \xrightarrow{\;g\;} \underset{\bullet}{(D,d_1,d_2)} \quad}$$

6.1.3.19   Definition of limit

A product of two objects $X, Y \in \text{Ob}()$ is a special case of a limit, namely, one in which the indexing category is $\underline{2}$. To handle arbitrary limits, we replace $\underline{2}$ with an arbitrary indexing category I, and use the following definition to generalize the category of spans, defined in (6.5).

Definition 6.1.3.20. Let $\mathcal{C}$ be a category, let I be a category. Let $I^\triangleleft$ be the left cone on I, and let $i\colon I \to I^\triangleleft$ be the inclusion. Suppose that $X\colon I \to \mathcal{C}$ is an I-shaped diagram in $\mathcal{C}$. The *slice category of $\mathcal{C}$ over X*, denoted $\mathcal{C}_{/X}$, is the category whose objects and morphisms are as follows:

$$\mathrm{Ob}(\mathcal{C}_{/X}) = \{S\colon I^\triangleleft \to \mathcal{C} \mid S \circ i = X\};$$
$$\mathrm{Hom}_{\mathcal{C}_{/X}}(S, S') = \{\alpha\colon S \to S' \mid \alpha \circ i = \mathrm{id}_X\}.$$

A *limit of X*, denoted $\lim_I X$ or $\lim X$, is a terminal object in $\mathcal{C}_{/X}$.

Remark 6.1.3.21. Perhaps the following diagram will be helpful for understanding limits. Given a functor $X\colon I \to \mathcal{C}$, what is its limit? The solid-arrow part of the figure is the data we start with, i.e., the category $\mathcal{C}$, the indexing category I, and the diagram $X\colon I \to \mathcal{C}$, as well as the part we automatically add, the cone $I^\triangleleft$ with the inclusion $I \xrightarrow{i} I^\triangleleft$. The category $\mathcal{C}_{/X}$ is found in the dotted arrow part: its objects are the dotted arrows $S\colon I^\triangleleft \to \mathcal{C}$ that make the following triangle commute, and its morphisms are the natural transformations $\alpha\colon S \to S'$x between them:

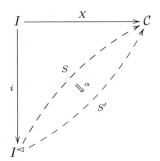

The limit of X is the initial object in this category.

Pullbacks The relevant indexing category for pullbacks is the cospan, $I = \underline{2}^\triangleright$, drawn as on the left:

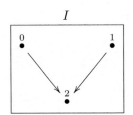

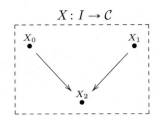

A I-shaped diagram in $\mathcal{C}$ is a functor $X\colon I \to \mathcal{C}$, which might be drawn as on the right (e.g., $X_0 \in \mathrm{Ob}(\mathcal{C})$).

An object S in the slice category $\mathcal{C}_{/X}$ is a commutative diagram $S\colon I^{\triangleleft} \to \mathcal{C}$ over X, which looks like the left-hand box:

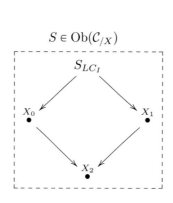

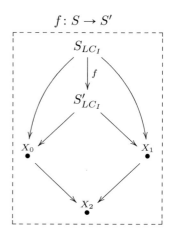

A morphism in $\mathcal{C}_{/X}$ is drawn as in the right-hand box. A terminal object in $\mathcal{C}_{/X}$ is precisely the gateway we want, i.e., the limit of X is the pullback $X_0 \times_{X_2} X_1$ (see Remark 6.1.1.9).

Remark 6.1.3.22. Let $\mathcal{C}$ be a category, and suppose given a functor $X\colon I \to \mathcal{C}$. Its limit is a certain functor $\lim X\colon I^{\triangleleft} \to \mathcal{C}$. The category $I^{\triangleleft}$ looks basically the same as I, except it has an extra cone point LC_I mapping to everything in I (see Definition 6.1.2.8). The functor $\lim X$ can be applied to this object in $I^{\triangleleft}$ to get an object in $\mathcal{C}$, and it is this object that people often refer to as the limit of X. We call it the *limit set* of X.

For example, if $I = \underline{2}$ then a functor $X\colon \underline{2} \to \mathcal{C}$ consists of two objects in $\mathcal{C}$, say X_1 and X_2. The left cone $\underline{2}^{\triangleleft}$ is the span category, so the limit of X is a span, in particular it is the product span $X_1 \leftarrow X_1 \times X_2 \to X_2$. But people often speak of the product as if it was just $X_1 \times X_2$, the cone point of the span.

Exercise 6.1.3.23.

Let **GrIn** be the graph-indexing category (see (5.8)).

a. What is **GrIn**$^{\triangleleft}$?

b. Let $G\colon$ **GrIn** $\to$ **Set** be the graph from Example 4.3.1.2. Give an example of an object in **Set**$_{/G}$.

◊

Solution 6.1.3.23.

a. As in Exercise 6.1.2.12,

$$\mathbf{GrIn}^{\triangleleft} \cong \boxed{\begin{array}{ccccc} E & \xrightarrow{\ f\ } & Ar & \overset{src}{\underset{tgt}{\rightrightarrows}} & Ve \\ \bullet & & \bullet & & \bullet \end{array}}$$

with the PED $_E[f, src] = {}_E[f, tgt]$.

b. An object in $\mathbf{Set}_{/G}$ is a functor $S\colon \mathbf{GrIn}^{\triangleleft} \to \mathbf{Set}$ such that $S(Ar) = \{f, g, h, i, j, k\}$ and $S(Ve) = \{v, w, x, y, z\}$, and with source and target maps as in Example 4.3.1.2. To specify the functor S, we need only specify what it does on the remaining object, $S(E)$, and on the remaining morphism $S(f)\colon S(E) \to S(Ar)$, and the necessary diagram must commute. So we could take $S(E) = \{1, 2, 3\}$; we are forced to have $S(f)(1) = S(f)(2) = S(f)(3) = i$. Indeed, since $src \circ f = tgt \circ f$, we must have $S(src)(S(f)(x)) = S(tgt)(S(f)(x))$, which means that $S(f)(x)$ must be a loop in G, and i is the only one.

♦

Exercise 6.1.3.24.

Let $\mathcal{C}$ be a category, and let $I = \underline{0}$ be the empty category. There is a unique functor $X\colon \underline{0} \to \mathcal{C}$.

a. What is the slice category $\mathcal{C}_{/X}$?

b. What is a limit of X?

◊

Solution 6.1.3.24.

a. The left cone of $\underline{0}$ is the terminal category $\underline{0}^{\triangleleft} = \underline{1}$, and since every diagram

commutes, we have an isomorphism $\mathrm{Fun}(\underline{1}, \mathcal{C}) \xrightarrow{\cong} \mathcal{C}_{/X}$. But by (5.14), we have an isomorphism $\mathcal{C} \xrightarrow{\cong} \mathrm{Fun}(\underline{1}, \mathcal{C})$, so in fact $\mathcal{C}_{/X} \cong \mathcal{C}$.

b. A limit of X is defined to be a terminal object in $\mathcal{C}_{/X}$, which is a terminal object in $\mathcal{C}$, if it exists. In other words, terminal objects in a category give us a canonical example of limits. This was hinted at in Exercise 3.2.3.5.

◆

Example 6.1.3.25. In the course of doing math, random-looking diagrams sometimes come up, for which one wants to take the limit. We have now constructed the limit for any shape diagram. For example, if we wanted to take the product of more than two, say, n, objects, we could use the diagram shape $I = \underline{n}$. A functor $X\colon \underline{n} \to \mathbf{Set}$ is n sets $X_1, X_2, \ldots, X_n$, and their limit is a functor $\lim X\colon \underline{n}^{\triangleleft} \to \mathbf{Set}$,

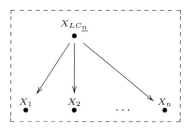

which, of course, is the product, $X_{LC_{\underline{n}}} = X_1 \times X_2 \times \cdots \times X_n$.

Example 6.1.3.26. We have now defined limits in any category, so we have defined limits in **Cat**. Let $[1]$ denote the category depicted

and let $\mathcal{C}$ be an arbitrary category. Naming two categories is the same thing as naming a functor $X\colon \underline{2} \to \mathbf{Cat}$; consider the functor $X(1) = [1], X(2) = \mathcal{C}$. The limit of X is a product of categories (see Example 6.1.1.17); it is denoted $[1] \times \mathcal{C}$. It turns out that $[1] \times \mathcal{C}$ looks like a $\mathcal{C}$-shaped prism. It consists of two panes, front and back, say, each having the precise shape as $\mathcal{C}$ (same objects, same arrows, same composition) as well as morphisms from the front pane to the back pane making all front-to-back squares commute. For example, if $\mathcal{C}$ was the category generated by the left-hand schema , then $\mathcal{C} \times [1]$ would be the category generated by the right-hand schema:

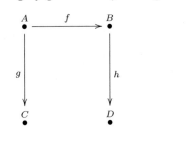
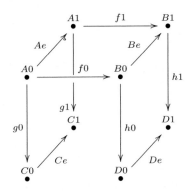

It turns out that a natural transformation $\alpha\colon F \to G$ between functors $F, G\colon \mathcal{C} \to \mathcal{D}$ is the same thing as a functor $\mathcal{C} \times [1] \to \mathcal{D}$ such that the front pane is sent via F and the back pane is sent via G. The components are captured by the front-to-back morphisms, and the naturality is captured by the commutativity of the front-to-back squares in $\mathcal{C} \times [1]$.
Exercise 6.1.3.27.

Recall that Section 3.4.6.5 described relative sets. In fact, Definition 3.4.6.6 basically defines a category of relative sets over any fixed set B. Let $B\colon \underline{1} \to \mathbf{Set}$ be the functor representing the object $B \in \mathrm{Ob}(\mathbf{Set})$.

a. What is the relationship between the slice category $\mathbf{Set}_{/B}$, as defined in Definition 6.1.3.20, and the category of relative sets over B?

b. What is the limit of the functor $B\colon \underline{1} \to \mathbf{Set}$?

◊

Solution 6.1.3.27.

a. They are identical: $\mathbf{Set}_{/B}$ is the category of relative sets over B.

b. The limit of the diagram $B\colon \underline{1} \to \mathbf{Set}$ is the terminal object in the category of "sets mapping to B." One can check that $B \xrightarrow{\mathrm{id}} B$ is the terminal object.

♦

Theorem 6.1.3.28. *Let I be a category and let $F\colon I \to \mathbf{Set}$ be a functor. Then its limit set $\lim_I F \in \mathrm{Ob}(\mathbf{Set})$ exists and one can find its elements as follows. An element of the set $\lim_I F$ is given by choosing an element of $x_i \in F(i)$ for each object $i \in \mathrm{Ob}(I)$ such that, for each $f\colon i \to i'$ one has $F(f)(x_i) = x_{i'}$.*

Proof. See [29].

□

Exercise 6.1.3.29.

Let I be the category given by the following schema:

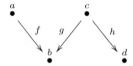

Let $X: I \to \mathbf{Set}$ be given on objects by $X(a) := \underline{2}, X(b) := \underline{1}, X(c) := \underline{3}, X(d) = \underline{2}$, and given (in sequence notation) on morphisms by $X(f) = (1,1), X(g) = (1,1,1), X(h) = (1,2,1)$. What is the limit $\lim_I X$.   ◊

Solution 6.1.3.29.

By Theorem 6.1.3.28, an element of the limit set $\lim_I X$ is given by choosing a tuple (x_a, x_b, x_c, x_d) such that

- $x_a \in X(a), x_b \in X(b), x_c \in X(c)$, and $x_d \in X(d)$, and

- $X(f)(x_a) = x_b, X(g)(x_b) = x_c$, and $X(h)(x_b) = x_d$.

Without the second condition, there are two ways to choose x_a, one way to choose x_b, three ways to choose x_c and two ways to choose x_d, giving twelve possible tuples. The second restriction reduces this number. We need $X(h)(x_c) = x_d$, so in fact x_d is determined by x_c; this reduces our choices to $2 * 3 = 6$. We also need $X(f)(x_a) = x_b$ and $X(g)(x_c) = x_b$, but because $X(b) = \underline{1}$, this is automatic. Thus the answer is $\underline{6}$.   ◆

6.1.3.30   Definition of colimit

The definition of colimits is appropriately dual to the definition of limits. Instead of looking at left cones, we look at right cones; instead of being interested in terminal objects, we are interested in initial objects.

Definition 6.1.3.31. Let $\mathcal{C}$ be a category, let I be a category; let $I^\rhd$ be the right cone on I, and let $i: I \to I^\rhd$ be the inclusion. Suppose that $X: I \to \mathcal{C}$ is an I-shaped diagram in $\mathcal{C}$. The *coslice category of $\mathcal{C}$ over X*, denoted $\mathcal{C}_{X/}$, is the category whose objects and morphisms are as follows:

$$\mathrm{Ob}(\mathcal{C}_{X/}) = \{S: I^\rhd \to \mathcal{C} \mid S \circ i = X\};$$
$$\mathrm{Hom}_{\mathcal{C}_{X/}}(S, S') = \{\alpha: S \to S' \mid \alpha \diamond i = \mathrm{id}_X\}.$$

A *colimit of* X, denoted $\operatorname{colim}_I X$ or $\operatorname{colim} X$, is an initial object in $\mathcal{C}_{X/}$.

Remark 6.1.3.32. Perhaps the following diagram will be helpful for understanding colimits. Given a functor $X: I \to \mathcal{C}$, what is its colimit? The solid-arrow part of the figure is the data we start with, i.e., the category $\mathcal{C}$, the indexing category I, and the diagram $X: I \to \mathcal{C}$, as well as the part we automatically add, the cone $I^{\triangleright}$ with the inclusion $I \overset{i}{\to} I^{\triangleright}$. The category $\mathcal{C}_{X/}$ is found in the dotted arrow part: its objects are the dotted arrows $S: I^{\triangleright} \to \mathcal{C}$ that make the following triangle commute, and its morphisms are the natural transformations $\alpha: S \to S'$ between them:

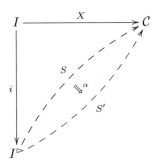

The colimit of X is the initial object in this category.

Pushouts   The relevant indexing category for pushouts is the span, $I = \underline{2}^{\triangleleft}$ drawn as on the left:

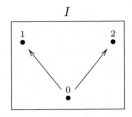 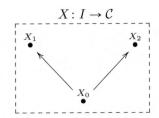

An I-shaped diagram in $\mathcal{C}$ is a functor $X: I \to \mathcal{C}$, which might be drawn as on the right (e.g., $X_0 \in \operatorname{Ob}(\mathcal{C})$).

An object S in the coslice category $\mathcal{C}_{X/}$ is a commutative diagram $S: I^{\triangleright} \to \mathcal{C}$ over

X, which looks like the left-hand box:

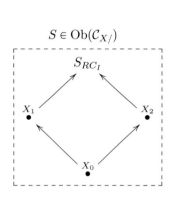

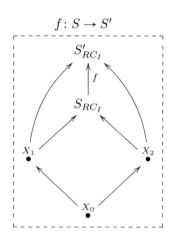

A morphism in $\mathcal{C}_{X/}$ is drawn as in right-hand box. An initial object in $\mathcal{C}_{X/}$ is precisely the gateway we want, i.e., the colimit of X is the pushout, $X_1 \sqcup_{X_0} X_2$.

Exercise 6.1.3.33.

Let **GrIn** be the graph-indexing category (see (5.8)).

a. What is **GrIn**$^{\triangleright}$?

b. Let $G \colon$ **GrIn** $\to$ **Set** be the graph from Example 4.3.1.2. Give an example of an object in **Set**$_{G/}$.

$\diamond$

Solution 6.1.3.33.

a. We have

$$\mathbf{GrIn}^{\triangleright} \cong \boxed{\begin{array}{ccccc} Ar & \xrightarrow[tgt]{src} & Ve & \xrightarrow{f} & C \\ \bullet & & \bullet & & \bullet \end{array}}$$

with the PED $_{Ar}[src, f] = {}_{Ar}[tgt, f]$.

b. An object in **Set**$_{G/}$ is a functor $S \colon$ **GrIn**$^{\triangleright} \to$ **Set** such that $S(Ar) = \{f, g, h, i, j, k\}$ and $S(Ve) = \{v, w, x, y, z\}$, and with source and target maps as in Example 4.3.1.2. To specify the functor S, we need only specify what it does on the remaining object, $S(C)$, and on the remaining morphism $S(f) \colon S(Ve) \to S(C)$, and the necessary diagram must commute. So we could take $S(C) = \{1, 2, 3\}$; we are forced to have $S(f)(v) = S(f)(w) = S(f)(x)$ and $S(f)(y) = S(f)(z)$. In other words, for each connected component of this graph (of which there are two), we can choose to which element of $S(C)$, of which there are three, we will send it. So we could take $S(f)(v) = 1$ and $S(f)(y) = 3$. We have now defined an object in $\mathbf{Set}_{G/}$.

♦

Exercise 6.1.3.34.

Let $\mathcal{C}$ be a category, and let $I = \underline{0}$ be the empty category. There is a unique functor $X : \underline{0} \to \mathcal{C}$.

a. What is the coslice category $\mathcal{C}_{X/}$?

b. What is a colimit of X (assuming it exists)?

◇

Solution 6.1.3.34.

a. The right cone of $\underline{0}$ is the terminal category $\underline{0}^{\triangleright} \cong \underline{1}$, and since every diagram

commutes, we have an isomorphism $\mathrm{Fun}(\underline{1}, \mathcal{C}) \xrightarrow{\cong} \mathcal{C}_{X/}$. But by (5.14), we have an isomorphism $\mathcal{C} \xrightarrow{\cong} \mathrm{Fun}(\underline{1}, \mathcal{C})$, so in fact $\mathcal{C} \cong \mathcal{C}_{X/}$.

b. A colimit of X is defined to be an initial object in $\mathcal{C}_{X/}$, which is an initial object in $\mathcal{C}$, if it exists. In other words, initial objects in a category give us a canonical example of colimits. This was hinted at in Exercise 3.3.3.4.

♦

Theorem 6.1.3.35. *Let I be a category and let $F : I \to \mathbf{Set}$ be a functor. Then its colimit set $\mathrm{colim}_I F \in \mathrm{Ob}(\mathbf{Set})$ exists and one can find its elements as follows. An element of the set $\mathrm{colim}_I F$ is given by choosing any $i \in \mathrm{Ob}(I)$ and any element of $x_i \in F(i)$, and then considering two such elements equivalent if there exists $f : i \to i'$ such that $X(f)(x_i) = x_{i'}$.*

Proof. See [29].

□

Exercise 6.1.3.36.

Let I be the category given by the following schema:

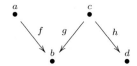

Let $X : I \to \mathbf{Set}$ be given on objects by $X(a) := \underline{2}, X(b) := \underline{2}, X(c) := \underline{4}, X(d) = \underline{3}$, and given (in sequence notation) on morphisms by $X(f) = (1,2), X(g) = (1,2,1), X(h) = (1,2,4)$. What is the colimit $\mathrm{colim}_I X$.   ◊

Solution 6.1.3.36.

We follow Theorem 6.1.3.35; to begin, we find the set of ways to choose an object $i \in \mathrm{Ob}(I)$ and an element $x_i \in X(i)$. This is the set of vertices in the following depiction of X:

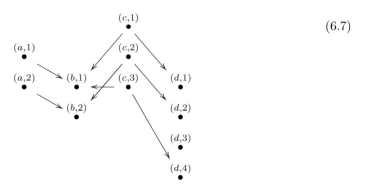

(6.7)

The equivalence relation is given by connections in this graph; e.g., $(a,2) \sim (b,2) \sim (c,3) \sim (d,4)$. The colimit is basically the set of connected components of this graph. There are three equivalence classes: the one containing $(a,1)$, the one containing $(a,2)$, and the one containing $(d,3)$. So we have $\mathrm{colim}_I X \cong \underline{3}$.   ♦

Remark 6.1.3.37. Definition 6.1.3.31 defined what it means to be a colimit in any category; however, in any particular category, some colimits may not exist. It is like defining the quotient of any two natural numbers $r, s \in \mathbb{N}$ by $r \div s = q$ if and only if $q * s = r$. We have defined what it means to be a quotient, but that doesn't mean the quotient of any two numbers exists, e.g. if $r = 7$ and $s = 2$.

The same goes for limits. A category $\mathcal{C}$ in which every diagram is guaranteed to have a limit is called *complete*. A category $\mathcal{C}$ in which every diagram is guaranteed to have a colimit is called *cocomplete*.

Example 6.1.3.38 (Cone as colimit). It turns out that **Cat** is cocomplete, meaning every diagram in $\mathcal{C}$ has a colimit. We give an example of a colimit in **Cat**.

Let $\mathcal{C}$ be a category, and recall from Example 6.1.3.26 the category $\mathcal{C} \times [1]$. The inclusion of the front pane is a functor $i_0 \colon \mathcal{C} \to \mathcal{C} \times [1]$. (Similarly, the inclusion of the back pane is a functor $i_1 \colon \mathcal{C} \to \mathcal{C} \times [1]$.) Finally, let $t \colon \mathcal{C} \to \underline{1}$ be the unique functor to the terminal category (see Exercise 5.1.2.40). We now have a diagram in **Cat** of the form

$$
\begin{array}{ccc}
\mathcal{C} & \xrightarrow{\;i_0\;} & \mathcal{C} \times [1] \\
{\scriptstyle t}\big\downarrow & & \\
\underline{1} & &
\end{array}
$$

The colimit (i.e., the pushout) of this diagram in **Cat** slurps down the entire front pane of $\mathcal{C} \times [1]$ to a point, and the resulting category is isomorphic to $\mathcal{C}^{\triangleleft}$. The diagrams in

(6.8) illustrate this phenomenon.

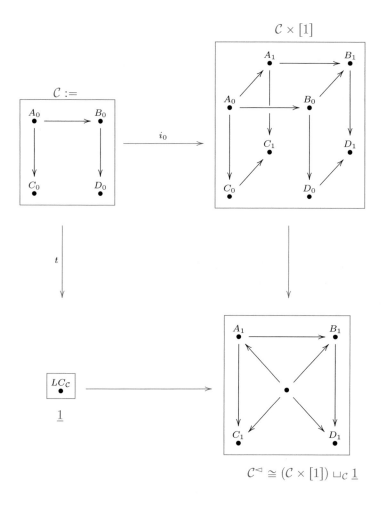

$$\mathcal{C}^{\lhd} \cong (\mathcal{C} \times [1]) \sqcup_{\mathcal{C}} \underline{1}$$

(6.8)

The category $\mathcal{C}$ is shown in the upper left-hand corner of (6.8). The left cone $\mathcal{C}^{\lhd}$ on $\mathcal{C}$ is obtained as a pushout in **Cat**. We first make a prism $\mathcal{C} \times [1]$ and then identify the front pane with a point. (Similarly, the pushout of an analogous diagram for i_1 would give $\mathcal{C}^{\rhd}$.)

Example 6.1.3.39. Consider the category **Top** of topological spaces. The (unfilled) circle is a topological space, which people often denote by S^1 (for one-dimensional sphere).

Topologically, it is equivalent to an oval, as shown in Figure 6.1. The filled-in circle, also called a two-dimensional disk, is denoted D^2. The inclusion of the circle into the disk, as its boundary, is continuous, so we have a morphism in **Top** of the form $i\colon S^1 \to D^2$. The terminal object in **Top** is the one-point space $\bullet$, so there is a unique morphism $t\colon S^1 \to \bullet$.

The pushout of the diagram $D^2 \xleftarrow{i} S^1 \xrightarrow{t} \bullet$ is isomorphic to the two-dimensional sphere (the exterior of a tennis ball), S^2. The reason is that we have slurped the entire bounding circle of D^2 to a point, which becomes, say, the south pole, and the interior area of D^2 becomes the surface area of the sphere. Mathematically, the category of topological spaces has the right morphisms to ensure that this intuitive picture is correct.

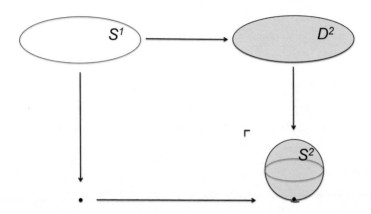

Figure 6.1 A pushout of topological spaces. A circle S^1 is both included as the boundary of a disk D^2 and sent to a single point $\bullet$. The resulting pushout is a 2-dimensional sphere S^2, formed by sewing the boundary circle of a disk all together into a single point.

Application 6.1.3.40. Consider the symmetric graph G_n consisting of a chain of n vertices,

Think of this as modeling a subway line. There are n-many graph homomorphisms $G_1 \to G_n$ given by the various vertices. One can create transit maps using colimits. For

example, the colimit of the left-hand diagram is the symmetric graph drawn at the right:

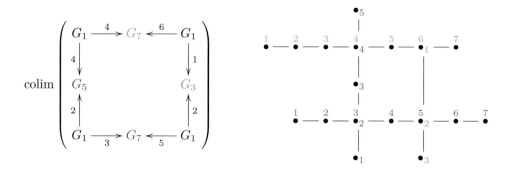

$$\Diamond\Diamond$$

6.2   Other notions in Cat

This section discusses some additional notions about categories. Section 6.2.1 explains a kind of duality for categories, in which arrows are flipped. Reversing the order in a preorder is an example of this duality, as is the similarity between the definitions of limit and colimit. Section 6.2.2 discusses the Grothendieck construction, which in some sense makes a histogram for a set-valued functor, and shows that this idea is useful for transforming databases into the kind of format (RDF) used in scraping data off web pages. Some ways of creating new categories from old are explained in Sections 6.2.3 and 6.2.4. Finally, Section 6.2.5 shows that precisely the same arithmetic statements that held for sets (see Section 3.4.3) hold for categories.

6.2.1   Opposite categories

In the early days of category theory, and still today, people would sometimes discuss two different kinds of functors between categories: *covariant functors* and *contravariant functors*. Covariant functors are what this book calls functors. The reader may have come across the idea of contravariance when considering Exercise 5.2.3.2,[9] which showed that a continuous mapping of topological spaces $f\colon X \to Y$ does not induce a morphism of orders on their open sets $\text{Open}(X) \to \text{Open}(Y)$; that is not required by the notion of continuity. Instead, a morphism of topological spaces $f\colon X \to Y$ induces a morphism of orders $\text{Open}(Y) \to \text{Open}(X)$, going backward. So we do not have a functor **Top** $\to$ **PrO**

---

[9]Similarly, see Exercise 5.2.4.4.

in this way, but it is quite close. It used to be said that Open is a *contravariant functor*
Top → **PrO**.

As important and common as contravariance is, one finds that keeping track of which
functors were covariant and which were contravariant is a big hassle. Luckily, there is a
simple work-around, which simplifies everything: the notion of opposite categories.

Definition 6.2.1.1. Let $\mathcal{C}$ be a category. The *opposite category* of $\mathcal{C}$, denoted $\mathcal{C}^{\mathrm{op}}$, has
the same objects as $\mathcal{C}$, i.e., $\mathrm{Ob}(\mathcal{C}^{\mathrm{op}}) = \mathrm{Ob}(\mathcal{C})$, and for any two objects c, c', one defines

$$\mathrm{Hom}_{\mathcal{C}^{\mathrm{op}}}(c, c') := \mathrm{Hom}_{\mathcal{C}}(c', c).$$

Example 6.2.1.2. If $n \in \mathbb{N}$ is a natural number and $\underline{n}$ the corresponding discrete category,
then $\underline{n}^{\mathrm{op}} = \underline{n}$. Recall the span category $I = \underline{2}^{\triangleleft}$ from Definition 6.1.1.8. Its opposite is
the cospan category $I^{\mathrm{op}} = \underline{2}^{\triangleright}$, from Definition 6.1.1.23.

Exercise 6.2.1.3.

Let $\mathcal{C}$ be the category from Example 6.1.3.18. Draw $\mathcal{C}^{\mathrm{op}}$.   ◊

Solution 6.2.1.3.

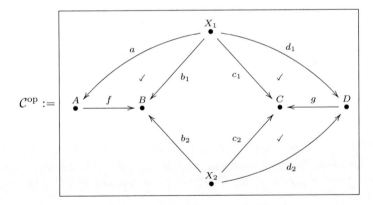

                                                                                                            ◆

Proposition 6.2.1.4. *Let $\mathcal{C}$ and $\mathcal{D}$ be categories. One has $(\mathcal{C}^{\mathrm{op}})^{\mathrm{op}} = \mathcal{C}$. Also one has a
canonical isomorphism $\mathrm{Fun}(\mathcal{C}, \mathcal{D}) \cong \mathrm{Fun}(\mathcal{C}^{\mathrm{op}}, \mathcal{D}^{\mathrm{op}})$. This implies that a functor $\mathcal{C}^{\mathrm{op}} \to \mathcal{D}$
can be identified with a functor $\mathcal{C} \to \mathcal{D}^{\mathrm{op}}$.*

Proof. This follows straightforwardly from the definitions.

                                                                                                            □

Exercise 6.2.1.5.

If $\mathcal{C}$ is a category and $c \in \mathrm{Ob}(\mathcal{C})$ is an initial object, does this imply that c is a terminal object in $\mathcal{C}^{\mathrm{op}}$?   ◇

Solution 6.2.1.5.

Yes. For any $x \in \mathrm{Ob}(\mathcal{C}^{\mathrm{op}}) = \mathrm{Ob}(\mathcal{C})$, we have $\mathrm{Hom}_{\mathcal{C}^{\mathrm{op}}}(x, c) = \mathrm{Hom}_{\mathcal{C}}(c, x) \cong \underline{1}$.   ◆

Exercise 6.2.1.6.

In Exercises 5.2.3.2, 5.2.4.3, and 5.2.4.4 there were questions about whether a certain function $\mathrm{Ob}(\mathcal{C}) \to \mathrm{Ob}(\mathcal{D})$ extended to a functor $\mathcal{C} \to \mathcal{D}$.

a. Does the function $\mathrm{Open}\colon \mathrm{Ob}(\mathbf{Top}) \to \mathrm{Ob}(\mathbf{PrO})$ extend to a functor $\mathrm{Open}\colon \mathbf{Top}^{\mathrm{op}} \to \mathbf{PrO}$?

b. Does the function $L\colon \mathrm{Ob}(J) \to \mathrm{Ob}(\mathbf{Prop})$ extend to a functor $L\colon J^{\mathrm{op}} \to \mathbf{Prop}$?

c. Does the function $R\colon \mathrm{Ob}(J) \to \mathrm{Ob}(\mathbf{Set})$ extend to a functor $R\colon J^{\mathrm{op}} \to \mathbf{Set}$?

                                                                                                                         ◇

Solution 6.2.1.6.

a. Yes.

b. No.

c. Yes.

                                                                                                                         ◆

Example 6.2.1.7 (Simplicial sets). Recall from Example 5.3.4.4 the category $\mathbf{\Delta}$ of linear orders $[n]$. For example, $[1]$ is the linear order $0 \leqslant 1$, and $[2]$ is the linear order $0 \leqslant 1 \leqslant 2$. Both $[1]$ and $[2]$ are objects of $\mathbf{\Delta}$. There are 6 morphisms from $[1]$ to $[2]$, which could be denoted

$$\mathrm{Hom}_{\mathbf{\Delta}}([1], [2]) = \{(0, 0), (0, 1), (0, 2), (1, 1), (1, 2), (2, 2)\}.$$

The category $\mathbf{\Delta}^{\mathrm{op}}$ turns out to be quite useful in algebraic topology. It is the indexing category for a combinatorial approach to the homotopy theory of spaces. That is, we can represent something like the category of spaces and continuous maps using the functor category $\mathrm{Fun}(\mathbf{\Delta}^{\mathrm{op}}, \mathbf{Set})$, which is called the *category of simplicial sets*.

This may seem very complicated compared to simplicial complexes (see Section 3.4.4.3). But simplicial sets have excellent formal properties that simplicial complexes do not. We do not go further with this here, but through the work of Dan Kan, André Joyal, Jacob Lurie, and many others, simplicial sets have allowed category theory to pierce deeply into the realm of topology, and vice versa.

6.2.2 Grothendieck construction

Let $\mathcal{C}$ be a database schema (or category), and let $J \colon \mathcal{C} \to \mathbf{Set}$ be an instance. We have been drawing this in table form, but there is another standard way of laying out the data in J, called the *resource descriptive framework*, or RDF. Developed for the World Wide Web, RDF is a useful format when one does not have a schema in hand. For example, when scraping information off a website, one does not know which schema will be best. In these cases information is stored in RDF triples, which are of the form

$$\langle \text{Subject, Predicate, Object} \rangle.$$

For example, one might see something like

Subject	Predicate	Object
A01	occurredOn	D13114
A01	performedBy	P44
A01	actionDescription	Told congress to raise the debt ceiling
D13114	hasYear	2013
D13114	hasMonth	January
D13114	hasDay	14
P44	FirstName	Barack
P44	LastName	Obama

(6.9)

This might be an RDF interpretation of the sentence "On January 14, 2013, Barack Obama told congress to raise the debt ceiling."

Category-theoretically, it is quite simple to convert a database instance $J \colon \mathcal{C} \to \mathbf{Set}$ into an RDF triple store. To do so, we use the *Grothendieck construction*, also known as the *category of elements*.

Definition 6.2.2.1. Let $\mathcal{C}$ be a category, and let $J \colon \mathcal{C} \to \mathbf{Set}$ be a functor. The *category of elements of J*, denoted $\int_{\mathcal{C}} J$, is defined as follows:

$$\mathrm{Ob}\left(\int_{\mathcal{C}} J\right) := \{(C, x) \mid C \in \mathrm{Ob}(\mathcal{C}), x \in J(C)\};$$

$$\mathrm{Hom}_{\int_{\mathcal{C}} J}((C, x), (C', x')) := \{f \colon C \to C' \mid J(f)(x) = x'\}.$$

There is a natural functor $\pi_J \colon \int_{\mathcal{C}} J \longrightarrow \mathcal{C}$. It sends each object $(C, x) \in \mathrm{Ob}(\int_{\mathcal{C}} J)$ to the object $C \in \mathrm{Ob}(\mathcal{C})$. And it sends each morphism $f \colon (C, x) \to (C', x')$ to the morphism $f \colon C \to C'$. We call π_J the *projection functor*.

Example 6.2.2.2. Let A be a set, and consider it as a discrete category. We saw in Exercise 5.3.3.4 that a functor $S \colon A \to \mathbf{Set}$ is the same thing as an A-indexed set, as discussed in Section 3.4.6.9. We follow Definition 3.4.6.11 and, for each $a \in A$, write $S_a := S(a)$.

What is the category of elements of a functor $S \colon A \to \mathbf{Set}$? The objects of $\int_A S$ are pairs (a, s), where $a \in A$ and $s \in S(a)$. Since A has nothing but identity morphisms, $\int_A S$ has nothing but identity morphisms, i.e., it is the discrete category on a set. In fact, that set is the disjoint union

$$\int_A S = \bigsqcup_{a \in A} S_a.$$

The functor $\pi_S \colon \int_A S \to A$ sends each element in S_a to the element $a \in A$.

One can see this as a kind of histogram. For example, let $A = \{\mathtt{BOS}, \mathtt{NYC}, \mathtt{LA}, \mathtt{DC}\}$, and let $S \colon A \to \mathbf{Set}$ assign

$$S_{\mathtt{BOS}} = \{\mathtt{Abby}, \mathtt{Bob}, \mathtt{Casandra}\},$$
$$S_{\mathtt{NYC}} = \varnothing,$$
$$S_{\mathtt{LA}} = \{\mathtt{John}, \mathtt{Jim}\},$$
$$S_{\mathtt{DC}} = \{\mathtt{Abby}, \mathtt{Carla}\}.$$

Then the category of elements of S would look like the (discrete) category at the top:

$$\int_A S = \quad (6.10)$$

We also see that the category of elements construction has converted an A-indexed set into a relative set over A, as in Definition 3.4.6.6.

The preceding example does not show how the Grothendieck construction transforms a database instance into an RDF triple store. The reason is that the database schema was A, a discrete category that specifies no connections between data (it simply collects the data into bins). So let's examine a more interesting database schema and instance. This is taken from Spivak [39].

Application 6.2.2.3. Consider the following schema, first encountered in Example 4.5.2.1:

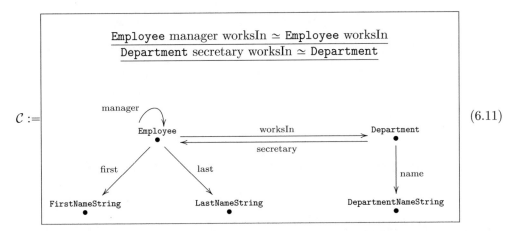

$$(6.11)$$

And consider the instance $J: \mathcal{C} \to \mathbf{Set}$, which we first encountered in (4.13) and (4.15):

Employee						Department		
ID	**first**	**last**	**manager**	**worksIn**		**ID**	**name**	**secretary**
101	David	Hilbert	103	q10		q10	Sales	101
102	Bertrand	Russell	102	x02		x02	Production	102
103	Emmy	Noether	103	q10				

FirstNameString
ID
Alan
Bertrand
Carl
David
Emmy

LastNameString
ID
Arden
Hilbert
Jones
Noether
Russell

DepartmentNameString
ID
Marketing
Production
Sales

The category of elements of $J: \mathcal{C} \to \mathbf{Set}$ looks like this:

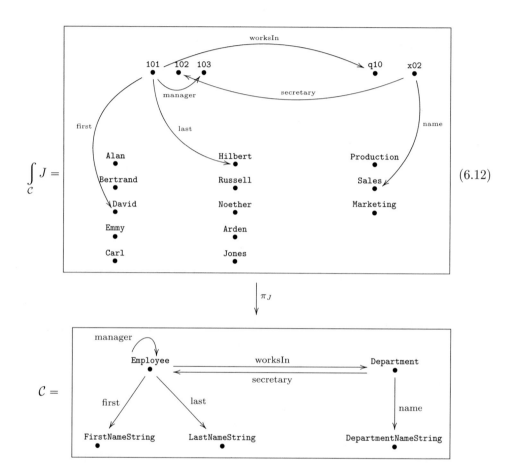

$$\int_{\mathcal{C}} J = \qquad\qquad\qquad\qquad\qquad (6.12)$$

In Diagram (6.12) of $\int_{\mathcal{C}} J$, ten arrows were omitted for ease of readability, for example, arrow $\overset{102}{\bullet} \xrightarrow{\text{first}} \overset{\text{Bertrand}}{\bullet}$ was omitted.

How do we see the category of elements $\int_{\mathcal{C}} J$ as an RDF triple store? For each arrow in $\int_{\mathcal{C}} J$, we take the triple consisting of the source vertex, the arrow name, and the target vertex. So the triple store would include triples such as $\langle 101 \; \text{worksIn} \; \text{q10} \rangle$ and $\langle \text{q10} \; \text{name} \; \text{Production} \rangle$. Note that if $\mathcal{C}$ were an olog, we could read off these triples (and concatenations of them) as English sentences. For example, the preceding two triples could be Englished as follows:

Employee 101 works in Department q10, which has as name Production.

◊◊

Exercise 6.2.2.4.

Devise a schema $\mathcal{C}$ for which you can imagine an instance $I\colon \mathcal{C} \to$ **Set** such that the category of elements $\int(I)$ is the triple store in (6.9).   ◊

Solution 6.2.2.4.

$$\mathcal{C} :=$$

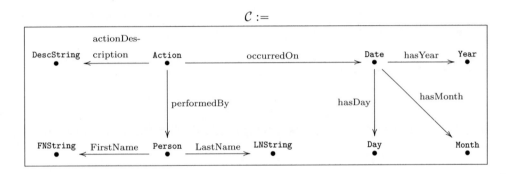

◆

Slogan 6.2.2.5.

> The Grothendieck construction takes structured, tabulated data and flattens it by throwing it all into one big space. The projection functor is then tasked with remembering which box each datum originally came from.

Exercise 6.2.2.6.

Recall from Section 4.1.2.10 that a finite state machine is a free monoid $(\mathrm{List}(\Sigma), [\,], +\!\!+\,)$ acting on a set X. Recall also that we can consider a monoid as a category $\mathcal{M}$ with one object, and we can consider a monoid action as a set-valued functor $F\colon \mathcal{M} \to$ **Set** (see Section 5.2.1.1). In the case of Figure 4.2 the monoid is $\mathrm{List}(a, b)$, which can be drawn as the schema

and the functor $F\colon \mathcal{M} \to$ **Set** is recorded in an action table in Example 4.1.3.1. What is $\int_{\mathcal{M}} F$? How does it relate to Figure 4.2?   ◊

Solution 6.2.2.6.

Figure 4.2 is a drawing of the category $\int_{\mathcal{M}} F$, or more precisely, of a presentation of it. The objects are drawn as boxes, $\mathrm{Ob}(\int_{\mathcal{M}} F) = \{\text{State } 0, \text{State } 1, \text{State } 2\}$, and the morphisms are generated by the six arrows drawn there.   ♦

6.2.3   Full subcategory

Definition 6.2.3.1. Let $\mathcal{C}$ be a category, and let $X \subseteq \mathrm{Ob}(\mathcal{C})$ be a set of objects in $\mathcal{C}$. The *full subcategory of $\mathcal{C}$ spanned by X* is the category, denoted $\mathcal{C}_{\mathrm{Ob}=X}$, with objects $\mathrm{Ob}(\mathcal{C}_{\mathrm{Ob}=X}) := X$ and with morphisms $\mathrm{Hom}_{\mathcal{C}_{\mathrm{Ob}=X}}(x, x') := \mathrm{Hom}_{\mathcal{C}}(x, x')$.

Example 6.2.3.2. The following are examples of full subcategories. For example, the category **Fin** of finite sets is the full subcategory of **Set** spanned by the finite sets.

- If $X = \{s \in \mathrm{Ob}(\mathbf{Set}) \mid s \text{ is finite}\}$, then $\mathbf{Fin} = \mathbf{Set}_{\mathrm{Ob}=X}$.

- If $X = \{P \in \mathrm{Ob}(\mathbf{PrO}) \mid P \text{ is a finite linear order}\}$, then $\mathbf{FLin} = \mathbf{PrO}_{\mathrm{Ob}=X}$.

- If $X = \{[n] \in \mathbf{FLin} \mid n \in \mathbb{N}\}$ (see Example 5.3.4.4), then $\mathbf{\Delta} = \mathbf{FLin}_{\mathrm{Ob}=X}$.

- If $X = \{M \in \mathrm{Ob}(\mathbf{Mon}) \mid \mathrm{M} \text{ is a group}\}$, then $\mathbf{Grp} = \mathbf{Mon}_{\mathrm{Ob}=X}$.

- If $X = \{\mathcal{C} \in \mathrm{Ob}(\mathbf{Cat}) \mid \mathcal{C} \text{ has one object}\}$, then $\mathbf{Mon} = \mathbf{Cat}_{\mathrm{Ob}=X}$.

- If $X = \{\underline{n} \in \mathrm{Ob}(\mathbf{Fin}) \mid n \in \mathbb{N}\}$, then there is an equivalence of categories $\mathbf{Fin} \simeq \mathbf{Fin}_{\mathrm{Ob}=X}$.

- If $X = \{(V, A, src, tgt) \in \mathrm{Ob}(\mathbf{Grph}) \mid A = \varnothing\}$, then $\mathbf{Set} \cong \mathbf{Grph}_{\mathrm{Ob}=X}$.

- If $X = \{\mathcal{C} \in \mathbf{Cat} \mid \mathcal{C} \text{ is discrete}\}$, then $\mathbf{Set} \cong \mathbf{Cat}_{\mathrm{Ob}=X}$.

Remark 6.2.3.3. A subcategory $\mathcal{C} \subseteq \mathcal{D}$ is (up to isomorphism) just a functor $i \colon \mathcal{C} \to \mathcal{D}$ that happens to be injective on objects and arrows. The subcategory is full if and only if i is a full functor in the sense of Definition 5.3.4.8.

Example 6.2.3.4. Let $\mathcal{C}$ be a category, let $X \subseteq \mathrm{Ob}(\mathcal{C})$ be a set of objects, and let $\mathcal{C}_{\mathrm{Ob}=X}$ denote the full subcategory of $\mathcal{C}$ spanned by X. We can realize this as a fiber product of categories. Indeed, recall that for any set, we can form the indiscrete category on that set (see Example 5.3.4.3). In fact, we have a functor $Ind \colon \mathbf{Set} \to \mathbf{Cat}$. Thus the function $X \to \mathrm{Ob}(\mathcal{C})$ can be converted into a functor between indiscrete categories $Ind(X) \to Ind(\mathrm{Ob}(\mathcal{C}))$. There is also a unique functor $\mathcal{C} \to Ind(\mathrm{Ob}(\mathcal{C}))$ sending each object to itself. Then the full subcategory of $\mathcal{C}$ spanned by X is the fiber product of categories,

$$
\begin{array}{ccc}
\mathcal{C}_{\text{Ob}=X} & \longrightarrow & \mathcal{C} \\
\downarrow & \lrcorner & \downarrow \\
Ind(X) & \longrightarrow & Ind(\text{Ob}(\mathcal{C}))
\end{array}
$$

Exercise 6.2.3.5.

Recall the sets $\underline{0}, \underline{1}, \underline{2} \in \text{Ob}(\textbf{Set})$ from Notation 2.1.2.21. Including all identities and all compositions, how many morphisms are there in the full subcategory $\textbf{Set}_{\text{Ob}=\{\underline{0},\underline{1},\underline{2}\}}$?

◊

Solution 6.2.3.5.

We can write a function $\underline{m} \to \underline{n}$ as a length m sequence of numbers i with $1 \leqslant i \leqslant n$. For example, there is function $\underline{3} \to \underline{5}$, that we would write $(4, 4, 2)$.

There are 11 morphisms in $\mathcal{C}$; every entry in the table is the set $\text{Hom}_{\mathcal{C}}(\underline{m}, \underline{n})$:

$\begin{array}{c} m \\ n \end{array}$	0	1	2
0	$\{(\,)\}$	$\varnothing$	$\varnothing$
1	$\{(\,)\}$	$\{(1)\}$	$\{(1,1)\}$
2	$\{(\,)\}$	$\{(1),(2)\}$	$\{(1,1),(1,2),(2,1),(2,2)\}$

♦

6.2.4 Comma categories

Category theory includes a highly developed and interoperable catalogue of materials (categories such as $[n]$, **GrIn**, **PrO**, etc.) and production techniques for making new categories from old. One such was the full subcategory idea in the previous section— given any category and any subset of objects, one can form a new category to restrict attention to the subset. Another is the comma category construction.

Definition 6.2.4.1. Let $\mathcal{A} \xrightarrow{F} \mathcal{C} \xleftarrow{G} \mathcal{B}$ be a cospan of categories. The *comma category of* $\mathcal{C}$ *morphisms from* F *to* G, denoted $(F \downarrow_{\mathcal{C}} G)$ or simply $(F \downarrow G)$, is the category with objects

$$
\text{Ob}(F \downarrow G) = \{(a, b, f) \mid a \in \text{Ob}(\mathcal{A}), b \in \text{Ob}(\mathcal{B}), f\colon F(a) \to G(b) \text{ in } \mathcal{C}\},
$$

and for any two objects (a, b, f) and (a', b', f') the set $\mathrm{Hom}_{(F \downarrow G)}((a, b, f), (a', b', f'))$ of morphisms $(a, b, f) \longrightarrow (a', b', f')$ is

$$\{(q, r) \mid q \colon a \to a' \text{ in } \mathcal{A}, \ r \colon b \to b' \text{ in } \mathcal{B}, \text{ such that } f' \circ F(q) = G(r) \circ f\}.$$

In diagram form,

$$
\mathrm{Hom}_{(F \downarrow G)}((a, b, f), (a', b', f')) := \left\{
\begin{array}{ccccc}
a & F(a) & \xrightarrow{\ f\ } & G(b) & b \\
q \downarrow & F(q) \downarrow & \checkmark & \downarrow G(r) & \downarrow r \\
a' & F(a') & \xrightarrow{\ f'\ } & G(b') & b'
\end{array}
\right\}
$$

There is a canonical functor $(F \downarrow G) \to \mathcal{A}$, called *left projecton*, sending (a, b, f) to a, and a canonical functor $(F \downarrow G) \to \mathcal{B}$, called *right projection*, sending (a, b, f) to b.

A cospan $\mathcal{A} \xrightarrow{F} \mathcal{C} \xleftarrow{G} \mathcal{B}$ is reversible, i.e., we can flip it to obtain $\mathcal{B} \xrightarrow{G} \mathcal{C} \xleftarrow{F} \mathcal{A}$. However, note that $(F \downarrow G)$ is different than (i.e., almost never equivalent to) $(G \downarrow F)$.

Slogan 6.2.4.2.

> When two categories $\mathcal{A}, \mathcal{B}$ can be interpreted in a common setting $\mathcal{C}$, the comma category integrates them by recording how to move from $\mathcal{A}$ to $\mathcal{B}$ inside $\mathcal{C}$.

Example 6.2.4.3. Let $\mathcal{C}$ be a category and $I \colon \mathcal{C} \to \mathbf{Set}$ a functor. This example shows that the comma category construction captures the notion of taking the category of elements $\int_{\mathcal{C}} I$ (see Definition 6.2.2.1).

Consider the set $\underline{1}$, the category $Disc(\underline{1})$, and the functor $F \colon Disc(\underline{1}) \to \mathbf{Set}$ sending the unique object to the set $\underline{1}$. We use the cospan $Disc(\underline{1}) \xrightarrow{F} \mathbf{Set} \xleftarrow{I} \mathcal{C}$. There is an isomorphism of categories

$$\int_{\mathcal{C}} I \cong (F \downarrow I).$$

Indeed, an object in $(F \downarrow I)$ is a triple (a, b, f), where $a \in \mathrm{Ob}(Disc(\underline{1})), b \in \mathrm{Ob}(\mathcal{C})$, and $f \colon F(a) \to I(b)$ is a morphism in $\mathbf{Set}$. There is only one object in $Disc(\underline{1})$, so this reduces to a pair (b, f), where $b \in \mathrm{Ob}(\mathcal{C})$ and $f \colon \{\odot\} \to I(b)$. The set of functions $\{\odot\} \to I(b)$ is isomorphic to $I(b)$ (see Exercise 2.1.2.20). So we have reduced $\mathrm{Ob}(F \downarrow I)$ to the set of pairs (b, x), where $b \in \mathrm{Ob}(\mathcal{C})$ and $x \in I(b)$; this is $\mathrm{Ob}(\int_{\mathcal{C}} I)$. Because there is only one function $\underline{1} \to \underline{1}$, a morphism $(b, x) \to (b', x')$ in $(F \downarrow I)$ boils down to a morphism $r \colon b \to b'$ such that the diagram

$$
\begin{array}{ccc}
\underline{1} & \xrightarrow{\ x\ } & I(b) \\
\big\| & & \big\downarrow I(r) \\
\underline{1} & \xrightarrow[\ x'\ ]{} & I(b')
\end{array}
$$

commutes. But such diagrams are in one-to-one correspondence with the diagrams defining morphisms in $\int_{\mathcal{C}} I$.

Exercise 6.2.4.4.

Let $\mathcal{C}$ be a category, and let $c, c' \in \mathrm{Ob}(\mathcal{C})$ be objects represented by the functors $c, c' \colon \underline{1} \to \mathcal{C}$. Consider the cospan $\underline{1} \xrightarrow{\ c\ } \mathcal{C} \xleftarrow{\ c'\ } \underline{1}$. What is the comma category $(c \downarrow c')$? ◊

Solution 6.2.4.4.

Its objects are in bijection with $\mathrm{Hom}_{\mathcal{C}}(c, c')$. Since the only morphism in $\underline{1}$ is identity, $(c \downarrow c')$ is discrete. ◆

Exercise 6.2.4.5.

Let $\mathcal{C}$ and $\mathcal{D}$ be categories, and let $! \colon \mathcal{C} \to \underline{1}$ and $! \colon \mathcal{D} \to \underline{!}$ be the unique functors to the terminal category. What is the comma category for $\mathcal{C} \xrightarrow{\ !\ } \underline{1} \xleftarrow{\ !\ } \mathcal{D}$? ◊

Solution 6.2.4.5.

There is an isomorphism to the product category,

$$
(\mathcal{C} \downarrow \mathcal{D}) \xrightarrow{\cong} \mathcal{C} \times \mathcal{D}.
$$

◆

Exercise 6.2.4.6.

Let $\mathcal{C}$ be a category.

a. If $c \in \mathcal{C}$ is an initial object, what is the comma category for the cospan $\underline{1} \xrightarrow{\ c\ } \mathcal{C} \xleftarrow{\ \mathrm{id}_{\mathcal{C}}\ } \mathcal{C}$?

b. If $d \in \mathcal{C}$ is a terminal object, what is the comma category for the cospan $\mathcal{C} \xrightarrow{\ \mathrm{id}_{\mathcal{C}}\ } \mathcal{C} \xleftarrow{\ d\ } \mathcal{C}$?

◊

Solution 6.2.4.6.

a. If $c \in \mathrm{Ob}(\mathcal{C})$ is initial, it is easy to check that there is an isomorphism $(c \downarrow \mathcal{C}) \cong \mathcal{C}$.

b. If $d \in \mathrm{Ob}(\mathcal{C})$ is terminal, it is easy to check that there is an isomorphism $(\mathcal{C} \downarrow d) \cong \mathcal{C}$.

$\blacklozenge$

6.2.5   Arithmetic of categories

Section 3.4.3 summarized some of the properties of products, coproducts, and exponentials for sets, showing that they lined up precisely with familiar arithmetic properties of natural numbers. We can do the same for categories.

In the following proposition, we denote the coproduct of two categories $\mathcal{A}$ and $\mathcal{B}$ by the notation $\mathcal{A} + \mathcal{B}$ rather than $\mathcal{A} \sqcup \mathcal{B}$. We also denote the functor category $\mathrm{Fun}(\mathcal{A}, \mathcal{B})$ by $\mathcal{B}^{\mathcal{A}}$. Finally, we use $\underline{0}$ and $\underline{1}$ to refer to the discrete category on 0 objects and on 1 object respectively.

Proposition 6.2.5.1. *The following isomorphisms exist for any small categories* $\mathcal{A}, \mathcal{B}$, *and* $\mathcal{C}$.

- $\mathcal{A} + \underline{0} \cong \mathcal{A}$.

- $\mathcal{A} + \mathcal{B} \cong \mathcal{B} + \mathcal{A}$.

- $(\mathcal{A} + \mathcal{B}) + \mathcal{C} \cong \mathcal{A} + (\mathcal{B} + \mathcal{C})$.

- $\mathcal{A} \times \underline{0} \cong \underline{0}$.

- $\mathcal{A} \times \underline{1} \cong \mathcal{A}$.

- $\mathcal{A} \times \mathcal{B} \cong \mathcal{B} \times \mathcal{A}$.

- $(\mathcal{A} \times \mathcal{B}) \times \mathcal{C} \cong \mathcal{A} \times (\mathcal{B} \times \mathcal{C})$.

- $\mathcal{A} \times (\mathcal{B} + \mathcal{C}) \cong (\mathcal{A} \times \mathcal{B}) + (\mathcal{A} \times \mathcal{C})$.

- $\mathcal{A}^{\underline{0}} \cong \underline{1}$.

- $\mathcal{A}^{\underline{1}} \cong \mathcal{A}$.

- $\underline{0}^{\mathcal{A}} \cong \underline{0}, \quad$ *if* $\mathcal{A} \neq \underline{0}$.

- $\underline{1}^{\mathcal{A}} \cong \underline{1}$.

- $\mathcal{A}^{\mathcal{B}+\mathcal{C}} \cong \mathcal{A}^{\mathcal{B}} \times \mathcal{A}^{\mathcal{C}}$.

- $(\mathcal{A}^{\mathcal{B}})^{\mathcal{C}} \cong \mathcal{A}^{\mathcal{B} \times \mathcal{C}}$.

- $(\mathcal{A} \times \mathcal{B})^{\mathcal{C}} \cong \mathcal{A}^{\mathcal{C}} \times \mathcal{B}^{\mathcal{C}}$.

Proof. These are standard results; see Mac Lane [29].

$\square$

Chapter 7

Categories at Work

The reader should now have an understanding of the basic notions of category theory: categories, functors, natural transformations, and universal properties. As well, we have discussed many sources of examples: orders, graphs, monoids, and databases. This chapter begins with the notion of *adjoint functors* (also known as *adjunctions*), which are like dictionaries translating back and forth between different categories.

7.1 Adjoint functors

How far can we take this dictionary analogy?

In the common understanding of dictionaries, we assume that two languages (say, French and English) are equally expressive and that a good dictionary will assist in an even exchange of ideas. But in category theory we often have two categories that are not on the same conceptual level. This is most clear in the case of *free-forgetful adjunctions*. Section 7.1.1 explores the sense in which each adjunction provides a dictionary between two categories that are not necessarily on an equal footing, so to speak.

7.1.1 Discussion and definition

Consider the category of monoids and the category of sets. A monoid $(M, e, \star)$ is a set with a unit element and a multiplication formula that is associative. A set is just a set. A dictionary between **Mon** and **Set** should not be required to set up an even exchange but rather an exchange that is appropriate to the structures at hand. It will be in the form of two functors, denoted $L\colon \mathbf{Set} \to \mathbf{Mon}$ and $R\colon \mathbf{Mon} \to \mathbf{Set}$. So we can translate back and forth, but to say what kind of exchange is appropriate will require more work.

An extended analogy will introduce the subject. A one-year-old can make repeatable noises, and an adult can make repeatable noises. One might say, "After all, talking is nothing but making repeatable noises." But the adult's repeatable noises are called words, they form sentences, and those sentences can cause nuclear wars. There is something more in adult language than simply repeatable sounds. In the same vein, a game of tennis can be viewed in terms of physics, the movement of trillions of atoms, but in so doing one won't see the game aspect. So we have here something analogous to two categories here: {repeated noises} and {meaningful words}. We are looking for adjoint functors to serve as the appropriate sort of dictionary.

To translate baby talk into adult language we would make every repeated noise a kind of word, thereby granting it meaning. We do not know what a given repeated noise should mean, but we give it a slot in our conceptual space while always pondering, "I wonder what she means by Koh...." On the other hand, to translate from meaningful words to repeatable noises is easy. We just hear the word as a repeated noise, which is how the baby probably hears it.

Adjoint functors often come in the form of "free" and "forgetful." Here we freely add Koh to our conceptual space without having any idea how it adheres to the rest of the child's noises or feelings. But it does not act like a sound to us, it acts like a word; we do not know what it means, but we figure it means something. Conversely, the translation going the other way is "forgetful," forgetting the meaning of the words and just hearing them as sounds. The baby hears our words and accepts them as mere sounds, not knowing that there is anything extra to get.

Sets are like the babies in the story: they are simple objects full of unconnected dots. Monoids are like the adults, forming words and performing actions. In the monoid each element means something and combines with other elements in certain ways. There are many different sets and many different monoids, just as there are many babies and many adults, but there are differences in how they interact, so we put them in different categories.

Applying free functor $L\colon \mathbf{Set} \to \mathbf{Mon}$ to a set X makes every element $x \in X$ a word, and these words can be strung together to form more complex words. (Section 4.1.1.12 discussed the free monoid functor L.) Since a set such as X carries no information about the meaning or structure of its various elements, the free monoid $F(X)$ does not relate different words in any way. To apply the forgetful functor $R\colon \mathbf{Mon} \to \mathbf{Set}$ to a monoid, even a structured one, is to simply forget that its elements are anything but mere elements of a set. It sends a monoid $(M, e, \star)$ to the set M.

Definition 7.1.1.1. Let $\mathcal{B}$ and $\mathcal{A}$ be categories.[1] An *adjunction between $\mathcal{B}$ and $\mathcal{A}$* is a pair of functors

$$L\colon \mathcal{B} \to \mathcal{A} \qquad \text{and} \qquad R\colon \mathcal{A} \to \mathcal{B}$$

together with a natural isomorphism[2] whose component for any objects $A \in \mathrm{Ob}(\mathcal{A})$ and $B \in \mathrm{Ob}(\mathcal{B})$ is

$$\alpha_{B,A}\colon \mathrm{Hom}_{\mathcal{A}}(L(B), A) \overset{\cong}{\longrightarrow} \mathrm{Hom}_{\mathcal{B}}(B, R(A)). \tag{7.1}$$

This isomorphism is called the *adjunction isomorphism* for the (L, R) adjunction, and for any morphism $f\colon L(B) \to A$ in $\mathcal{A}$, we refer to $\alpha_{B,A}(f)\colon B \to R(A)$ as *the adjunct of f*.[3]

The functor L is called the *left adjoint* and the functor R is called the *right adjoint*. We may say that *L is the left adjoint of R* or that *R is the right adjoint of L*.[4] We often denote this setup

$$L\colon \mathcal{B} \rightleftarrows \mathcal{A} \colon R \tag{7.2}$$

Proposition 7.1.1.2. *Let $L\colon \mathbf{Set} \to \mathbf{Mon}$ be the functor sending $X \in \mathrm{Ob}(\mathbf{Set})$ to the free monoid $L(X) := (\mathrm{List}(X), [\,], \text{++})$, as in Definition 4.1.1.15. Let $R\colon \mathbf{Mon} \to \mathbf{Set}$ be the functor sending each monoid $\mathcal{M} := (M, e, \star)$ to its underlying set $R(\mathcal{M}) := M$. Then L is left adjoint to R.*

Proof. This is precisely the content of Proposition 4.1.4.9.

$\square$

Example 7.1.1.3. We need to ground the discussion in some concrete mathematics. In Proposition 7.1.1.2 we provided an adjunction between sets and monoids. A set X gets transformed into a monoid by considering lists in X; a monoid $\mathcal{M}$ gets transformed into a set by forgetting the multiplication law. So we have a functor for translating each way,

$$L\colon \mathbf{Set} \to \mathbf{Mon}, \qquad\qquad R\colon \mathbf{Mon} \to \mathbf{Set},$$

but an adjunction is more than that: it includes a guarantee about the relationship between these two functors. What is the relationship between L and R? Consider an arbitrary monoid $\mathcal{M} = (M, e, \star)$.

---

[1]Throughout this definition, notice that B's come before A's, especially in (7.1), which might be confusing. It was a stylistic choice to match with the **B**abies and **A**dults discussion.

[2]The natural isomorphism α (see Proposition 5.3.2.12) is between two functors $\mathcal{B}^{\mathrm{op}} \times \mathcal{A} \to \mathbf{Set}$, namely, the functor $(B, A) \mapsto \mathrm{Hom}_{\mathcal{A}}(L(B), A)$ and the functor $(B, A) \mapsto \mathrm{Hom}_{\mathcal{B}}(B, R(A))$.

[3]Conversely, for any $g\colon B \to R(A)$ in $\mathcal{B}$, we refer to $\alpha_{B,A}^{-1}(g)\colon L(B) \to A$ as *the adjunct of g*.

[4]The left adjoint does not have to be called L, nor does the right adjoint have to be called R, of course.

If we want to pick out three elements of the set M, that is the same thing as giving a function $\{a, b, c\} \to M$. But that function exists in the category of sets; in fact it is an element of $\mathrm{Hom}_{\mathbf{Set}}(\{a, b, c\}, M)$. But since $M = R(\mathcal{M})$ is the underlying set of the monoid, we can view the current paragraph in the light of adjunction (7.1) by saying the set

$$\mathrm{Hom}_{\mathbf{Set}}(\{a, b, c\}, R(\mathcal{M})).$$

classifies all the ways to choose three elements out of the underlying set of monoid $\mathcal{M}$. It was constructed completely from within the context of sets and functions.

Now, what does (7.1) mean? The equation

$$\mathrm{Hom}_{\mathbf{Mon}}(L(\{a, b, c\}), \mathcal{M}) \cong \mathrm{Hom}_{\mathbf{Set}}(\{a, b, c\}, R(\mathcal{M}))$$

tells us that somehow we can classify all the ways to choose three elements from M, while staying in the context of monoids and monoid homomorphisms. In fact, it tells us how to do so, namely, as $\mathrm{Hom}_{\mathbf{Mon}}(\mathrm{List}(\{1, 2, 3\}), \mathcal{M})$. Exercise 7.1.1.4 looks at that. The answer can be extracted from the proof of Proposition 4.1.4.9.

Exercise 7.1.1.4.

Let $X = \{a, b, c\}$, and let $\mathcal{M} = (\mathbb{N}, 1, *)$ be the multiplicative monoid of natural numbers (see Example 4.1.3.2). Let $g \colon X \to \mathbb{N}$ be the function given by $g(a) = 7, g(b) = 2, g(c) = 2$, and let $\beta_{X, \mathcal{M}} \colon \mathrm{Hom}_{\mathbf{Set}}(X, R(\mathcal{M})) \to \mathrm{Hom}_{\mathbf{Mon}}(L(X), \mathcal{M})$ be as in the proof of Proposition 4.1.4.9.

Consider the list $[b, b, a, c] \in L(X)$. What is $\beta_{X, \mathcal{M}}(g)([b, b, a, c])$? ◊

Solution 7.1.1.4.

By definition, we have

$$\beta_{X, \mathcal{M}}(g)([b, b, a, c]) := g(b) * g(b) * g(a) * g(c) = 2 * 2 * 7 * 2 = 56.$$

♦

Let us look once more at the adjunction between adults and babies. Using the notation of Definition 7.1.1.1, $\mathcal{A}$ is the adult category of meaningful words, and $\mathcal{B}$ is the baby category of repeated noises. The left adjoint turns every repeated sound into a meaningful word (having free meaning), and the right adjoint forgets the meaning of any word and considers it merely as a sound.

At the risk of taking this simple analogy too far, let's look at the heart of the issue: how to conceive of the isomorphism (7.1) of hom-sets. Once we have freely given a slot to each of the baby's repeated sounds, we try to find a mapping from the lexicon $L(B)$ of these new words to the adult lexicon A of meaningful words; these are mappings in the adult category $\mathcal{A}$ of the form $L(B) \to A$. And (stretching it) the baby tries to find a mapping (which we might see as emulation) from her set B of repeatable sounds to the set $R(A)$ of the sounds the adult seems to repeat. If there were a global system for making these transformations, that would establish (7.1) and hence the adjunction.

Note that the directionality of the adjunction makes a difference. If $L\colon \mathcal{B} \to \mathcal{A}$ is left adjoint to $R\colon \mathcal{A} \to \mathcal{B}$, there is no reason to think that L is also a right adjoint. In the case of babies and adults, we see that it would make little sense to look for a mapping in the category of meaningful words from the adult lexicon to the wordifications of baby sounds $\mathrm{Hom}_{\mathcal{A}}(A, L(B))$, because there is unlikely to be a good candidate for most of the words. That is, to which of the child's repeated noises would we assign the concept "weekday"?

Again, this is simply an analogy and should not be taken to seriously. The next example shows mathematically that the directionality of an adjunction is not arbitrary.

Example 7.1.1.5. Let $L\colon \mathbf{Set} \to \mathbf{Mon}$ and $R\colon \mathbf{Mon} \to \mathbf{Set}$ be the free and forgetful functors from Proposition 7.1.1.2. We know that L is left adjoint to R; however L is *not* right adjoint to R. In other words, we can show that the necessary natural isomorphism cannot exist.

Let $X = \{a, b\}$, and let $\mathcal{M} = \underline{1}$ be the trivial monoid. Then the necessary natural isomorphism would need to give a bijection

$$\mathrm{Hom}_{\mathbf{Mon}}(\mathcal{M}, L(X)) \cong^{?} \mathrm{Hom}_{\mathbf{Set}}(\{1\}, X).$$

But the left-hand side has one element, because $\mathcal{M}$ is the initial object in $\mathbf{Mon}$ (see Example 6.1.3.7), whereas the right-hand side has two elements. Therefore, no isomorphism can exist.

Example 7.1.1.6. Preorders have underlying sets, giving rise to a functor $U\colon \mathbf{PrO} \to \mathbf{Set}$. The functor U has both a left adjoint and a right adjoint. The left adjoint of U is $D\colon \mathbf{Set} \to \mathbf{PrO}$, sending a set X to the discrete preorder on X (the preorder with underlying set X, having the fewest possible $\leqslant$'s). The right adjoint of U is $I\colon \mathbf{Set} \to \mathbf{PrO}$, sending a set X to the indiscrete preorder on X (the preorder with underlying set X, having the most possible $\leqslant$'s). See Example 4.4.4.5.

Exercise 7.1.1.7.

Let $U\colon \mathbf{Grph} \to \mathbf{Set}$ denote the functor sending a graph to its underlying set of vertices. This functor has both a left and a right adjoint.

a. What functor $\mathbf{Set} \to \mathbf{Grph}$ is the left adjoint of U?

b. What functor $\mathbf{Set} \to \mathbf{Grph}$ is the right adjoint of U?

$\diamond$

Solution 7.1.1.7.

a. The discrete graph functor $Disc$: **Set** → **Grph** is left adjoint to U.

b. The indiscrete graph (or complete graph) functor Ind: **Set** → **Grph** (see Exercise 4.3.1.6) is right adjoint to U.

♦

Example 7.1.1.8. Here are some other adjunctions:

- Ob: **Cat** → **Set** has a left adjoint $Disc$: **Set** → **Cat** given by the discrete category.

- Ob: **Cat** → **Set** has a right adjoint Ind: **Set** → **Cat** given by the indiscrete category.

- The underlying graph functor **Cat** → **Grph** has a left adjoint **Grph** → **Cat** given by the free category.

- The inclusion **Grp** → **Mon** has a right adjoint **Mon** $\xrightarrow{\text{Core}}$ **Grp**, called the *core*, that sends a monoid to its subgroup of invertible elements.

- The functor **PrO** → **Grph**, given by drawing edges for ⩽'s, has a left adjoint given by existence of paths.

- The forgetful functor from partial orders to preorders has a left adjoint given by quotienting out the cliques (see Exercise 4.4.1.15).

- Given a set A, the functor $(- \times A)$: **Set** → **Set** has a right adjoint $\mathrm{Hom}(A, -)$ (this was called currying in Section 3.4.2).

Exercise 7.1.1.9.

Let $\underline{1}$ denote the terminal category. There is a unique functor !: **Set** → $\underline{1}$.

a. Does ! have a left adjoint? If so, what is it; if not, why not?

b. Does ! have a right adjoint? If so, what is it; if not, why not?

◊

Solution 7.1.1.9.

a. Yes. To give a functor $U: \underline{1} \to \mathbf{Set}$ is to give a set U, but which one? For any set $X \in \mathrm{Ob}(\mathbf{Set})$, we need a bijection $\mathrm{Hom}_{\underline{1}}(1,1) \cong \mathrm{Hom}_{\mathbf{Set}}(U, X)$. In other words, $U = \varnothing$ is the only choice.

b. Yes. We need to give a set $U: \underline{1} \to \mathbf{Set}$ such that for any set $X \in \mathrm{Ob}(\mathbf{Set})$, we have a bijection $\mathrm{Hom}_{\underline{1}}(1,1) \cong \mathrm{Hom}_{\mathbf{Set}}(X, U)$. In other words, any singleton set, such as $U = \{\odot\}$, will work.

♦

Exercise 7.1.1.10.

The discrete category functor $Disc: \mathbf{Set} \to \mathbf{Cat}$ has a left adjoint $p: \mathbf{Cat} \to \mathbf{Set}$. In this exercise you will work out how to unpack this idea and begin to deduce how p must behave.

a. For an arbitrary object $X \in \mathrm{Ob}(\mathbf{Set})$ and an arbitrary object $\mathcal{C} \in \mathrm{Ob}(\mathbf{Cat})$, write the adjunction in the style of (7.2), appropriately filling in all the variables (e.g., decide whether $\mathcal{B} = \mathbf{Cat}$ or $\mathcal{B} = \mathbf{Set}$, etc.).

b. For X and $\mathcal{C}$ as in part (a), write the adjunction isomorphism in the style of (7.1), appropriately filling in all the variables.

c. Let $\mathcal{C}$ be the free category on the graph G

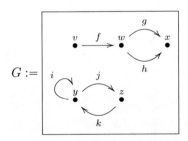

and let $X = \{1, 2, 3\}$. How many elements does the set $\mathrm{Hom}_{\mathbf{Cat}}(\mathcal{C}, Disc(X))$ have?

d. What can you do to an arbitrary category $\mathcal{C} \in \mathrm{Ob}(\mathbf{Cat})$ to make a set $p(\mathcal{C})$ such that the adjunction isomorphism holds? That is, how does the functor $p: \mathbf{Cat} \to \mathbf{Set}$ behave on objects?

◊

Solution 7.1.1.10.

a. The adjunction looks like this:

$$p\colon \mathbf{Cat} \rightleftarrows \mathbf{Set} : Disc\ .$$

b. The adjunction isomorphism for $\mathcal{C}$ and X looks like this:

$$\alpha_{\mathcal{C},X} \colon \mathrm{Hom}_{\mathbf{Set}}(p(\mathcal{C}), X) \overset{\cong}{\longrightarrow} \mathrm{Hom}_{\mathbf{Cat}}(\mathcal{C}, Disc(X)).$$

c. Suppose $F\colon \mathcal{C} \to Disc(X)$ is a functor. The only morphisms in $Disc(X)$ are identities, so F must send every arrow in $\mathcal{C}$ to an identity, meaning that v, w, x must all be sent to the same element of X, i.e., $F(v) = F(w) = F(x)$, and similarly $F(y) = F(z)$. But this is the only criterion for F. In other words, there are $\mathrm{Hom}_{\mathbf{Cat}}(\mathcal{C}, Disc(X)) \cong \{1, 2, 3\}^{\{v,y\}} \cong \underline{9}$ choices for F.

d. Given a category $\mathcal{C}$, we find that for any functor $F\colon \mathcal{C} \to Disc(X)$, if two objects $c, c' \in \mathrm{Ob}(\mathcal{C})$ are connected by a morphism $c \to c'$ or $c' \to c$, then $F(c) = F(c')$. Let $K = \mathrm{Ob}(\mathcal{C})/\sim$ be the quotient by the equivalence relation generated by declaring $c \sim c'$ if there is a morphism between them. To give such a functor is essentially the same as giving a function $K \to X$.

   To make this precise, recall that we have sets and functions

$$\mathrm{Hom}_{\mathcal{C}} \overset{dom}{\underset{cod}{\rightrightarrows}} \mathrm{Ob}(\mathcal{C})\ ,$$

   constituting the underlying graph of $\mathcal{C}$, and taking the coequalizer, we get the set $p(\mathcal{C})$ of "islands in $\mathcal{C}$." In the case of the category $\mathcal{C}$ in part (c), there were two islands: one containing v and one containing y.

                                                                                            ♦

   The following proposition says that all adjoints to a given functor are isomorphic to each other.

Proposition 7.1.1.11. *Let $\mathcal{C}$ and $\mathcal{D}$ be categories, let $F\colon \mathcal{C} \to \mathcal{D}$ be a functor, and let $G, G'\colon \mathcal{D} \to \mathcal{C}$ also be functors. If both G and G' are right adjoint (resp. if both are left adjoint) to F, then there is a natural isomorphism $\phi\colon G \overset{\cong}{\to} G'$.*

Proof. Suppose that both G and G' are right adjoint to F (the case of G and G' being left adjoint is similarly proved). We first give a formula for the components of $\phi\colon G \to G'$

and its inverse $\psi\colon G' \to G$. Given an object $d \in \mathrm{Ob}(\mathcal{D})$, we use $c = G(d)$ to obtain two natural isomorphisms, one from each adjunction:

$$\mathrm{Hom}_{\mathcal{C}}(G(d), G(d)) \cong \mathrm{Hom}_{\mathcal{D}}(F(G(d)), d) \cong \mathrm{Hom}_{\mathcal{C}}(G(d), G'(d)).$$

The identity morphism $\mathrm{id}_{G(d)}$ is then sent to some morphism $G(d) \to G'(d)$, which we take to be the component ϕ_d. Similarly, we use $c' = G'(d)$ to obtain two natural isomorphisms, one from each adjunction:

$$\mathrm{Hom}_{\mathcal{C}}(G'(d), G'(d)) \cong \mathrm{Hom}_{\mathcal{D}}(F(G'(d)), d) \cong \mathrm{Hom}_{\mathcal{C}}(G'(d), G(d)).$$

Again, the identity element $\mathrm{id}_{G'(d)}$ is sent to some morphism $G'(d) \to G(d)$, which we take to be the d-component ψ_d. The naturality of the adjunction isomorphisms implies that ϕ and ψ are natural transformations, and it is straightforward to check that they are mutually inverse.

$\square$

7.1.1.12 Quantifiers as adjoints

One of the simplest places where adjoints show up is between preimages and the logical quantifiers $\exists$ and $\forall$, ideas first discussed in Notation 2.1.1.1. The setting in which to discuss this is that of sets and their power preorders. That is, if X is a set, then recall from Section 4.4.2 that the power-set $\mathbb{P}(X)$ has a natural ordering by inclusion of subsets.

Given a function $f\colon X \to Y$ and a subset $V \subseteq Y$ the preimage is $f^{-1}(V) := \{x \in X \mid f(x) \in V\}$. If $V' \subseteq V$, then $f^{-1}(V') \subseteq f^{-1}(V)$, so in fact $f^{-1}\colon \mathbb{P}(Y) \to \mathbb{P}(X)$ can be considered a functor (where of course we are thinking of preorders as categories). The quantifiers $\exists$ and $\forall$ appear as adjoints of f^{-1}.

Let's begin with the left adjoint of $f^{-1}\colon \mathbb{P}(Y) \to \mathbb{P}(X)$. It is a functor $L_f\colon \mathbb{P}(X) \to \mathbb{P}(Y)$. Choose an object $U \subseteq X$ in $\mathbb{P}(X)$. It turns out that

$$L_f(U) = \{y \in Y \mid \exists x \in f^{-1}(y) \text{ such that } x \in U\}.$$

And the right adjoint $R_f\colon \mathbb{P}(X) \to \mathbb{P}(Y)$, when applied to U, is

$$R_f(U) = \{y \in Y \mid \forall x \in f^{-1}(y), x \in U\}.$$

In fact, the functor L_f is generally denoted $\exists_f\colon \mathbb{P}(X) \to \mathbb{P}(Y)$, and R_f is generally denoted $\forall_f\colon \mathbb{P}(X) \to \mathbb{P}(Y)$.

$$\mathbb{P}(X) \underset{\forall_f}{\overset{\exists_f}{\underset{\longleftarrow}{\rightrightarrows}}} \mathbb{P}(Y).$$

The next example shows why this notation is apt.

Example 7.1.1.13. In logic or computer science the quantifiers $\exists$ and $\forall$ are used to ask whether any or all elements of a set have a certain property. For example, one may have a set U of natural numbers and want to know whether any or all are even or odd. Let $Y = \{\mathsf{even}, \mathsf{odd}\}$, and let

$$p\colon \mathbb{N} \to Y$$

be the function that assigns to each natural number its parity (even or odd). Because the elements of $\mathbb{P}(\mathbb{N})$ and $\mathbb{P}(Y)$ are ordered by inclusion of subsets, we can construe these orders as categories (by Proposition 5.2.1.13). What is new is that we have adjunctions between these categories:

$$\mathbb{P}(\mathbb{N}) \underset{\forall_p}{\overset{\exists_p}{\rightleftharpoons}} \overset{p^{-1}}{\longleftarrow} \mathbb{P}(Y).$$

Given a subset $U \subseteq \mathbb{N}$, i.e., an object $U \in \mathrm{Ob}(\mathbb{P}(\mathbb{N}))$, we investigate the objects $\exists_p(U), \forall_p(U)$. These are both subsets of $\{\mathsf{even}, \mathsf{odd}\}$. The set $\exists_p(U)$ includes the element even if there exists an even number in U; it includes the element odd if there exists an odd number in U. Similarly, the set $\forall_p(U)$ includes the element even if every even number is in U, and it includes odd if every odd number is in U.

Let's use the definition of adjunction to ask whether every element of $U \subseteq \mathbb{N}$ is even. Let $V = \{\mathsf{even}\} \subseteq Y$. Then $f^{-1}(V) \subseteq \mathbb{N}$ is the set of even numbers, and there is a morphism $U \to f^{-1}(V)$ in the preorder $\mathbb{P}(\mathbb{N})$ if and only if every element of U is even. Therefore, the adjunction isomorphism $\mathrm{Hom}_{\mathbb{P}(\mathbb{N})}(U, f^{-1}(V)) \cong \mathrm{Hom}_{\mathbb{P}(Y)}(\exists_p U, V)$ says that $\exists_p U \subseteq \{\mathsf{even}\}$ if and only if every element of U is even.

Exercise 7.1.1.14.

The national scout jamboree is a gathering of Boy Scouts from troops across the United States. Let S be the set of Boy Scouts in the U.S., and let T be the set of Boy Scout troops in the U.S. Let $t\colon S \to T$ be the function that assigns to each Boy Scout his troop. Let $U \subseteq S$ be the set of Boy Scouts in attendance at this year's jamboree.

a. What is the meaning of the object $\exists_t U$

b. What is the meaning of the object $\forall_t U$?

$\diamond$

Solution 7.1.1.14.

a. The object $\exists_t U \in \mathbb{P}(T)$ is a set of troops; which one? It is the set of troops being represented at the jamboree, i.e., the set of troops x having at least one scout in attendance. "Welcome to the jamboree, troop x."

b. The object $\forall_t U \in \mathbb{P}(T)$ is a set of troops; which one? It is the set of troops x for which every member is in attendance at the jamboree. "Way to go, troop x!"

♦

Exercise 7.1.1.15.

Let X be an arbitrary set and $U \subseteq X$ a subset.

a. Find a set Y and a function $f\colon X \to Y$ such that $\exists_f U$ tells you whether U is nonempty.

b. What is the meaning of $\forall_f U$ for your choice of Y and f?

◊

Solution 7.1.1.15.

a. If this is going to work for any X, we should probably try to find a universal choice for Y and f. Let's take $Y = \{\odot\}$, so there is a unique function $!\colon X \to \{\odot\}$. We find that $\varnothing = \exists_! U$ means "it is not true that something exists in U," and $\{\odot\} = \exists_! U$ means "something exists in U."

b. We find that $\varnothing = \forall_! U$ means "it is not true that everything is in U," and $\{\odot\} = \forall_! U$ means "everything is in U." In other words, $\forall_!$ tells us whether a subset $U \subseteq X$ is actually equal to X or not.

♦

In fact, the idea of quantifiers as adjoints is part of a larger story. Suppose we think of elements of a set X as bins, or storage areas. An element of $\mathbb{P}(X)$ can be construed as an injection $U \hookrightarrow X$, i.e., an assignment of a bin to each element of U, with at most one element of U in each bin. Relaxing the injectivity restriction, we may consider arbitrary sets U and assignments $U \to X$ of a bin to each element $u \in U$. Given a function $f\colon X \to Y$, we can generalize $\exists_f$ and $\forall_f$ to functors denoted Σ_f and Π_f, which will parameterize disjoint unions and products (respectively) over $y \in Y$. This is discussed in Section 7.1.4.

7.1.2 Universal concepts in terms of adjoints

This section explores how universal concepts, i.e., initial objects and terminal objects, colimits and limits, are easily phrased in the language of adjoint functors. We say that a functor $F\colon \mathcal{C} \to \mathcal{D}$ *is a left adjoint* or *has a right adjoint* if there exists a functor $G\colon \mathcal{D} \to \mathcal{C}$ such that F is a left adjoint of G. Proposition 7.1.1.11 showed that if F is a left adjoint of some functor G, then it is isomorphic to every other left adjoint of G, and G is isomorphic to every other right adjoint of F.

Example 7.1.2.1. Let C be a category and $t\colon C \to \underline{1}$ the unique functor to the terminal category. Then t has a right adjoint if and only if C has a terminal object, and t has a left adjoint if and only if C has an initial object. The proofs are dual, so let's focus on the first.

The functor t has a right adjoint $R\colon \underline{1} \to C$ if and only if for every object $c \in \mathrm{Ob}(C)$ there is an isomorphism

$$\mathrm{Hom}_{C}(c, r) \cong \mathrm{Hom}_{\underline{1}}(t(c), 1),$$

where $r = R(1)$. But $\mathrm{Hom}_{\underline{1}}(t(c), 1)$ has one element. Thus t has a right adjoint iff $\mathrm{Hom}_{C}(c, r)$ has one element for each $c \in \mathrm{Ob}(C)$. This is the definition of r being a terminal object.

When colimits and limits were defined in Definitions 6.1.3.31 and 6.1.3.20, it was for individual I-shaped diagrams $X\colon I \to C$. Using adjoints we can define the limit of every I-shaped diagram in C at once.

Let $t\colon I \to \underline{1}$ denote the unique functor to the terminal category. Suppose given an object $c \in \mathrm{Ob}(C)$, represented by the functor $c\colon \underline{1} \to C$. Then $c \circ t\colon I \to C$ is the *constant functor at* c, sending each object in I to the same C-object, c, and every morphism in I to id_{c}. Thus composing with t induces a functor $C \cong \mathrm{Fun}(\underline{1}, C) \to \mathrm{Fun}(I, C)$, denoted $\Delta_{t}\colon C \to \mathrm{Fun}(I, C)$. It sends each object c to the associated constant functor $c \circ t$.

Suppose we want to take the colimit or limit of X. We are given an object X of $\mathrm{Fun}(I, C)$, and we want back an object of C. We could hope, and it turns out to be true, that the adjoints of Δ_{t} are the limit and colimit. Indeed, let $\Sigma_{t}\colon \mathrm{Fun}(I, C) \to C$ denote the left adjoint of Δ_{t}, and let $\Pi_{t}\colon \mathrm{Fun}(I, C) \to C$ denote the right adjoint of Δ_{t}. Then Σ_{t} is the functor that takes colimits, and Π_{t} is the functor that takes limits.

A generalization of colimits and limits is given in Section 7.1.4. But for now, let's consider a concrete example.

Example 7.1.2.2. Let $C = \mathbf{Set}$, and let $I = \underline{3}$. The category $\mathrm{Fun}(\underline{3}, \mathbf{Set})$ is the category of $\{1, 2, 3\}$-indexed sets, e.g., $(\mathbb{Z}, \mathbb{N}, \mathbb{Z}) \in \mathrm{Ob}(\mathrm{Fun}(\underline{3}, \mathbf{Set}))$ is an object of it. We will obtain the limit, i.e., the product of these three sets $\underline{3} \to \mathbf{Set}$ using adjoints.

In fact, the limit will be right adjoint to a functor $\Delta_{t}\colon \mathbf{Set} \to \mathrm{Fun}(\underline{3}, \mathbf{Set})$, defined as follows. Given a set $c \in \mathrm{Ob}(\mathbf{Set})$, represented by a functor $c\colon \underline{1} \to \mathbf{Set}$, and define $\Delta_{t}(c)$ to be the composite $c \circ t\colon \underline{3} \to \mathbf{Set}$; it is the constant functor. That is, $\Delta_{t}(c)\colon \underline{3} \to \mathbf{Set}$ is the $\{1, 2, 3\}$-indexed set (c, c, c).

To say that Δ_{t} has a right adjoint called $\Pi_{t}\colon \mathrm{Fun}(\underline{3}, \mathbf{Set}) \to \mathbf{Set}$ and that Π_{t} takes limits should mean that the definition of right adjoint provides the formula that yields the appropriate limit. Fix a functor $D\colon \underline{3} \to \mathbf{Set}$, so $D(1), D(2)$, and $D(3)$ are sets. We know from Example 6.1.3.25 that the limit, $\lim D$, of D is supposed to be the product $D(1) \times D(2) \times D(3)$. For example, if $D = (\mathbb{Z}, \mathbb{N}, \mathbb{Z})$, then $\lim D = \mathbb{Z} \times \mathbb{N} \times \mathbb{Z}$. How does this fact arise in the definition of adjoint?

The definition of Π_t being the right adjoint to Δ_t says that for any $c \in \mathrm{Ob}(\mathbf{Set})$ and $D \in \mathrm{Fun}(\underline{3}, \mathbf{Set})$, there is a natural isomorphism of sets,

$$\alpha_{c,D} \colon \mathrm{Hom}_{\mathrm{Fun}(\underline{3},\mathbf{Set})}(\Delta_t(c), D) \cong \mathrm{Hom}_{\mathbf{Set}}(c, \Pi_t(D)). \tag{7.3}$$

The domain of $\alpha_{c,D}$ has elements $f \in \mathrm{Hom}_{\mathrm{Fun}(\underline{3},\mathbf{Set})}(\Delta_t(c), D)$ that look like the left-hand drawing, but having these three maps is equivalent to having the right-hand diagram:

The isomorphism $\alpha_{c,D}$ in (7.3) says that choosing the three functions $f(1), f(2), f(3)$ is the same thing as choosing a function $c \to \Pi_t(D)$. This is basically the universal property for limits: there is a unique function $\ell \colon c \to D(1) \times D(2) \times D(3)$, so this product is isomorphic to Π_t. I have not given a formal proof here but hopefully enough for the interested reader to work it out.

7.1.3 Preservation of colimits or limits

One useful fact about adjunctions is that left adjoints preserve all colimits, and right adjoints preserve all limits.

Proposition 7.1.3.1. *Let* $L \colon \mathcal{B} \rightleftarrows \mathcal{A} \colon R$ *be an adjunction. For any indexing category* I *and functor* $D \colon I \to \mathcal{B}$, *if* D *has a colimit in* $\mathcal{B}$, *then there is a unique isomorphism*

$$L(\mathrm{colim}\, D) \cong \mathrm{colim}(L \circ D).$$

Similarly, for any $I \in \mathrm{Ob}(\mathbf{Cat})$ *and functor* $D \colon I \to \mathcal{A}$, *if* D *has a limit in* $\mathcal{A}$, *then there is a unique isomorphism*

$$R(\lim D) \cong \lim(R \circ D).$$

Proof. The proof is simple if one knows the Yoneda lemma (Section 7.2.1.14). See Mac Lane [29] for details. □

Example 7.1.3.2. Since Ob: **Cat** → **Set** is both a left adjoint and a right adjoint, it must preserve both limits and colimits. This means that if one wants to know the set of objects in the fiber product of some categories, one can simply take the fiber product of the set of objects in those categories,

$$\mathrm{Ob}(\mathcal{A} \times_{\mathcal{C}} \mathcal{B}) \cong \mathrm{Ob}(\mathcal{A}) \times_{\mathrm{Ob}(\mathcal{C})} \mathrm{Ob}(\mathcal{B}).$$

While the right-hand side might look daunting, it is just a fiber product in **Set**, which is quite understandable (see Definition 3.2.1.1).

This is greatly simplifying. If one thinks through what defines a limit in **Cat**, one encounters notions of slice categories and terminal objects in them. These slice categories are in **Cat** so they involve several categories and functors, and it is difficult for a beginner. Knowing that the objects are given by a simple fiber product makes the search for limits in **Cat** much simpler.

For example, if $[n]$ is the linear order category of length n, then $[n] \times [m]$ has $(n + 1)(m + 1)$ objects because $[n]$ has $n + 1$ objects and $[m]$ has $m + 1$ objects.

Example 7.1.3.3. The path preorder functor $L\colon$ **Grph** → **PrO** given by existence of paths (see Exercise 5.1.2.13) is left adjoint to the functor $R\colon$ **PrO** → **Grph** given by replacing $\leqslant$'s by arrows. This means that L preserves colimits. So taking the union of graphs G and H results in a graph whose path poset $L(G \sqcup H)$ is the union of the path posets of G and H. But this is not so for products, i.e., we do not expect to have an isomorphism $L(G \times H) \cong^? L(G) \times L(H)$.

As an example, let $G = H = $. Then $L(G) = L(H) = [1]$, the linear order of length 1. But the product $G \times H$ in **Grph** looks like the graph

$$\begin{matrix} (a,a) & & (a,b) \\ \bullet & & \bullet \\ & \searrow & \\ (b,a) & & (b,b) \\ \bullet & & \bullet \end{matrix}$$

Its preorder $L(G \times H)$ does not have $(a, a) \leqslant (a, b)$, whereas this is the case in the preorder $L(G) \times L(H)$. So $L(G \times H) \not\cong L(G) \times L(H)$. The left adjoint preservers all colimits, but not necessarily limits.

7.1.4   Data migration

As we saw in Sections 5.2.2 and 5.2.2.6, a database schema is a category $\mathcal{C}$, and an instance is a functor $I\colon \mathcal{C} \to$ **Set**.

Notation 7.1.4.1. Let $\mathcal{C}$ be a category. The category $\mathrm{Fun}(\mathcal{C}, \mathbf{Set})$ of functors from $\mathcal{C}$ to **Set**, i.e., the category of instances on $\mathcal{C}$, is denoted $\mathcal{C}$–**Set**.

This section discusses what happens to the resulting instances when different schemas are connected by a functor, say, $F\colon \mathcal{C} \to \mathcal{D}$. It turns out that three adjoint functors emerge: $\Delta_F\colon \mathcal{D}$–**Set** $\to \mathcal{C}$–**Set**, $\Sigma_F\colon \mathcal{C}$–**Set** $\to \mathcal{D}$–**Set**, and $\Pi_F\colon \mathcal{C}$–**Set** $\to \mathcal{D}$–**Set**, where Δ_F is adjoint to both of them:

$$\Sigma_F\colon \mathcal{C}\text{–}\mathbf{Set} \rightleftarrows \mathcal{D}\text{–}\mathbf{Set} :\Delta_F \qquad\qquad \Delta_F\colon \mathcal{D}\text{–}\mathbf{Set} \rightleftarrows \mathcal{C}\text{–}\mathbf{Set} :\Pi_F.$$

Interestingly, many of the basic database operations are captured by these three functors. For example, Δ_F handles the job of duplicating or deleting tables as well as duplicating or deleting columns in a single table. The functor Σ_F handles taking unions, and the functor Π_F handles joining tables together, matching columns, or selecting the rows with certain properties (e.g., everyone whose first name is Mary).

This section is challenging, and it can be safely skipped, resuming at Section 7.2. For those who want to pursue it, there is an open source implementation of these ideas and more, called FQL,[5] which stands for *functorial query language* (not to be confused with Facebook query language).

7.1.4.2   Pullback: Δ

Given a functor $F\colon \mathcal{C} \to \mathcal{D}$ and a functor $I\colon \mathcal{D} \to \mathbf{Set}$, we can compose them to get a functor $I \circ F\colon \mathcal{C} \to \mathbf{Set}$. In other words, the presence of F provides a way to convert $\mathcal{D}$-instances into $\mathcal{C}$-instances. In fact, this conversion is functorial, meaning that a morphism of $\mathcal{D}$-instances $\alpha\colon I \to I'$ is sent to a morphism of $\mathcal{C}$-instances. This can be seen by whiskering (see Definition 5.3.2.16):

$$\mathcal{C} \xrightarrow{\ F\ } \mathcal{D} \overset{I}{\underset{I'}{\Downarrow \alpha}} \mathbf{Set}$$

We denote the resulting functor $\Delta_F\colon \mathcal{D}$–**Set** $\to \mathcal{C}$–**Set** and call it *pullback along F*.

An example of this was given in Example 5.3.2.15, which showed how a monoid homomorphism $F\colon \mathcal{M}' \to \mathcal{M}$ could add functionality to a finite state machine. More generally, we can use pullbacks to reorganize data, copying and deleting tables and columns.

Remark 7.1.4.3. Given a functor $F\colon \mathcal{C} \to \mathcal{D}$, which we think of as a schema translation, the functor $\Delta_F\colon \mathcal{D}$–**Set** $\to \mathcal{C}$–**Set** goes the opposite way. The reasoning is simple to

---

[5]FQL is available on the Internet. See http://categoricaldata.net/fql.html.

explain (we are composing functors) but something about it often seems strange at first. The rough idea of this contravariance is captured by the role-reversal in the following slogan:

Slogan 7.1.4.4.

> If I get my information from you, then your information becomes my information.

Consider the following functor $F: \mathcal{C} \to \mathcal{D}$: [6]

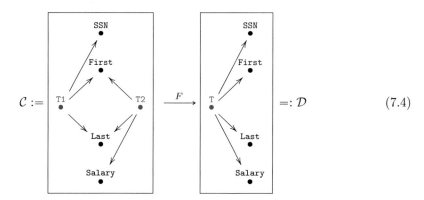

$$(7.4)$$

Recall how to read schemas. In schema $\mathcal{C}$ there are leaf tables SSN, First, Last, Salary, which represent different kinds of basic data. More interestingly, there are two *fact tables*. The first is called T1, and it relates SSN, First, and Last. The second is called T2, and it relates First, Last, and Salary.

The functor $F: \mathcal{C} \to \mathcal{D}$ relates $\mathcal{C}$ to a schema $\mathcal{D}$ which has a single fact table relating all four attributes: SSN, First, Last, and Salary. We are interested in $\Delta_F: \mathcal{D}\text{--}\mathbf{Set} \to \mathcal{C}\text{--}\mathbf{Set}$. Suppose given the following database instance $I: \mathcal{D} \to \mathbf{Set}$ on $\mathcal{D}$:

T				
ID	**SSN**	**First**	**Last**	**Salary**
XF667	115-234	Bob	Smith	$250
XF891	122-988	Sue	Smith	$300
XF221	198-877	Alice	Jones	$100

---

[6]This example was taken from Spivak [38].

SSN
ID
115-234
118-334
122-988
198-877
342-164

First
ID
Adam
Alice
Bob
Carl
Sam
Sue

Last
ID
Jones
Miller
Pratt
Richards
Smith

Salary
ID
$100
$150
$200
$250
$300

How does one get the instance $\Delta_F(I)\colon \mathcal{C} \to \mathbf{Set}$? The formula was given: compose I with F. In terms of tables, it is like duplicating table T as T1 and T2 but deleting a column from each in accordance with the definition of $\mathcal{C}$ in (7.4). Here is the result, $\Delta_F(I)$, in table form:

T1			
ID	**SSN**	**First**	**Last**
XF667	115-234	Bob	Smith
XF891	122-988	Sue	Smith
XF221	198-877	Alice	Jones

T2			
ID	**First**	**Last**	**Salary**
XF221	Alice	Jones	$100
XF667	Bob	Smith	$250
XF891	Sue	Smith	$300

SSN
ID
115-234
118-334
122-988
198-877
342-164

First
ID
Adam
Alice
Bob
Carl
Sam
Sue

Last
ID
Jones
Miller
Pratt
Richards
Smith

Salary
ID
$100
$150
$200
$250
$300

Exercise 7.1.4.5.

Consider the schemas

$$[1] = \boxed{\overset{0}{\bullet} \xrightarrow{\ f\ } \overset{1}{\bullet}} \quad \text{and} \quad [2] = \boxed{\overset{0}{\bullet} \xrightarrow{\ g\ } \overset{1}{\bullet} \xrightarrow{\ h\ } \overset{2}{\bullet}}$$

and the functor $F\colon [1] \to [2]$ given by sending $0 \mapsto 0$ and $1 \mapsto 2$.

a. How many possibilities are there for $F(f)$?

b. Suppose $I \colon [2] \to \mathbf{Set}$ is given by the following tables:

0	
ID	**g**
Am	To be verb
Baltimore	Place
Carla	Person
Develop	Action verb
Edward	Person
Foolish	Adjective
Green	Adjective

1	
ID	**h**
Action verb	Verb
Adjective	Adjective
Place	Noun
Person	Noun
To be verb	Verb

2
ID
Adjective
Noun
Verb

Write the two tables associated to the [1]-instance $\Delta_F(I) \colon [1] \to \mathbf{Set}$.

◊

Solution 7.1.4.5.

a. Only one possibility: $F(f) = h \circ g$.

b.

0	
ID	**f**
Am	Verb
Baltimore	Noun
Carla	Noun
Develop	Verb
Edward	Noun
Foolish	Adjective
Green	Adjective

♦

7.1.4.6　Left pushforward: Σ

Let $F \colon \mathcal{C} \to \mathcal{D}$ be a functor. The functor $\Delta_F \colon \mathcal{D}\text{–}\mathbf{Set} \to \mathcal{C}\text{–}\mathbf{Set}$ has a left adjoint, $\Sigma_F \colon \mathcal{C}\text{–}\mathbf{Set} \to \mathcal{D}\text{–}\mathbf{Set}$. The rough idea is that Σ_F performs parameterized colimits. Given an instance $I \colon \mathcal{C} \to \mathbf{Set}$, we get an instance on $\mathcal{D}$ that acts as follows. For each object $d \in \mathrm{Ob}(\mathcal{D})$, the set $\Sigma_F(I)(d)$ is the colimit (think of union) of some diagram in $\mathcal{C}$.

Left pushforwards (also known as left Kan extensions) are discussed at length in Spivak [38]; here we examine some examples from that paper.

Example 7.1.4.7. We again use the functor $F\colon \mathcal{C} \to \mathcal{D}$ from (7.4):

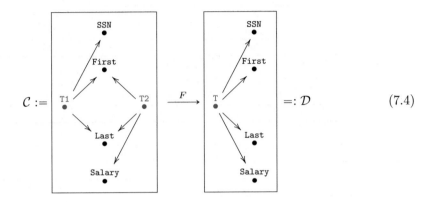

$$\mathcal{C} := \qquad \xrightarrow{\quad F \quad} \qquad =: \mathcal{D} \qquad\qquad (7.4)$$

We apply the left pushforward $\Sigma_F\colon \mathcal{C}\text{--}\mathbf{Set} \to \mathcal{D}\text{--}\mathbf{Set}$ to the following instance $I\colon \mathcal{C} \to \mathbf{Set}$:

T1			
ID	**SSN**	**First**	**Last**
T1-001	115-234	Bob	Smith
T1-002	122-988	Sue	Smith
T1-003	198-877	Alice	Jones

T2			
ID	**First**	**Last**	**Salary**
T2-001	Alice	Jones	$100
T2-002	Sam	Miller	$150
T2-004	Sue	Smith	$300
T2-010	Carl	Pratt	$200

SSN
ID
115-234
118-334
122-988
198-877
342-164

First
ID
Adam
Alice
Bob
Carl
Sam
Sue

Last
ID
Jones
Miller
Pratt
Richards
Smith

Salary
ID
$100
$150
$200
$250
$300

The functor $F\colon \mathcal{C} \to \mathcal{D}$ sends both tables T1 and T2 to table T. Applying Σ_F takes what was in T1 and T2 and puts the union in T. The result, $\Sigma_F I\colon \mathcal{D} \to \mathbf{Set}$, is as follows:

T				
ID	**SSN**	**First**	**Last**	**Salary**
T1-001	115-234	Bob	Smith	T1-001.Salary
T1-002	122-988	Sue	Smith	T1-002.Salary
T1-003	198-877	Alice	Jones	T1-003.Salary
T2-001	T2-A101.SSN	Alice	Jones	$100
T2-002	T2-A102.SSN	Sam	Miller	$150
T2-004	T2-004.SSN	Sue	Smith	$300
T2-010	T2-A110.SSN	Carl	Pratt	$200

SSN
ID
115-234
118-334
122-988
198-877
342-164
T2-001.SSN
T2-002.SSN
T2-004.SSN
T2-010.SSN

First
ID
Adam
Alice
Bob
Carl
Sam
Sue

Last
ID
Jones
Miller
Pratt
Richards
Smith

Salary
ID
$100
$150
$200
$250
$300
T1-001.Salary
T1-002.Salary
T1-003.Salary

As one can see, no set salary information for any data comes from table T1, nor does any set SSN information come form table T2. But the definition of adjoint, given in Definition 7.1.1.1, yields the universal response: freely add new variables that take the place of missing information. It turns out that this idea already has a name in logic, *Skolem variables*, and a name in database theory, *labeled nulls*.

Exercise 7.1.4.8.

Consider the functor $F: \underline{3} \to \underline{2}$ given by the sequence $(1, 2, 2)$.

a. Write an instance $I: \underline{3} \to \mathbf{Set}$.

b. Given the description "Σ_F performs a parameterized colimit," make an educated guess about what $\Sigma_F(I): \underline{2} \to \mathbf{Set}$ is. Give your answer in the form of two sets that are made up from the three sets you already wrote.

◊

Solution 7.1.4.8.

a.

1
ID
Science
Math
English
Gym
Social studies
Recess

2
ID
Peter
David
Robert
Bryan
Michelle
Sam
Samantha
Brendan

3
ID
Cool
Slick
Bad
Awesome
Sweet
Rockin
Sick

b.

1
ID
Science
Math
English
Gym
Social studies
Recess

2
ID
Peter
David
Robert
Bryan
Michelle
Sam
Samantha
Brendan
Cool
Slick
Bad
Awesome
Sweet
Rockin
Sick

♦

Here is the actual formula for computing left pushforwards. Suppose that $F\colon \mathcal{C} \to \mathcal{D}$ is a functor, and let $I\colon \mathcal{C} \to \mathbf{Set}$ be a set-valued functor on $\mathcal{C}$. Then $\Sigma_F(I)\colon \mathcal{D} \to \mathbf{Set}$ is defined as follows. Given an object $d \in \mathrm{Ob}(\mathcal{D})$, we first form the comma category (see Definition 6.2.4.1) for the cospan

$$\mathcal{C} \xrightarrow{F} \mathcal{D} \xleftarrow{d} \underline{1}$$

and denote it $(F \downarrow d)$. There is a canonical projection functor $\pi\colon (F \downarrow d) \to \mathcal{C}$, which we can compose with $I\colon \mathcal{C} \to \mathbf{Set}$ to obtain a functor $(F \downarrow d) \to \mathbf{Set}$. We are ready to define $\Sigma_F(I)(d)$ to be its colimit,

$$\Sigma_F(I)(d) := \operatorname*{colim}_{(F \downarrow d)} I \circ \pi.$$

$\Sigma_F(I)\colon \mathcal{D} \to \mathbf{Set}$ has been defined on objects $d \in \mathrm{Ob}(\mathcal{D})$. Morphisms are treated here only briefly; see Spivak [38] for details. Given a morphism $g\colon d \to d'$, there is an induced functor $(F \downarrow g)\colon (F \downarrow d) \to (F \downarrow d')$ and a commutative diagram of categories:

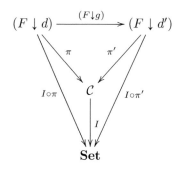

By the universal property for colimits, this induces the required function

$$\operatorname*{colim}_{(F\downarrow d)} I \circ \pi \xrightarrow{\ \Sigma_F(I)(g)\ } \operatorname*{colim}_{(F\downarrow d')} I \circ \pi'.$$

7.1.4.9   Right pushforward: Π

Let $F\colon \mathcal{C} \to \mathcal{D}$ be a functor. Section 7.1.4.6 explained that the functor $\Delta_F\colon \mathcal{D}\text{--}\mathbf{Set} \to \mathcal{C}\text{--}\mathbf{Set}$ has a left adjoint. The present section explains that Δ_F has a right adjoint, $\Pi_F\colon \mathcal{C}\text{--}\mathbf{Set} \to \mathcal{D}\text{--}\mathbf{Set}$ as well. The rough idea is that Π_F performs parameterized limits. Given an instance $I\colon \mathcal{C} \to \mathbf{Set}$, we get an instance on $\mathcal{D}$ that acts as follows. For each object $d \in \mathrm{Ob}(\mathcal{D})$, the set $\Pi_F(I)(d)$ is the limit (think of fiber product) of some diagram in $\mathcal{C}$.

Right pushforwards (also known as right Kan extensions) are discussed at length in Spivak [38]; here we look at some examples from that paper.

Example 7.1.4.10. We again use the functor $F \colon \mathcal{C} \to \mathcal{D}$ from (7.4) and Example 7.1.4.7. We apply the right pushforward Π_F to instance $I \colon \mathcal{C} \to \mathbf{Set}$ from that example.[7]

The instance $\Pi_F(I)$ puts data in all five tables in $\mathcal{D}$. In T it puts pairs (t_1, t_2), where t_1 is a row in T1, and t_2 is a row in T2, for which the first and last names agree. It copies the leaf tables exactly, so they are not displayed here; the following is the table T for $\Pi_F(I)$:

T				
ID	**SSN**	**First**	**Last**	**Salary**
T1-002T2-A104	122-988	Sue	Smith	$300
T1-003T2-A101	198-877	Alice	Jones	$100

From T1 and T2 there are only two ways to match first and last names.

---

[7]Repeated for convenience,

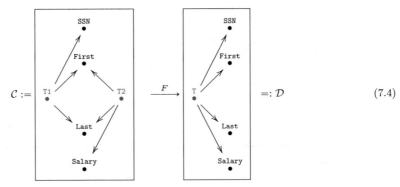

$$\mathcal{C} := \quad \xrightarrow{\ F\ } \quad =: \mathcal{D} \qquad (7.4)$$

$I \colon \mathcal{C} \to \mathbf{Set}$ is

T1			
ID	**SSN**	**First**	**Last**
T1-001	115-234	Bob	Smith
T1-002	122-988	Sue	Smith
T1-003	198-877	Alice	Jones

T2			
ID	**First**	**Last**	**Salary**
T2-001	Alice	Jones	$100
T2-002	Sam	Miller	$150
T2-004	Sue	Smith	$300
T2-010	Carl	Pratt	$200

SSN
ID
115-234
118-334
122-988
198-877
342-164

First
ID
Adam
Alice
Bob
Carl
Sam
Sue

Last
ID
Jones
Miller
Pratt
Richards
Smith

Salary
ID
$100
$150
$200
$250
$300

Exercise 7.1.4.11.

Consider the functor $F \colon \underline{3} \to \underline{2}$ given by the sequence $(1, 2, 2)$.

a. Write an instance $I \colon \underline{3} \to \textbf{Set}$.

b. Given the description "Π_F performs a parameterized limit," make an educated guess about what $\Pi_F(I) \colon \underline{2} \to \textbf{Set}$ is. Give your answer in the form of two sets that are made up from the three sets you already wrote down.

◇

Solution 7.1.4.11.

a.

1
ID
Science
Math
English
Gym
Social studies
Recess

2
ID
John
Deb
Liz

3
ID
Cool
Slick
Awesome
Sweet

b.

1
ID
Science
Math
English
Gym
Social studies
Recess

2
ID
(John,Cool)
(John,Slick)
(John,Awesome)
(John,Sweet)
(Deb,Cool)
(Deb,Slick)
(Deb,Awesome)
(Deb,Sweet)
(Liz,Cool)
(Liz,Slick)
(Liz,Awesome)
(Liz,Sweet)

♦

Here is the actual formula for computing right pushforwards. Suppose that $F\colon \mathcal{C} \to \mathcal{D}$ is a functor, and let $I\colon \mathcal{C} \to \mathbf{Set}$ be a set-valued functor on $\mathcal{C}$. Then $\Pi_F(I)\colon \mathcal{D} \to \mathbf{Set}$ is defined as follows. Given an object $d \in \mathrm{Ob}(\mathcal{D})$, we first form the comma category (see Definition 6.2.4.1) for the cospan

$$\underline{1} \xrightarrow{d} \mathcal{D} \xleftarrow{F} \mathcal{C}$$

and denote it $(d \downarrow F)$. There is a canonical projection functor $\pi\colon (d \downarrow F) \to \mathcal{C}$, which we can compose with $I\colon \mathcal{C} \to \mathbf{Set}$ to obtain a functor $(d \downarrow F) \to \mathbf{Set}$. We are ready to define $\Pi_F(I)(d)$ to be its limit,

$$\Pi_F(I)(d) := \lim_{(d \downarrow F)} I \circ \pi.$$

$\Pi_F(I)\colon \mathcal{D} \to \mathbf{Set}$ has been defined on objects $d \in \mathrm{Ob}(\mathcal{D})$, and morphisms are treated only briefly; see Spivak [38] for details. Given a morphism $g\colon d \to d'$, there is an induced functor $(g \downarrow F)\colon (d' \downarrow F) \to (d \downarrow F)$ and a commutative diagram of categories:

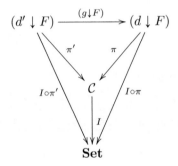

By the universal property for limits, this induces the required function

$$\lim_{(d \downarrow F)} I \circ \pi \xrightarrow{\Pi_F(I)(g)} \lim_{(d' \downarrow F)} I \circ \pi'.$$

Proposition 7.1.4.12. *Left adjoints are closed under composition, as are right adjoints. That is, given adjunctions,*

$$\mathcal{C} \underset{R}{\overset{L}{\rightleftarrows}} \mathcal{D} \underset{R'}{\overset{L'}{\rightleftarrows}} \mathcal{E}$$

their composite is also an adjunction:

$$\mathcal{C} \underset{R \circ R'}{\overset{L' \circ L}{\rightleftarrows}} \mathcal{E}.$$

Proof. This is a straightforward calculation. For any objects $c \in \mathrm{Ob}(\mathcal{C})$ and $e \in \mathrm{Ob}(\mathcal{E})$ we have adjunction isomorphisms:

$$\mathrm{Hom}_{\mathcal{E}}(L'(L(c)), e) \cong \mathrm{Hom}_{\mathcal{D}}(L(c), R'(e)) \cong \mathrm{Hom}_{\mathcal{C}}(c, R(R'(e)))$$

whose composite is the required adjunction isomorphism. It is natural in our choice of objects c and e.

$\square$

Example 7.1.4.13 (Currying via Δ, Σ, Π). This example shows how currying (as in Sections 3.4.2 and 7.1.1.8) arises out of a certain combination of data migration functors.

Let A, B, and C be sets. Consider the unique functor $a \colon A \to \underline{1}$ and consider B and C as functors $\underline{1} \xrightarrow{\ B\ } \mathbf{Set}$ and $\underline{1} \xrightarrow{\ C\ } \mathbf{Set}$ respectively.

$$A \xrightarrow{\ a\ } \underline{1} \underset{C}{\overset{B}{\rightrightarrows}} \mathbf{Set}$$

Note that $\underline{1}\text{–}\mathbf{Set} \cong \mathbf{Set}$, and we elide the difference.

We know that Σ_a is left adjoint to Δ_a and that Δ_a is left adjoint to Π_a, so by Proposition 7.1.4.12, the composite $\Sigma_a \circ \Delta_a$ is left adjoint to $\Pi_a \Delta_a$. The goal is to see currying arise out of the adjunction isomorphism

$$\mathrm{Hom}_{\mathbf{Set}}(\Sigma_a \Delta_a(B), C) \cong \mathrm{Hom}_{\mathbf{Set}}(B, \Pi_a \Delta_a(C)). \tag{7.5}$$

By definition, $\Delta_a(B) \colon A \to \mathbf{Set}$ assigns to each element $a \in A$ the set B. Since Σ_A takes disjoint unions, we have a bijection

$$\Sigma_a(\Delta_a(B)) = \left(\bigsqcup_{a \in A} B\right) \cong A \times B.$$

Similarly, $\Delta_a(C) \colon A \to \mathbf{Set}$ assigns to each element $a \in A$ the set C. Since Π_A takes products, we have a bijection

$$\Pi_a(\Delta_a(C)) = \left(\prod_{a \in A} C\right) \cong C^A.$$

The currying isomorphism $\mathrm{Hom}_{\mathbf{Set}}(A \times B, C) \cong \mathrm{Hom}_{\mathbf{Set}}(B, C^A)$ falls out of (7.5).

7.2   Categories of functors

For any two categories $\mathcal{C}$ and $\mathcal{D}$,[8] Section 5.3.2.1 discussed the category $\mathrm{Fun}(\mathcal{C}, \mathcal{D})$ of functors and natural transformations between them. This section discusses functor categories a bit more and gives some important applications in mathematics (sheaves) that extend to the real world.

7.2.1   Set-valued functors

Let $\mathcal{C}$ be a category. We have been denoted by $\mathcal{C}$–**Set** the functor category $\mathrm{Fun}(\mathcal{C}, \mathbf{Set})$. Here is a nice result about these categories.

Proposition 7.2.1.1. *Let $\mathcal{C}$ be a category. The category $\mathcal{C}$–**Set** is closed under colimits and limits. That is, for any category I and functor $D: I \to \mathcal{C}$–**Set**, both the limit and the colimit of D exist in $\mathcal{C}$–**Set**.*

Sketch of proof. We rely on the fact that the category **Set** is complete and cocomplete (see Remark 6.1.3.37), i.e., that it has all limits and colimits (see Theorems 6.1.3.28 and 6.1.3.35 for constructions). Let J be an indexing category and $D: J \to \mathcal{C}$–**Set** a functor. For each object $c \in \mathrm{Ob}(\mathcal{C})$, we have a functor $D_c: J \to \mathbf{Set}$ defined by $D_c(j) = D(j)(c)$. Define a functor $L: \mathcal{C} \to \mathbf{Set}$ by $L(c) = \lim_J D_c$, and note that for each $f: c \to c'$ in $\mathcal{C}$ there is an induced function $L(f): L(c) \to L(c')$. One can check that L is a limit of J, because it satisfies the relevant universal property.

The dual proof holds for colimits.

$\square$

Application 7.2.1.2. When taking in data about a scientific subject, one often finds that how one thinks about the problem changes over time. We understand this phenomenon in the language of databases in terms of a series of schemas $\mathcal{C}_1, \mathcal{C}_2, \ldots, \mathcal{C}_{n+1}$, perhaps indexed chronologically. The problem is that previously-collected data is held in what may be outdated schemas, and we want to work with it in our current understanding. By finding appropriate functors between these schemas, or possibly with the help of auxiliary schemas, we can make a chain of categories and functors

$$\mathcal{C}_1 \xleftarrow{F_1} \mathcal{D}_1 \xrightarrow{G_1} \mathcal{E}_1 \xrightarrow{H_1} \mathcal{C}_2 \xleftarrow{F_2} \mathcal{D}_2 \xrightarrow{G_2} \mathcal{E}_2 \xrightarrow{H_2} \cdots \xrightarrow{G_n} \mathcal{E}_n \xrightarrow{H_n} \mathcal{C}_{n+1}.$$

We can then use the data migration functors Δ_F, Π_G, and Σ_H to move data from category $\mathcal{C}_1$ to category $\mathcal{C}_{n+1}$ using projections, joins, and unions in any combination. Theorems

---

[8]Technically $\mathcal{C}$ has to be small but, as mentioned in Remark 5.1.1.2), we do not worry about that distinction in this book.

about sequences of Δ's, Π's, and Σ's can help us understand how such a transformation will behave, before we spend the resources to enact it.

$\Diamond\Diamond$

Exercise 7.2.1.3.

By Proposition 7.2.1.1, the category $\mathcal{C}$–**Set** is closed under taking colimits and limits. By Exercises 6.1.3.24 and 6.1.3.34, this means in particular, that $\mathcal{C}$–**Set** has an initial object and a terminal object.

a. Let $A \in \mathrm{Ob}(\mathcal{C}$–**Set**$)$ be the initial object, considered as a functor $A \colon \mathcal{C} \to$ **Set**. For any $c \in \mathrm{Ob}(\mathcal{C})$, what is the set $A(c)$?

b. Let $Z \in \mathrm{Ob}(\mathcal{C}$–**Set**$)$ be the terminal object, considered as a functor $Z \colon \mathcal{C} \to$ **Set**. For any $c \in \mathrm{Ob}(\mathcal{C})$, what is the set $Z(c)$?

$\Diamond$

Solution 7.2.1.3.

a. If A is initial, then for any object $c \in \mathrm{Ob}(\mathcal{C})$, we have $A(c) = \varnothing$. In other words, the initial instance makes every table empty.

b. If A is terminal, then for any object $c \in \mathrm{Ob}(\mathcal{C})$, we have $A(c) \cong \{\smiley\}$. In other words, the terminal instance puts one row in every table.

♦

Proposition 7.2.1.1 says that we can add or multiply database instances together. In fact, database instances on $\mathcal{C}$ form a *topos*, which means that just about every consideration we made for sets holds for instances on any schema. Perhaps the simplest schema is $\mathcal{C} = \boxed{\bullet}$, on which the relevant topos $\boxed{\bullet}$–**Set** is indeed equivalent to **Set**. But schemas can be arbitrarily complex categories, and it is impressive that all these set-theoretic notions make sense in such generality. Here is a table that compares these domains:

Dictionary between **Set** and $\mathcal{C}$–**Set**	
Concept in Set	**Concept in $\mathcal{C}$–Set**
Set	Object in $\mathcal{C}$–**Set**
Function	Morphism in $\mathcal{C}$–**Set**
Element	Representable functor
Empty set	Initial object
Natural numbers	Natural numbers object
Image	Image
(Co)limits	(Co)limits
Exponential objects	Exponential objects
"Familiar" arithmetic	"Familiar" arithmetic
Power-sets 2^X	Power objects Ω^X
Characteristic functions	Characteristic morphisms
Surjections, injections	Epimorphisms, monomorphisms

Thus elements of a set are akin to representable functors in $\mathcal{C}$–**Set**, which are defined in Section 7.2.1.6. We briefly discuss monomorphisms and epimorphisms first in general (Definition 7.2.1.4) and then in $\mathcal{C}$–**Set** (Proposition 7.2.1.5).

Definition 7.2.1.4 (Monomorphism, epimorphism). Let $\mathcal{S}$ be a category, and let $f\colon X \to Y$ be a morphism. We say that f is a *monomorphism* if it has the following property. For all objects $A \in \mathrm{Ob}(\mathcal{S})$ and morphisms $g, g'\colon A \to X$ in $\mathcal{S}$,

$$A \underset{g'}{\overset{g}{\rightrightarrows}} X \xrightarrow{\ f\ } Y,$$

if $f \circ g = f \circ g'$, then $g = g'$.

We say that $f\colon X \to Y$ is an *epimorphism* if it has the following property. For all objects $B \in \mathrm{Ob}(\mathcal{S})$ and morphisms $h, h'\colon Y \to B$ in $\mathcal{S}$,

$$X \xrightarrow{\ f\ } Y \underset{h'}{\overset{h}{\rightrightarrows}} B,$$

if $h \circ f = h' \circ f$, then $h = h'$.

In the category of sets, monomorphisms are the same as injections, and epimorphisms are the same as surjections (see Proposition 3.4.5.8). The same is true in $\mathcal{C}$–**Set**: one can check table by table that a morphism of instances is mono or epi.

Proposition 7.2.1.5. *Let C be a category, let $X, Y\colon C \to$ **Set** be objects in C–**Set**, and let $f\colon X \to Y$ be a morphism in C–**Set**. Then f is a monomorphism (resp. an epimorphism) if and only if for every object $c \in \mathrm{Ob}(C)$, the function $f(c)\colon X(c) \to Y(c)$ is injective (resp. surjective).*

Sketch of proof. We first show that if f is mono (resp. epi), then so is $f(c)$, for all $c \in \mathrm{Ob}(C)$. Considering c as a functor $c\colon \underline{1} \to C$, this result follows from the fact that Δ_c preserves limits and colimits, hence monos and epis.

We now check that if $f(c)$ is mono for all $c \in \mathrm{Ob}(C)$, then f is mono. Suppose that $g, g'\colon A \to X$ are morphisms in C–**Set** such that $f \circ g = f \circ g'$. Then for every c, we have $f \circ g(c) = f \circ g'(c)$, which implies by hypothesis that $g(c) = g'(c)$. But the morphisms in C–**Set** are natural transformations, and if two natural transformations g, g' have the same components, then they are the same.

A similar argument works to show the analogous result for epimorphisms.

$\square$

7.2.1.6 Representable functors

Given a category C, there are certain functors $C \to$ **Set** that come with the package, i.e., that are not arbitrary from a mathematical perspective as database instances usually are. In fact, there is a certain instance corresponding to each object in C. So if C is a database schema, then for every table $c \in \mathrm{Ob}(C)$ there is a certain database instance associated to it. These instances, i.e., set-valued functors, are called representable functors (see Definition 7.2.1.7). The idea is that if a database schema is a conceptual layout of types (e.g., as an olog), then each type c has an instance associated to it, standing for "the generic thing of type c with all its generic attributes."

Definition 7.2.1.7. Let C be a category, and let $c \in \mathrm{Ob}(C)$ be an object. The functor $\mathrm{Hom}_C(c, -)\colon C \to$ **Set**, sending $d \in \mathrm{Ob}(C)$ to the set $\mathrm{Hom}_C(c, d)$ and acting similarly on morphisms $d \to d'$, is said to be *represented by c*. If a functor $F\colon C \to$ **Set** is isomorphic to $\mathrm{Hom}_C(c, -)$, we say that F is *a representable functor*. To shorten notation we sometimes write

$$Y_c := \mathrm{Hom}_C(c, -).$$

Example 7.2.1.8. Given a category C and an object $c \in \mathrm{Ob}(C)$, we get a representable functor Y_c. If we think of C as a database schema and c as a table, then what does the representable functor $Y_c\colon C \to$ **Set** look like in terms of databases? It turns out that the following procedure will generate it.

Begin by writing a new row, say, "☺," in the ID column of table c. For each foreign key column $f\colon c \to c'$, add a row in the ID column of table c' called "$f(☺)$" and record that result, "$f(☺)$," in the f column of table c. Repeat as follows: for each table d, identify all rows r that have a blank cell in column $g\colon d \to e$. Add a new row called "$g(r)$" to table e and record that result, "$g(r)$," in the (r, g) cell of table d.

Here is a concrete example. Let $\mathcal{C}$ be the following schema:

$$\mathcal{C} := \begin{array}{c} \overset{A}{\bullet} \overset{f}{\longrightarrow} \overset{B}{\bullet} \underset{g_2}{\overset{g_1}{\rightrightarrows}} \overset{C}{\bullet} \overset{i}{\longrightarrow} \overset{D}{\bullet} \\ h \downarrow \\ \overset{E}{\bullet} \end{array}$$

Then $Y_B\colon \mathcal{C} \to \mathbf{Set}$ is given by "morphisms from B to $-$," i.e., it is the following instance:

A
ID || f

B			
ID || g_1	g_2	h
☺ || $g_1(☺)$	$g_2(☺)$	$h(☺)$

C	
ID || i
$g_1(☺)$ || $i(g_1(☺))$
$g_2(☺)$ || $i(g_2(☺))$

D
ID
$i(g_1(☺))$
$i(g_2(☺))$

E
ID
$h(☺)$

To create Y_B we began with a single element in table B and followed the arrows, putting new entries wherever they were required. One might call this the *schematically implied reference spread* or SIRS of the element ☺ in table B. Notice that the table at A is empty, because there are no morphisms $B \to A$ in $\mathcal{C}$.

Representable functors Y_c yield database instances that are as free as possible, subject to having the initial row ☺ in table c. We saw this before (as Skolem variables) when studying the left pushforward Σ. Indeed, suppose $c \in \mathrm{Ob}(\mathcal{C})$ is an object represented by the functor $c\colon \underline{1} \to \mathcal{C}$. A database instance on $\underline{1}$ is the same thing as a set X. The left pushforward $\Sigma_c(X)$ has the same kinds of Skolem variables as Y_c does. In fact, if $X = \{☺\}$ is a one-element set, then we get the representable functor

$$Y_c \cong \Sigma_c\{☺\}.$$

Exercise 7.2.1.9.

Consider the schema for graphs,

$$\textbf{GrIn} := \boxed{\begin{array}{ccc} Ar & \overset{src}{\underset{tgt}{\rightrightarrows}} & Ve \\ \bullet & & \bullet \end{array}}$$

a. Write the representable functor $Y_{Ar} \colon \textbf{GrIn} \to \textbf{Set}$ as two tables.

b. Write the representable functor Y_{Ve} as two tables.

$\Diamond$

Solution 7.2.1.9.

a. This was done in Exercise 5.3.3.7, although not with the most natural names. Here we rewrite $Y_{Ar} = \text{Hom}_{\textbf{GrIn}}(Ar, -)$ as

Ar		
ID	**src**	**tgt**
☺	$src(☺)$	$tgt(☺)$

Ve
ID
$src(☺)$
$tgt(☺)$

b. Here is $Y_{Ve} = \text{Hom}_{\textbf{GrIn}}(Ve, -)$ with "natural names":

Ar		
ID	**src**	**tgt**

Ve
ID
☺

(The left-hand table is empty because there are no morphisms $Ve \to Ar$ in **GrIn**.)

◆

Exercise 7.2.1.10.

Consider the loop schema

$$\mathcal{L}oop := \boxed{\begin{array}{c} f \\ \circlearrowright_s \\ \bullet \end{array}}$$

Express the representable functor $Y_s \colon \mathcal{L}oop \to \textbf{Set}$ in table form.

$\Diamond$

Solution 7.2.1.10.

We have $Y_s = \text{Hom}_{\mathcal{L}oop}(s, -) \colon \mathcal{L}oop \to \mathbf{Set}$. On objects, of which there is only $\text{Ob}(\mathcal{L}oop) = \{s\}$, we have $Y_s(s) = \{f^n \mid n \in \mathbb{N}\}$. The morphism $f \colon s \to s$ acts on $Y_s(s)$ by composing. Here is Y_s in table form:

s	
ID	**f**
☺	$f(☺)$
$f(☺)$	$f^2(☺)$
$f^2(☺)$	$f^3(☺)$
$f^3(☺)$	$f^4(☺)$
$f^4(☺)$	$f^5(☺)$
⋮	⋮

◆

Let B be a box in an olog, say, ⌜a person⌝, and recall that an aspect of B is an outgoing arrow, such as ⌜a person⌝ $\xrightarrow{\text{has as height in inches}}$ ⌜an integer⌝. The following slogan explains representable functors in those terms.

Slogan 7.2.1.11.

> *The functor represented by ⌜a person⌝ simply leaves a placeholder, like ⟨person's name here⟩ or ⟨person's height here⟩, for every aspect of ⌜a person⌝.*
>
> *In general, there is a representable functor for every type in an olog. The representable functor for type T simply encapsulates the most generic or abstract example of type T, by leaving a placeholder for each of its attributes.*

Exercise 7.2.1.12.

Recall from Definition 7.2.1.7 that a functor $F \colon \mathcal{C} \to \mathbf{Set}$ is said to be represented by c if there is a natural isomorphism $F \cong \text{Hom}_{\mathcal{C}}(c, -)$.

a. There is a functor $\text{Ob} \colon \mathbf{Cat} \to \mathbf{Set}$ (see Exercise 5.1.2.41) sending a category $\mathcal{C}$ to its set $\text{Ob}(\mathcal{C})$ of objects, and sending a functor to its on-objects part. This functor is representable by some category. Name a category A that represents Ob.

b. There is a functor $\text{Hom} \colon \mathbf{Cat} \to \mathbf{Set}$ (see Exercise 5.1.2.42) sending a category $\mathcal{C}$ to the set $\text{Hom}_{\mathcal{C}}$ of all morphisms in $\mathcal{C}$ and sending a functor to its on-morphisms part. This functor is representable by a category. Name a category B that represents Hom.

◊

Solution 7.2.1.12.

a. The functor Ob is represented by the category $\underline{1}$. That is, there is a natural isomorphism of sets,
$$\mathrm{Ob}(\mathcal{C}) \cong \mathrm{Hom}_{\mathbf{Cat}}(\underline{1}, \mathcal{C}).$$

b. The functor Hom is represented by the free arrow category $[1] = \boxed{\bullet \longrightarrow \bullet}$. That is, there a natural isomorphism of sets
$$\mathrm{Hom}_{\mathcal{C}} \cong \mathrm{Hom}_{\mathbf{Cat}}([1], \mathcal{C}).$$

♦

Exercise 7.2.1.13.

Let $\mathcal{C}$ be a category, let $c, c' \in \mathrm{Ob}(\mathcal{C})$ be objects, and let $Y_c, Y_{c'} : \mathcal{C} \to \mathbf{Set}$ be the associated representable functors. Given $f : c \to c'$, we want to construct a morphism $Y_f : Y_{c'} \to Y_c$ in $\mathrm{Fun}(\mathcal{C}\text{–}\mathbf{Set})$. Of course, Y_f is supposed to be a natural transformation, so we need to provide a component $(Y_f)_d$ for every object $d \in \mathrm{Ob}(\mathcal{C})$.

a. What must the domain and codomain of $(Y_f)_d$ be? (Simplify your answer using Definition 7.2.1.7.)

b. Can you make sense of the statement, "Define $(Y_f)_d$ by precomposition"?

c. If $h : d \to e$ is a morphism in $\mathcal{C}$, draw the naturality square for Y_f. Does it commute?

◊

Solution 7.2.1.13.

a. We have $(Y_f)_d : Y_{c'}(d) \to Y_c(d)$. But by definition, this is $(Y_f)_d : \mathrm{Hom}_{\mathcal{C}}(c', d) \to \mathrm{Hom}_{\mathcal{C}}(c, d)$.

b. Given an element $g \in \mathrm{Hom}_{\mathcal{C}}(c', d)$, we can precompose with f to get a morphism $c \xrightarrow{f} c' \xrightarrow{g} d$, so let's define $(Y_f)_d(g) = g \circ f$.

c. The naturality square is as follows

$$
\begin{array}{ccc}
Y_{c'}(d) & \xrightarrow{Y_{c'}(h)} & Y_{c'}(e) \\
{\scriptstyle (Y_f)_d} \downarrow & & \downarrow {\scriptstyle (Y_f)_e} \\
Y_c(d) & \xrightarrow[Y_c(h)]{} & Y_c(e)
\end{array}
$$

and it commutes because, for any element $g \in Y_{c'}(d)$, the composition $c \xrightarrow{f} c' \xrightarrow{g} d \xrightarrow{h} e$ is associative. More explicitly, going down then right we have $(Y_f)_d(g) = g \circ f$ and $Y_c(h)(g \circ f) = h \circ (g \circ f)$. Going right then down we have $Y_{c'}(h)(g) = h \circ g$ and $(Y_f)_e(h \circ g) = (h \circ g) \circ f$. To reiterate, the associativity of composition in $\mathcal{C}$ insures that this square commutes.

◆

7.2.1.14 Yoneda's lemma

One of the most powerful tools in category theory is Yoneda's lemma. It is often considered by students to be quite abstract, but grounding it in databases may help.

Suppose that $I \colon \mathcal{C} \to \mathbf{Set}$ is an arbitrary database instance, let $c \in \mathrm{Ob}(\mathcal{C})$ be an object, and let $f \colon c \to c'$ be any outgoing arrow. Because I is a functor, we know that for every row $r \in I(c)$ in table c, a value has been recorded in the f column. The value in the (r, f) cell refers to some row in table c'. That is, each row in table c induces SIRS throughout the database as freely as possible (see Example 7.2.1.8). The instance Y_c consists entirely of a single row ☺ in table c and its SIRS. The idea is that for any row $r \in I(c)$ in arbitrary instance I, there exists a unique map $Y_c \to I$ sending ☺ to r.

Proposition 7.2.1.15 (Yoneda's lemma, part 1). *Let $\mathcal{C}$ be a category, $c \in \mathrm{Ob}(\mathcal{C})$ an object, and $I \colon \mathcal{C} \to \mathbf{Set}$ a set-valued functor. There is a natural bijection*

$$\mathrm{Hom}_{\mathcal{C}\text{-}\mathbf{Set}}(Y_c, I) \xrightarrow{\ \cong\ } I(c).$$

Proof. See Mac Lane [29].

□

Example 7.2.1.16. Consider the category $\mathcal{C}$ drawn as follows:

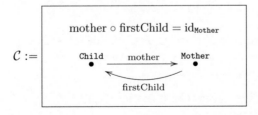

There are two representable functors, $Y_{\texttt{Child}}$ and $Y_{\texttt{Mother}}$. The former, $Y_{\texttt{Child}} \colon \mathcal{C} \to \mathbf{Set}$, is shown here:

Child (Y_{Child})	
ID	**mother**
☺	mother(☺)
firstChild(mother(☺))	mother(☺)

Mother (Y_{Child})	
ID	**firstChild**
mother(☺)	firstChild(mother(☺))

The representable functor Y_{Child} is the freest instance possible, starting with one element in the Child table and satisfying the constraints. The latter, Y_{Mother} is the freest instance possible, starting with one element in the Mother table and satisfying the constraints. Since mother∘firstChild=id$_{\text{Mother}}$, this instance has just one row in each table:

Child (Y_{Mother})	
ID	**mother**
firstChild(☺)	☺

Mother (Y_{Mother})	
ID	**firstChild**
☺	firstChild(☺)

Here is an arbitrary instance $I\colon \mathcal{C} \to$ **Set**:

Child (I)	
ID	**mother**
Amy	Ms. Adams
Bob	Ms. Adams
Carl	Ms. Jones
Deb	Ms. Smith

Mother (I)	
ID	**firstChild**
Ms. Adams	Bob
Ms. Jones	Carl
Ms. Smith	Deb

Yoneda's lemma (7.2.1.15) is about the set of natural transformations $Y_{\text{Child}} \to I$. Recall from Definition 5.3.1.2 that a search for natural transformations can get tedious. Yoneda's lemma makes the calculation quite trivial. In this case there are exactly four such natural transformations, $\text{Hom}_{\mathcal{C}\text{-}\mathbf{Set}}(Y_{\text{Child}}, I) \cong I(\text{Child}) \cong \underline{4}$, and they are completely determined by where ☺ goes. In some sense the symbol ☺ in Y_{Child} *represents* childness in this database.

Exercise 7.2.1.17.

Consider the schema $\mathcal{C}$ and instance $I\colon \mathcal{C} \to \mathbf{Set}$ from Example 7.2.1.16. Let Y_{Child} be the representable functor, and write (☺ ↦ Amy) for the unique natural transformation $Y_{\text{Child}} \to I$ sending ☺ to Amy, and so on.

a. What is (☺ ↦ Amy)$_{\text{Child}}$(firstChild(mother(☺)))?[9]

---

[9]There is a lot of clutter here. Note that "firstChild(mother(☺))" is a row in the Child table of Y_{Child}. Assuming that the math follows the meaning, if ☺ points to Amy, where should firstChild(mother(☺)) point?

b. What is $(\odot \mapsto \text{Bob})_{\texttt{Child}}(\text{firstChild}(\text{mother}(\odot)))$?

c. What is $(\odot \mapsto \text{Carl})_{\texttt{Child}}(\text{firstChild}(\text{mother}(\odot)))$?

d. What is $(\odot \mapsto \text{Amy})_{\texttt{Mother}}(\text{mother}(\odot))$?

e. In parts (a)–(d), what information does the first subscript (`Child`, `Child`, `Child`, `Mother`) give you about the answer?

$\diamond$

Solution 7.2.1.17.

The math works out as expected.

a. $(\odot \mapsto \text{Amy})_{\texttt{Child}}(\text{firstChild}(\text{mother}(\odot)))$ is the first child of the mother of Amy, namely Bob.

b. $(\odot \mapsto \text{Bob})_{\texttt{Child}}(\text{firstChild}(\text{mother}(\odot))) = \text{Bob}$.

c. $(\odot \mapsto \text{Carl})_{\texttt{Child}}(\text{firstChild}(\text{mother}(\odot))) = \text{Carl}$.

d. $(\odot \mapsto \text{Amy})_{\texttt{Mother}}(\text{mother}(\odot)) = \text{Ms. Adams}$.

e. Abstractly, is the component of the natural transformation. Practically, it tells us what kind of answer we are expecting: the first three (Bob, Bob, Carl) are children and the last (Ms. Adams) is a mother.

$\blacklozenge$

Section 7.2.1.6 showed that a representable functor $\mathcal{C} \to \textbf{Set}$ is a mathematically generated database instance for an abstract thing of type $T \in \text{Ob}(\mathcal{C})$. It creates placeholders for every attribute that things of type T are supposed to have.

Slogan 7.2.1.18.

> *Yoneda's lemma says the following. Specifying an actual thing of type T is the same as filling in all placeholders found in the generic thing of type T.*

Yoneda's lemma is considered by many category theorists to be the most important tool in the subject. While its power is probably unclear to students whose sole background in category theory comes from this book, Yoneda's lemma is indeed extremely useful for reasoning. It allows us to move the notion of functor application into the realm of morphisms between functors (i.e., morphisms in $\mathcal{C}$–**Set**, which are natural transformations). This keeps everything in one place—it is all in the morphisms—and thus more interoperable.

Example 7.2.1.19. Example 4.1.1.27 discussed the cyclic monoid $\mathcal{M}$ generated by the symbol Q and subject to the relation $Q^7 = Q^4$, depicted as

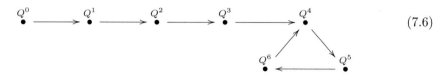

$$(7.6)$$

Here is the mathematical foundation for this picture. Since $\mathcal{M}$ is a category with one object, ▲, there is a unique representable functor (up to isomorphism) $Y := Y_▲ \colon \mathcal{M} \to$ **Set**. Any functor $\mathcal{M} \to$ **Set** can be thought of as a set with an $\mathcal{M}$ action (see Section 5.2.1.1). In the case of Y, the required set is

$$Y(▲) = \mathrm{Hom}_{\mathcal{M}}(▲, ▲) \cong \{Q^0, Q^1, Q^2, Q^3, Q^4, Q^5, Q^6\},$$

and the action is pretty straightforward (it is called the *principal action*). For example, $Q^5 \circlearrowleft Q^2 = Q^4$. We might say that (7.6) is a picture of this principal action of $\mathcal{M}$.

However, we can go one step further. Given the functor $Y \colon \mathcal{M} \to$ **Set**, we can take its category of elements, $\int_{\mathcal{M}} Y$ (see Section 6.2.2). The category $\int_{\mathcal{M}} Y$ has objects $Y(▲) \in \mathrm{Ob}(\mathbf{Set})$, i.e., the set of dots in (7.6), and it has a unique morphism $Q^i \to Q^j$ for every path of length $\leqslant 6$ from Q^i to Q^j in that picture. So the drawing of $\mathcal{M}$ in (7.6) is actually the category of elements of $\mathcal{M}$'s unique representable functor.

Exercise 7.2.1.20.

Let $\mathcal{C}$ be a category, let $c \in \mathrm{Ob}(\mathcal{C})$ be an object, and let $I \in \mathrm{Ob}(\mathcal{C}\text{–}\mathbf{Set})$ be in instance of $\mathcal{C}$. Consider c also as a functor $c \colon \underline{1} \to \mathcal{C}$ and recall the pullback functor $\Delta_c \colon \mathcal{C}\text{–}\mathbf{Set} \to \mathbf{Set}$ and its left adjoint $\Sigma_c \colon \mathbf{Set} \to \mathcal{C}\text{–}\mathbf{Set}$ (see Section 7.1.4).

a. What is the set $\Delta_c(I)$?

b. What is $\mathrm{Hom}_{\mathbf{Set}}(\{\odot\}, \Delta_c(I))$?

c. What is $\mathrm{Hom}_{\mathcal{C}\text{–}\mathbf{Set}}(\Sigma_c(\{\odot\}), I)$?

d. How does $\Sigma_c(\{\odot\})$ compare to Y_c, the functor represented by c, as objects in $\mathcal{C}\text{–}\mathbf{Set}$?

◊

Solution 7.2.1.20.

These ideas were mentioned on page 405.

a. We have $\Delta_c(I) \colon \underline{1} \to \mathbf{Set}$, which we can consider as a set. It is $\Delta_c(I) = I(c)$.

b. For any set X, we have a bijection $\text{Hom}_{\textbf{Set}}(\{\odot\}, X) \cong X$ (see Exercise 2.1.2.20). So $\text{Hom}_{\textbf{Set}}(\{\odot\}, \Delta_c(I)) \cong I(c)$.

c. Since Σ_c is left adjoint to Δ_c (see Section 7.1.4.6), we must have

$$\text{Hom}_{\mathcal{C}\text{-}\textbf{Set}}(\Sigma_c(\{\odot\}), I) \cong \text{Hom}_{\textbf{Set}}(\{\odot\}, \Delta_c(I)) \cong I(c).$$

d. Since for any $I \in \text{Ob}(\mathcal{C}\text{-}\textbf{Set})$, we have a natural isomorphism

$$\text{Hom}_{\mathcal{C}\text{-}\textbf{Set}}(Y_c, I) \cong I(c) \cong \text{Hom}_{\mathcal{C}\text{-}\textbf{Set}}(\Sigma_c(\{\odot\}), I),$$

it seems that Y_c and $\Sigma_c(\{\odot\})$ are similar. In fact, we could say that Y_c and $\Sigma_c(\{\odot\})$ represent the same functor $\mathcal{C}\text{-}\textbf{Set} \to \textbf{Set}$. Yoneda's embedding Proposition 7.2.1.21 implies that they are isomorphic,

$$Y_c \cong \Sigma_c(\{\odot\}).$$

$\blacklozenge$

Proposition 7.2.1.21 (Yoneda's lemma, part 2). *Let $\mathcal{C}$ be a category. The assignment $c \mapsto Y_c$ from Proposition 7.2.1.15 extends to a functor $Y\colon \mathcal{C}^{\text{op}} \to \mathcal{C}\text{-}\textbf{Set}$, and this functor is fully faithful.*

In particular, if $c, c' \in \text{Ob}(\mathcal{C})$ are objects and there is an isomorphism $Y_c \cong Y_{c'}$ in $\mathcal{C}\text{-}\textbf{Set}$, then there is an isomorphism $c \cong c'$ in $\mathcal{C}$.

Proof. See Mac Lane [29]. $\square$

Exercise 7.2.1.22.

The distributive law for addition of natural numbers says $c \times (a + b) = c \times a + c \times b$. Following is a proof of the distributive law using category-theoretic reasoning. Annotate anything shown in red with a justification for why it is true.

Proposition (Distributive law). *For any natural numbers $a, b, c \in \mathbb{N}$, the distributive law holds:*

$$c(a + b) = ca + cb.$$

Sketch of proof. To finish, justify things shown in red.
Let A, B, C be finite sets, and let X be another finite set.

$$\begin{aligned}
\text{Hom}_{\textbf{Set}}(C \times (A + B), X) &\cong \text{Hom}_{\textbf{Set}}(A + B, X^C) \\
&\cong \text{Hom}_{\textbf{Set}}(A, X^C) \times \text{Hom}_{\textbf{Set}}(B, X^C) \\
&\cong \text{Hom}_{\textbf{Set}}(C \times A, X) \times \text{Hom}_{\textbf{Set}}(C \times B, X) \\
&\cong \text{Hom}_{\textbf{Set}}((C \times A) + (C \times B), X).
\end{aligned}$$

By the appropriate application of Yoneda's lemma, we see that there is an isomorphism

$$C \times (A + B) \cong (C \times A) + (C \times B)$$

in **Fin**. The result about natural numbers follows.   □

                                                                                         ◊

Solution 7.2.1.22.

There are six red things. The first four are arithmetic of categories, justified by the Proposition 6.2.5.1. In that language we have

$$X^{C \times (A+B)} \cong (X^C)^{A+B} \cong (X^C)^A \times (X^C)^B \cong X^{C \times A} \times X^{C \times B} \cong X^{(C \times A)+(C \times B)}.$$

So now we have an isomorphism

$$\mathrm{Hom}_{\mathbf{Set}}(C \times (A + B), X) \cong \mathrm{Hom}_{\mathbf{Set}}(C \times A + C \times B, X)$$

for any set X. This isomorphism is natural in X, but this very important issue is not elaborated here (see Mac Lane [29] for such details). Thus the representable functor $Y_{C \times (A+B)}$ is isomorphic to the representable functor $Y_{C \times A + C \times B}$. By applying Yoneda's Proposition 7.2.1.21, we have an isomorphism $C \times (A + B) \cong (C \times A) + (C \times B)$. This is the fifth red thing.

Thus if $a, b, c \in \mathbb{N}$ are such that $A \cong \underline{a}, B \cong \underline{b}$, and $C \cong \underline{c}$, it follows that $c(a + b) = ca + cb$, justifying the sixth red thing.   ♦

7.2.1.23   The subobject classifier $\Omega \in \mathrm{Ob}(\mathcal{C}\text{–}\mathbf{Set})$

If $\mathcal{C}$ is a category, then the functor category $\mathcal{C}$–**Set** is a special kind of category, called a *topos*. Note that when $\mathcal{C} = \underline{1}$ is the terminal category, then we have an isomorphism $\underline{1}$–**Set** $\cong$ **Set**, so the category of sets is a special case of a topos. What is interesting about toposes (or topoi) is that they generalize many properties of **Set**. This short section investigates only one such property, namely, that $\mathcal{C}$–**Set** has a subobject classifier, denoted $\Omega \in \mathrm{Ob}(\mathcal{C}\text{–}\mathbf{Set})$. In the case $\mathcal{C} = \underline{1}$ the subobject classifier is $\{True, False\} \in \mathrm{Ob}(\mathbf{Set})$ (see Definition 3.4.4.9).

As usual, we consider the matter of subobject classifiers by grounding the discussion in terms of databases. The analogue of $\{True, False\}$ for an arbitrary database can be quite complex—it encodes the whole story of relational database instances for that schema.

Definition 7.2.1.24. Let $\mathcal{C}$ be a category, let $\mathcal{C}$–**Set** denote its category of instances, and let $1_{\mathcal{C}} \in \mathrm{Ob}(\mathcal{C}$–**Set**$)$ denote the terminal object. A *subobject classifier for $\mathcal{C}$*–**Set** is an object $\Omega_{\mathcal{C}} \in \mathrm{Ob}(\mathcal{C}$–**Set**$)$ and a morphism $t\colon 1_{\mathcal{C}} \to \Omega_{\mathcal{C}}$ with the following property. For any monomorphism $f\colon I \to J$ in $\mathcal{C}$–**Set**, there exists a unique morphism $char(f)\colon J \to \Omega_{\mathcal{C}}$ such that the following diagram is a pullback in $\mathcal{C}$–**Set**:

$$
\begin{array}{ccc}
I & \xrightarrow{\;\;!\;\;} & 1_{\mathcal{C}} \\
{\scriptstyle f}\downarrow & \lrcorner & \downarrow{\scriptstyle t} \\
J & \xrightarrow[char(f)]{} & \Omega_{\mathcal{C}}
\end{array}
$$

That is, for any instance J there is a bijection

$$\mathrm{Hom}_{\mathcal{C}\text{–}\mathbf{Set}}(J, \Omega) \cong \{I \in \mathrm{Ob}(\mathcal{C}\text{–}\mathbf{Set}) \mid I \subseteq J\}.$$

In terms of databases, what this means is that for every schema $\mathcal{C}$, there is some special instance $\Omega_{\mathcal{C}} \in \mathrm{Ob}(\mathcal{C}$–**Set**$)$ that somehow classifies subinstances of anything. When the schema is the terminal category, $\mathcal{C} = \underline{1}$, instances are sets and according to Definition 3.4.4.9 the subobject classifier is $\Omega_{\underline{1}} = \{True, False\}$. One might think that the subobject classifier for $\mathcal{C}$–**Set** should just consist of a two-element set table by table, i.e., that for every $c \in \mathrm{Ob}(\mathcal{C})$, we should have $\Omega_{\mathcal{C}} =^{?} \{True, False\}$, but this is not correct.

In fact, for any object $c \in \mathrm{Ob}(\mathcal{C})$, there is a way to figure out what $\Omega_{\mathcal{C}}(c)$ has to be. We know by Yoneda's lemma (Proposition 7.2.1.15) that $\Omega_{\mathcal{C}}(c) = \mathrm{Hom}_{\mathcal{C}\text{–}\mathbf{Set}}(Y_c, \Omega_{\mathcal{C}})$, where Y_c is the functor represented by c. There is a bijection between $\mathrm{Hom}_{\mathcal{C}\text{–}\mathbf{Set}}(Y_c, \Omega_{\mathcal{C}})$ and the set of subinstances of Y_c. Thus we have

$$\Omega_{\mathcal{C}}(c) = \{I \in \mathrm{Ob}(\mathcal{C}\text{–}\mathbf{Set}) \mid I \subseteq Y_c\}. \tag{7.7}$$

How should $\Omega_{\mathcal{C}}\colon \mathcal{C} \to \mathbf{Set}$ behave on morphisms? By Exercise 7.2.1.13, each morphism $f\colon c \to d$ in $\mathcal{C}$ induces a morphism $Y_f\colon Y_d \to Y_c$, and the map $\Omega_{\mathcal{C}}(f)\colon \Omega_{\mathcal{C}}(c) \to \Omega_{\mathcal{C}}(d)$ sends a subinstance $A \subseteq Y_c$ to the pullback

$$
\begin{array}{ccc}
Y_f^{-1}(A) & \longrightarrow & A \\
\downarrow & \lrcorner & \downarrow \\
Y_d & \xrightarrow[Y_f]{} & Y_c
\end{array}
\tag{7.8}
$$

That is, $\Omega_{\mathcal{C}}(f)(A) = Y_f^{-1}(A)$.

We have now fully described $\Omega_{\mathcal{C}}$ as a functor, but the description is very abstract. Here is an example of a subobject classifier.

Example 7.2.1.25. Consider the following category $\mathcal{C} \cong [3]$:

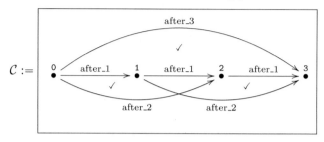

To write $\Omega_{\mathcal{C}}$, we need to understand the representable functors $Y_c \in \mathrm{Ob}(\mathcal{C}\text{–}\mathbf{Set})$, for $c = 0, 1, 2, 3$, as well as their subobjects. Here is Y_0 as an instance:

0 (Y_0)			
ID	**after_1**	**after_2**	**after_3**
☺	after_1(☺)	after_2(☺)	after_3(☺)

1 (Y_0)			
ID	**after_1**	**after_2**
after_1(☺)	after_2(☺)	after_3(☺)

2 (Y_0)	
ID	**after_1**
after_2(☺)	after_3(☺)

3 (Y_0)
ID
after_3(☺)

What are the subinstances of this? There is the empty subinstance $\varnothing \subseteq Y_0$ and the identity subinstance $Y_0 \subseteq Y_0$. But there are three more as well. Note that if we want to keep the ☺ row of table 0, then we have to keep everything. But if we throw away the ☺ row of table 0, we can still keep the rest and get a subinstance. If we want to keep the after_1(☺) row of table 1, then we have to keep its images in tables 2 and 3. But we could throw away both the ☺ row of table 0 and the after_1(☺) row of table 1 and still keep the rest. And so on. In other words, there are five subobjects of Y_0, i.e., elements of $\Omega_{\mathcal{C}}(0)$, but they are hard to name. We arbitrarily name them by $\Omega_{\mathcal{C}}(0) := \{yes, \; wait \; 1, \; wait \; 2, \; wait \; 3, \; never\}$.

The same analysis holds for the other tables of $\Omega_{\mathcal{C}}$. For example, we denote the three subinstances of Y_2 by $\Omega_{\mathcal{C}}(2) = \{yes, \; wait \; 1, \; never\}$. In sum, the database instance $\Omega_{\mathcal{C}}$ is:

0 $(\Omega_{\mathcal{C}})$			
ID	**after_1**	**after_2**	**after_3**
yes	*yes*	*yes*	*yes*
wait 1	*yes*	*yes*	*yes*
wait 2	*wait 1*	*yes*	*yes*
wait 3	*wait 2*	*wait 1*	*yes*
never	*never*	*never*	*never*

1 $(\Omega_{\mathcal{C}})$		
ID	**after_1**	**after_2**
yes	*yes*	*yes*
wait 1	*yes*	*yes*
wait 2	*wait 1*	*yes*
never	*never*	*never*

2 ($\Omega_\mathcal{C}$)	
ID	**after_1**
yes	*yes*
wait 1	*yes*
never	*never*

3 ($\Omega_\mathcal{C}$)
ID
yes
never

The morphism $1 \to \Omega_\mathcal{C}$ picks out the *yes* row of every table.

Now that we have constructed $\Omega_\mathcal{C} \in \mathrm{Ob}(\mathcal{C}\text{–}\mathbf{Set})$, we are ready to use it. What makes $\Omega_\mathcal{C}$ special is that for any instance $X \colon \mathcal{C} \to \mathbf{Set}$, the subinstances if X are in one-to-one correspondence with the instance morphisms $X \to \Omega_\mathcal{C}$. Consider the following arbitrary instance X, where the blue rows denote a subinstance $A \subseteq X$.

0 (X)			
ID	**after_1**	**after_2**	**after_3**
a_1	b_1	c_1	d_1
a_2	b_2	c_1	d_1
a_3	b_2	c_1	d_1
a_4	b_3	c_2	d_2
a_5	b_5	c_3	d_1

1 (X)		
ID	**after_1**	**after_2**
b_1	c_1	d_1
b_2	c_1	d_1
b_3	c_2	d_2
b_4	c_1	d_1
b_5	c_3	d_1

2 (X)	
ID	**after_1**
c_1	d_1
c_2	d_2
c_3	d_1

3 (X))
ID
d_1
d_2

$$(7.9)$$

This blue subinstance $A \subseteq X$ corresponds to a natural transformation $char(A) \colon X \to \Omega_\mathcal{C}$. That is, for each $c \in \mathrm{Ob}(\mathcal{C})$, all the rows in the c table of X are sent to the rows in the c table of $\Omega_\mathcal{C}$, as they would be for any natural transformation. The way $char(A)$ works is as follows. For each table i and row $x \in X(i)$, find the first column f in which the entry is blue (i.e., $f(x) \in A$), and send x to the corresponding element of $\Omega_\mathcal{C}(i)$. For example, $char(A)(0)$ sends a_1 to *wait 2* and sends a_4 to *never*, and $char(A)(2)$ sends c_1 to *yes* and sends c_2 to *never*.

Exercise 7.2.1.26.

a. Write the blue subinstance $A \subseteq X$ shown in (7.9) as an instance of $\mathcal{C}$, i.e., as four tables.

b. This subinstance $A \subseteq X$ corresponds to a map $\ell := char(A) \colon X \to \Omega_\mathcal{C}$. For all $c \in \mathrm{Ob}(\mathcal{C})$, we have a function $\ell(c) \colon X(c) \to \Omega_\mathcal{C}(c)$. With $c = 1$, write out $\ell(1) \colon X(1) \to \Omega_\mathcal{C}(1)$.

◊

Solution 7.2.1.26.

a.

0 (X)			
ID	**after_1**	**after_2**	**after_3**

1 (X)		
ID	**after_1**	**after_2**
b_2	c_1	d_1
b_4	c_1	d_1

2 (X)	
ID	**after_1**
c_1	d_1

3 $(X))$
ID
d_1

b.

char$(A)(1)$	
$X(1)$	$\Omega_{\mathcal{C}}(1)$
b_1	*wait 1*
b_2	*yes*
b_3	*never*
b_4	*yes*
b_5	*wait 2*

♦

Exercise 7.2.1.27.

Let $\mathcal{L}oop$ be the loop schema

$$\mathcal{L}oop = \boxed{\;\overset{f}{\circlearrowright}\,_s\;\bullet\;}$$

a. What is the subobject classifier $\Omega_{\mathcal{L}oop} \in \mathrm{Ob}(\mathcal{L}oop\text{–}\mathbf{Set})$? (Write it out in table form.)

b. In Exercise 7.2.1.10 you computed the representable functor Y_s. How does $\Omega_{\mathcal{L}oop}$ compare to Y_s?

c. Consider the discrete dynamical system X and its subset $W \subseteq X$:

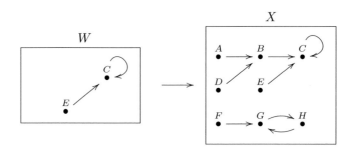

What is the morphism $char(W)\colon X \to \Omega_{\mathcal{L}oop}$ that corresponds to this subobject?

$\diamond$

Solution 7.2.1.27.

a. We see in (7.7) the formula for $\Omega_{\mathcal{L}oop}(s)$:

$$\Omega_{\mathcal{L}oop}(s) = \{I \in \mathrm{Ob}(\mathcal{L}oop\text{–}\mathbf{Set}) \mid I \subseteq Y_s\},$$

where $Y_s = \mathrm{Hom}_{\mathcal{L}oop}(s, -)$ is the representable functor at s. Recall from Exercise 7.2.1.10 that $Y_s(s)$ can be expressed in table form as

\multicolumn{2}{c}{s}	
ID	**f**
f^0	f^1
f	f^2
f^2	f^3
f^3	f^4
f^4	f^5
$\vdots$	$\vdots$

What is the set of subobjects of this? Which subsets of $Y_s(s) = \{f^0, f^1, f^2, \ldots\}$ are closed under composition with f? If a subset is closed under composition with f and contains f^k, then it contains f^{k+1}. So a subobject $A \subseteq Y_s$ is completely characterized by the least element (if it has one) in $A(s)$. We can write such a subobject as $A_k \subseteq Y_s$ for $k \in \mathbb{N} \sqcup \{\infty\}$, where $A_k(s) = \{f^i \mid i \in \mathbb{N}, \ i \geqslant k\} \subseteq Y_s(s)$. In particular, $A_\infty = \varnothing$ is the initial instance. We elide the difference between $A_k\colon \mathcal{L}oop \to \mathbf{Set}$ and $A_k(s) \in \mathrm{Ob}(\mathbf{Set})$.

Now we need to understand the function $\Omega_{\mathcal{L}oop}(f)\colon \Omega_{\mathcal{L}oop}(s) \to \Omega_{\mathcal{L}oop}(s)$. The function $Y_f\colon Y_s \to Y_s$ is simply "composition with f," sending f^k to f^{k+1}. So diagram (7.8), applied to s, says that for a subset A_k, the set $\Omega_{\mathcal{L}oop}(f)(A_k)$ is the pullback in the following diagram (left), which is abbreviated in the right-hand diagram:

In other words, we have

$$\Omega_{\mathcal{L}oop}(f)(A_k) = \begin{cases} A_\infty & \text{if } k = \infty, \\ A_{k-1} & \text{if } 1 \leqslant k < \infty, \\ A_0 & \text{if } k = 0. \end{cases}$$

At this point perhaps it is redundant, but we provide the requested table:

$\Omega_{\mathcal{L}oop}(s)$	
ID	**f**
A_0	A_0
A_1	A_0
A_2	A_1
A_3	A_2
$\vdots$	$\vdots$
A_∞	A_∞

b. The tables $Y_s(s)$ and $\Omega_{\mathcal{L}oop}(s)$ in part (a) are not isomorphic, even though they have basically the same number of rows (infinitely many). For example, Y_s has no fixed points under the action of f, whereas $\Omega_{\mathcal{L}oop}$ has two.

c. The idea is that $char(W)(s) \colon X(s) \to \Omega_{\mathcal{L}oop}(s)$ will send each element of $x \in X(s)$ to "the number k such that applying f^k to x will put it into W." So we have

$char(W)(s)$	
$X(s)$	$\Omega_{\mathcal{L}oop}(s)$
A	A_2
B	A_1
C	A_0
D	A_2
E	A_0
F	A_∞
G	A_∞
H	A_∞

Exercise 7.2.1.28.

Let **GrIn** = $\boxed{\begin{array}{c} Ar \xrightarrow[tgt]{src} Ve \\ \bullet \qquad \bullet \end{array}}$ be the indexing category for graphs.

a. Write the subobject classifier $\Omega_{\mathbf{GrIn}} \in \text{Ob}(\mathbf{GrIn}\text{–}\mathbf{Set})$ in tabular form, i.e., as two tables.

b. Draw $\Omega_{\mathbf{GrIn}}$ as a graph.

c. Let G be the following graph and $G' \subseteq G$ the blue part.

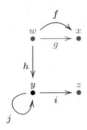

Write $G \in \text{Ob}(\mathbf{GrIn}\text{–}\mathbf{Set})$ in tabular form.

d. Write the components of the natural transformation $char(G') \colon G \to \Omega_{\mathbf{GrIn}}$.

◊

Solution 7.2.1.28.

a. Recall from Exercise 7.2.1.9 that Y_{Ar} and Y_{Ve} are as follows:

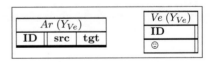

There are five subobjects of Y_{Ar}. We name them Arrow, Endpoints, Source, Target, and Nothing:

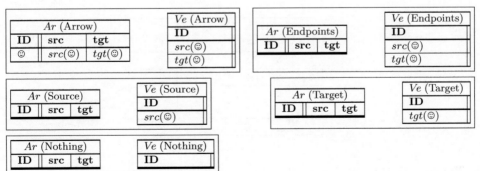

There are two subobjects of Y_{Ve}, namely, $\emptyset \subseteq Y_{Ve}$ and $Y_{Ve} \subseteq Y_{Ve}$. We are ready to write $\Omega_{\mathbf{GrIn}}$.

$\Omega_{\mathbf{GrIn}}(Ar)$		
ID	**src**	**tgt**
Arrow	Vertex	Vertex
Endpoints	Vertex	Vertex
Source	Vertex	Nothing
Target	Nothing	Vertex
Nothing	Nothing	Nothing

$\Omega_{\mathbf{GrIn}}(Ve)$
ID
Vertex
Nothing

b.

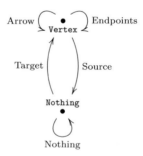

c. This is review and preparation for part (d).

$Ar\,(G)$		
ID	**src**	**tgt**
f	w	x
g	w	x
h	w	y
i	y	z
j	y	y

$Ve\,(G)$
ID
w
x
y
z

d. The natural transformation $char(G')\colon G \to \Omega_{\mathbf{GrIn}}$ has a component function for each object in $\mathrm{Ob}(\mathbf{GrIn}) = \{Ar, Ve\}$:

$char(G')_{Ar}$	
$G(Ar)$	$\Omega_{\mathbf{GrIn}}(Ar)$
f	Endpoints
g	Arrow
h	Source
i	Target
j	Nothing

$char(G')_{Ve}$	
$G(Ve)$	$\Omega_{\mathbf{GrIn}}(Ve)$
w	Vertex
x	Vertex
y	Nothing
z	Vertex

◆

7.2.2 Database instances in other categories

So far we have focused on the category $\mathcal{C}$–$\mathbf{Set} = \mathrm{Fun}(\mathcal{C}, \mathbf{Set})$ of set-valued functors $\mathcal{C} \to \mathbf{Set}$ for arbitrary categories, or database schemas, $\mathcal{C}$. What if we allow the target category $\mathbf{Set}$ to change?

7.2.2.1 Representations of groups

The classical mathematical subject of *representation theory* is the study of $\mathrm{Fun}(G, \mathbf{Vect})$, where G is a group and $\mathbf{Vect}$ is the category of vector spaces (over, say, $\mathbb{R}$). Every such functor $F\colon G \to \mathbf{Vect}$ is called a *representation of G*. Since G is a category with one object ▲, the functor F provides a single vector space $V = F(▲)$ together with an action of G on it.

We can think of this in terms of databases if we have a presentation of G in terms of generators and relations. The schema corresponding to G has one table, and this table has a column for each generator (see Section 4.1.3). Giving a representation F is the same as giving an instance on the schema, with some properties that stem from the fact that the target category is $\mathbf{Vect}$ rather than $\mathbf{Set}$. There are many possibilities for expressing such data.

One possibility is if we could draw V, say, if V were one-, two-, or three-dimensional. If so, let P be the chosen picture of V, e.g., P is the standard drawing of a Cartesian coordinate plane $V = \mathbb{R}^2$. Then every column of the table would consist entirely of the picture P instead of a set of rows. Touching a point in the ID column $\mathbb{R}^2$ would result in a point being drawn in the $\mathbb{R}^2$ corresponding to the other column, in accordance with the G action. Each column would, of course, respect addition and scalar multiplication.

Another possibility is to use the fact that there is a functor $U\colon \mathbf{Vect} \to \mathbf{Set}$, so the instance $F\colon G \to \mathbf{Vect}$ could be converted to an ordinary instance $U \circ F\colon G \to \mathbf{Set}$.

We would have an ordinary set of rows. This set would generally be infinite, but it would be structured by addition and scalar multiplication. For example, assuming V is finite-dimensional, one could find a few rows that generated the rest.

A third possibility is to use monads, which would allow the table to have only as many rows as V has dimensions. This yields a considerable saving of space. See Section 7.3. In all these possibilities, the usual tabulated format of databases has been slightly altered to accommodate the extra information in a vector space.

7.2.2.2   Representations of quivers

Representation theory also studies representations of quivers. A *quiver* is just the free category (see Example 5.1.2.33) on a graph. If P is a graph with free category $\mathcal{P}$, then a representation of the quiver $\mathcal{P}$ is a functor $F\colon \mathcal{P} \to \mathbf{Vect}$. Such a representation consists of a vector space at each vertex of P and a linear transformation for each arrow. All the discussion in Section 7.2.2.1 works in this setting, except that there is more than one table.

7.2.2.3   Other target categories

One can imagine the value of using target categories other than **Set** or **Vect** for databases.

Application 7.2.2.4. Geographic data consists of maps of the earth together with various functions on it. For example, for any point on the earth one may want to know the average of temperatures recorded in the past ten years or the precise temperature at this moment. Earth can be considered as a topological space, E. Similarly, temperatures on earth reside on a continuum, say, the space T of real numbers $[-100, 200]$. Thus the temperature record is a continuous function $E \to T$.

Other records such as precipitation, population density, elevation, and so on, can all be considered as continuous functions from E to some space. Agencies like the U.S. Geological Survey hold databases of such information. By modeling them on functors $\mathcal{C} \to \mathbf{Top}$, they may be able to employ mathematical tools such as persistent homology (see Weinberger [44]) to find interesting invariants of the data.

◇◇

Application 7.2.2.5. Application 7.2.2.4 discussed using topological database instances to model geographical data. Other scientific disciplines could use the same kind of tool. For example, in studying the mechanics of materials, one may want to consider the material as a topological space M and measure values such as energy as a continuous map $M \to E$. Such observations could be modeled by databases with target category **Top** or **Vect** rather than **Set**.

◇◇

7.2.3   Sheaves

Let X be a topological space (see Example 5.2.3.1), such as a sphere. Section 7.2.2.3 discussed continuous functions out of X and their use in science (e.g., recording temperatures on the earth as a continuous map $X \to [-100, 200]$). Sheaves allow us to consider the local-global nature of such maps, taking into account reparable discrepancies in data-gathering tools.

Application 7.2.3.1. Suppose that X is the topological space corresponding to the earth, and let *region* mean an open subset $U \subseteq X$. Suppose that we cover X with 10,000 regions $U_1, U_2, \ldots, U_{10000}$, such that some of the regions overlap in a nonempty subregion (e.g., $U_5 \cap U_9 \neq \varnothing$). For each i, j, let $U_{i,j} = U_i \cap U_j$.

For each region $U_i \subseteq X$, we have a temperature-recording device, which gives a function $T_i \colon U_i \to [-100, 200]$. If $U_i \cap U_j \neq \varnothing$, then two different recording devices give us temperature data for the intersection $U_{i,j}$. Suppose we find that they do not give precisely the same data but that there is a translation formula between their results. For example, T_i might register 3° warmer than T_j registers, throughout the region $U_i \cap U_j$.

Roughly speaking, a consistent system of translation formulas is called a *sheaf*. It does not demand a universal true temperature function but only a consistent translation system between them.

◊◊

Definitions 7.2.3.2 and 7.2.3.5 make the notion of sheaf precise, but it is developed slowly at first.

For every region U, we can record the value of some function (say, temperature) throughout U. Although this record might consist of a mountain of data (a temperature for each point in U), it can be thought of as one thing. That is, it is one element in the set of "value assignments throughout U". A sheaf holds the set of "value assignments throughout U" for each region U as well as how a "value assignment throughout U" restricts to a "value assignment throughout V" for any subset $V \subseteq U$.

Definition 7.2.3.2. Let X be a topological space, let $\mathrm{Open}(X)$ denote its partial order of open sets, and let $\mathrm{Open}(X)^{\mathrm{op}}$ be the opposite category. A *presheaf on X* is a functor $\mathcal{O} \colon \mathrm{Open}(X)^{\mathrm{op}} \to \mathbf{Set}$. For every open set $U \subseteq X$, we refer to the set $\mathcal{O}(U)$ as the *set of value assignments throughout U of $\mathcal{O}$*. If $V \subseteq U$ is an open subset, it corresponds to an arrow in $\mathrm{Open}(X)$, and applying the functor $\mathcal{O}$ yields a function called the *restriction map from U to V* and denoted $\rho_{V,U} \colon \mathcal{O}(U) \to \mathcal{O}(V)$. Given $a \in \mathcal{O}(U)$, we may denote $\rho_{V,U}(a)$ by $a|_V$; it is called *the restriction of a to V*.

The *category of presheaves on X* is simply $\mathrm{Open}(X)^{\mathrm{op}}\text{–}\mathbf{Set}$ (see Definition 5.3.3.1).

Exercise 7.2.3.3.

a. Find four overlapping open subsets that cover the square $X := [0,3] \times [0,3] \subseteq \mathbb{R}^2$. Write a label for each open set as well as a label for each overlap (two-fold, three-fold, etc.). You now have labeled n open sets. What is your n?

b. Draw the preorder Open(X). For each of the n open sets, draw a dot with the appropriate label. Then draw an arrow from one dot to another when the first refers to an open subset of the second. This is Open(X).

c. Make up and write formulas $R_1 \colon X \to \mathbb{R}$ and $R_2 \colon X \to \mathbb{R}$ with $R_1(x) \leqslant R_2(x)$ for all $x \in X$, expressing a range of temperatures $R_1(p) \leqslant Temp(p) \leqslant R_2(p)$ that an imaginary experiment shows can exist at each point p in the square. What is the temperature range at $p = (2,1) \in X$?

d. Make a presheaf $\mathcal{O} \colon \text{Open}(X)^{\text{op}} \to \mathbf{Set}$ as follows. For each of your open sets, say, $A \in \text{Open}(X)$, put

$$\mathcal{O}(A) := \{Temp \colon A \to \mathbb{R} \mid \forall a \in A, \ R_1(a) \leqslant Temp(a) \leqslant R_2(a)\}.$$

Call one of your n open sets A. What is $\mathcal{O}(A)$? Then choose some $A' \subseteq A$; what is $\mathcal{O}(A')$, and what is the restriction map $\rho_{A',A} \colon \mathcal{O}(A) \to \mathcal{O}(A')$ in this case? Do you like the name "value assignment throughout A" for an element of $\mathcal{O}(A)$?

$\Diamond$

Solution 7.2.3.3.

a. Inside the 3×3 grid, I drew four 2×2 squares, which I denoted $NE = [1,3] \times [1,3]$ (for NorthEast), $NW := [0,2] \times [1,3]$, $SE = [1,3] \times [0,2]$, and $SW = [0,2] \times [0,2]$. The two-fold intersections are denoted

$$N := NE \cap NW, \qquad S := SE \cap SW, \qquad E := NE \cap SE,$$
$$W := NW \cap SW, \qquad C := NW \cap SE = NE \cap SW.$$

All the other n-fold intersections are one of these, so my value of n is 9. (Other values are possible.)

b.

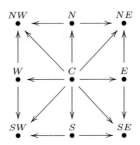

c. Let $R_1 = x + y$ and $R_2 = 2x + y + 1$. We have $3 \leqslant Temp(2,1) \leqslant 6$.

d. Let $A = N = [1,2] \times [2,3]$ be the North subset. We have

$$\mathcal{O}(N) = \{Temp\colon N \to \mathbb{R} \mid \forall (x,y) \in N, \ x + y \leqslant Temp(x,y) \leqslant 2x + y + 1\}.$$

We have $C \subseteq N$, so let $A' = C = [1,2] \times [1,2]$, so

$$\mathcal{O}(C) = \{Temp\colon C \to \mathbb{R} \mid \forall (x,y) \in C, \ x + y \leqslant Temp(x,y) \leqslant 2x + y + 1\}.$$

Clearly, there is a function $\mathcal{O}(N) \to \mathcal{O}(C)$, because if we have a $Temp$ function that is defined throughout N, we can restrict it to a $Temp$ function that is defined throughout C, and the conditions on it (namely, $x + y \leqslant Temp(x,y) \leqslant 2x + y + 1$) are the same. I think "value assignment throughout A" is a good name for this concept.

◆

Before moving to a definition of sheaves, we need to clarify the notion of covering. Suppose that U is a region and $V_1, \ldots, V_n$ are subregions (i.e., for each $1 \leqslant i \leqslant n$, we have $V_i \subseteq U$). Then we say that the V_i *collectively cover* U if every point in U is in V_i for some i. Another way to say this is that the natural function $\sqcup_i V_i \to U$ is surjective.

Example 7.2.3.4. Let $X = \mathbb{R}$ be the space of real numbers, and define the following open subsets: $U = (5,10), V_1 = (5,7), V_2 = (6,9), V_3 = (8,10).$[10] Then V_1, V_2, V_3 collectively cover of U. It has overlaps $V_{12} = V_1 \cap V_2 = (6,7), V_{13} = V_1 \cap V_3 = \varnothing, V_{23} = V_2 \cap V_3 = (8,9)$.

Given a presheaf $\mathcal{O}\colon \mathrm{Open}(X)^{\mathrm{op}} \to \mathbf{Set}$, we have sets and functions as in the following

---

[10]Parentheses are used to denote open intervals of real numbers. For example, $(6,9)$ denotes the set $\{x \in \mathbb{R} \mid 6 < x < 9\}$.

diagram

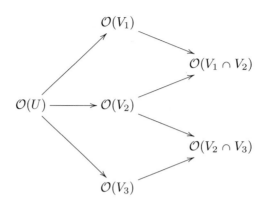

A presheaf $\mathcal{O}$ on X tells us what value assignments throughout U can exist for each U. Suppose we have a value assignment $a_1 \in \mathcal{O}(V_1)$ throughout V_1 and another value assignment $a_2 \in \mathcal{O}(V_2)$ throughout V_2, and suppose they agree as value assignments throughout $V_1 \cap V_2$, i.e., $a_1|_{V_1 \cap V_2} = a_2|_{V_1 \cap V_2}$. In this case we should have a unique value assignment $b \in \mathcal{O}(V_1 \cup V_2)$ throughout $V_1 \cup V_2$ that agrees on the V_1 part with a_1 and agrees on the V_2 part with a_2; i.e., $b|_{V_1} = a_1$ and $b|_{U_2} = a_2$. The condition that such equations hold for every covering is the sheaf condition.

For example, the elements of $\mathcal{O}(U)$ might be functions $h \colon U \to \mathbb{R}$, each of which we imagine as a curve defined on the interval $U = (5, 10)$. The sheaf condition says that if one is given a curve-snippet over $(5, 7)$, a curve-snippet over $(6, 9)$, and a curve snippet over $(8, 10)$, and these all agree on overlap intervals $(6, 7)$ and $(8, 9)$, then they can be put together to form a curve over all of U.

Definition 7.2.3.5. Let X be a topological space, let $\mathrm{Open}(X)$ be its partial order of open sets, and let $\mathcal{O} \colon \mathrm{Open}(X)^{\mathrm{op}} \to \mathbf{Set}$ be a presheaf. Given an open set $U \subseteq X$ and a cover $V_1, \ldots, V_n$ of U, the following condition is called the *sheaf condition* for that cover.

Sheaf condition Given a sequence $a_1, \ldots, a_n$, where each $a_i \in \mathcal{O}(V_i)$ is a value assignment throughout V_i, suppose that for all i, j, we have $a_i|_{V_i \cap V_j} = a_j|_{V_i \cap V_j}$; then there is a unique value assignment $b \in \mathcal{O}(U)$ such that $b|_{V_i} = a_i$.

The presheaf $\mathcal{O}$ is called a *sheaf* if it satisfies the sheaf condition for every cover.

Remark 7.2.3.6. Application 7.2.3.1 said that sheaves help us patch together information from different sources. Even if different temperature-recording devices T_i and T_j registered different temperatures on an overlapping region $U_i \cap U_j$, they could be patched together if given a consistent translation system between their results. What is actually needed is a set of isomorphisms

$$p_{i,j} \colon T_i|_{U_{i,j}} \xrightarrow{\cong} T_j|_{U_{i,j}}$$

that translate between them, and that these $p_{i,j}$'s act in concert with one another. This (when precisely defined) is called *descent data*. The way it interacts with the definition of sheaf given in Definitions 7.2.3.2 and 7.2.3.5 is buried in the restriction maps ρ for the overlaps as subsets $U_{i,j} \subseteq U_i$ and $U_{i,j} \subseteq U_j$ (see Grothendieck and Raynaud [18] for details).

Application 7.2.3.7. Consider outer space as a topological space X. Different amateur astronomers record observations of what they see in X on a given night. Let $C = [390, 700]$ denote the set of wavelengths in the visible light spectrum (written in nanometers). Given an open subset $U \subseteq X$, let $\mathcal{O}(U)$ denote the set of functions $U \to C$. The presheaf $\mathcal{O}$ satisfies the sheaf condition; this is the taken-for-granted fact that we can patch together different observations of space.

Figure 7.1 (see page 469) shows three views of the night sky. Given a telescope position to obtain the first view, one moves the telescope right and a little down to obtain the second, and one moves it down and left to obtain the third. These are value assignments $a_1 \in \mathcal{O}(V_1), a_2 \in \mathcal{O}(V_2)$, and $a_3 \in \mathcal{O}(V_3)$ throughout subsets $V_1, V_2, V_3 \subseteq X$ (respectively). These subsets V_1, V_2, V_3 cover some (strangely shaped) subset $U \subseteq X$. Because the restriction of a_1 to $V_1 \cap V_2$ is equal to the restriction of a_2 to $V_1 \cap V_2$, and so on, the sheaf condition says that these three value assignments glue together to form a single value assignment throughout U, as shown in Figure 7.2 (see page 470).

◊◊

Exercise 7.2.3.8.

Find an application of sheaves in your own domain of expertise. ◊

Solution 7.2.3.8.

Suppose a sociologist assigns to each open set U on earth the set $\mathcal{E}(U)$ of all ways that the inhabitants of U could feasibly be employed. This forms a sheaf. If $V \subseteq U$ is a subset, then any method to employ everyone in U gives a method by which to employ everyone in V. And if we know how to employ everyone in V_1 and everyone in V_2, and if our methods agree on $V_1 \cap V_2$, then we know a way to employ everyone in $V_1 \cup V_2$. ♦

Application 7.2.3.9. Suppose we have a sheaf for temperatures on earth. For every region U, we have a set of theoretically possible temperature assignments throughout U. For example, we may know that if it is warm in Texas, warm in Arkansas, and warm in Kansas, then it cannot be cold in Oklahoma. With such a sheaf $\mathcal{O}$ in hand, one can use facts about the temperature in one region U to predict the temperature in another region V.

The mathematics is as follows. Suppose given regions $U, V \subseteq X$ and a subset $A \subseteq \mathcal{O}(U)$ corresponding to what we know about the temperature assignment throughout U. We take the following fiber product:

$$
\begin{array}{ccc}
(\rho_{U,X})^{-1}(A) & \longrightarrow \mathcal{O}(X) \xrightarrow{\ \rho_{V,X}\ } \mathcal{O}(V) \\
\downarrow \quad\ \lrcorner & \downarrow {\scriptstyle \rho_{U,X}} \\
A & \longrightarrow \mathcal{O}(U)
\end{array}
$$

The image of the top composite $\mathrm{im}((\rho_{U,X})^{-1}(A) \to \mathcal{O}(V))$ is a subset of $\mathcal{O}(V)$ telling us which temperature assignments are possible throughout V, given our knowledge A about the temperature throughout U.

We can imagine the same type of prediction systems for other domains as well, such as the energy of various parts of a material.   ◊◊

Example 7.2.3.10. Exercises 5.2.4.3 and 5.2.4.4 discussed the idea of laws being dictated or respected throughout a jurisdiction. If X is earth, to every jurisdiction $U \subseteq X$ we assign the set $\mathcal{O}(U)$ of laws that are dictated to hold throughout U. Given a law on U and a law on V, we can see if they amount to the same law on $U \cap V$. For example, on U a law might say, "no hunting near rivers" and on V a law might say, "no hunting in public areas." It happens that on $U \cap V$ all public areas are near rivers, and vice versa, so the laws agree there. These laws patch together to form a single rule about hunting that is enforced throughout the union $U \cup V$, respected by all jurisdictions within it.

7.2.3.11   Sheaf of ologged concepts

Definition 7.2.3.5 defines what should be called a sheaf of sets. We can discuss sheaves of groups or even sheaves of categories. Here is an application of the latter.

Recall the notion of simplicial complexes (see Section 3.4.4.3). They look like this:

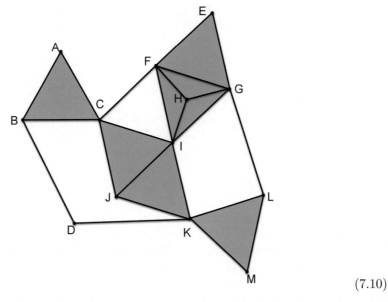

$$(7.10)$$

Given such a simplicial complex X, we can imagine each vertex $v \in X_0$ as an entity with a worldview (e.g., a person) and each simplex as the common worldview shared by its vertices. To model this, we assign to each vertex $v \in X$ an olog $\mathcal{O}(v)$, corresponding to the worldview held by that entity, and to each simplex $u \in X_n$, we assign an olog $\mathcal{O}(u)$ corresponding to a *common ground* worldview. Recall that X is a subset of $\mathbb{P}(X_0)$; it is a preorder and its elements (the simplices) are ordered by inclusion. If u, v are simplices with $u \subseteq v$, then we want a map of ologs (i.e., a schema morphism) $\mathcal{O}(v) \to \mathcal{O}(u)$. In this way the model says that any idea shared among the people in v is shared among the people in u. Thus we have a functor $\mathcal{O} \colon X \to \mathbf{Sch}$ (where we forget the distinction between ologs and databases for notational convenience).

To every simplicial complex (indeed every ordered set) one can associate a topological space; in fact, we have a functor $Alx \colon \mathbf{PrO} \to \mathbf{Top}$, called the Alexandrov functor. Applying $Alx(X^{\mathrm{op}})$, we have a space denoted $\mathcal{X}$. One can visualize $\mathcal{X}$ as X, but the open sets include unions of simplices. There is a unique sheaf of categories on $\mathcal{X}$ that behaves like $\mathcal{O}$ on simplices of X.

Example 7.2.3.12. Imagine two groups of people G_1 and G_2 each making observations about the world. Suppose there is some overlap $H = G_1 \cap G_2$. Then it may happen that there is a conversation including G_1 and G_2, and both groups are talking about something (though using different words). H says, "You guys are talking about the same things, you just use different words." In this case there is an observation being made throughout $G_1 \cup G_2$ that agrees with both those on G_1 and those on G_2.

7.2.3.13   Time

One can use sheaves to model objects in time; Goguen [17] gave an approach to this. For an approach that more closely fits the flow of this book, let $\mathcal{C}$ be a database schema. The lifespan of information about the world is generally finite; that is, what was true yesterday is not always the case today. Thus we can associate to each interval U of time the information that we deem to hold throughout U. This is sometimes called the *valid time* of the data.

If data is valid throughout U and we have a subset $V \subseteq U$, then of course it is valid throughout V. And the sheaf condition holds too. If some information is valid throughout U, and some other information is valid throughout U', and if these two things restrict to the same information on the overlap $U \cap V$, then they can be glued together to form information that is valid throughout the union $U \cup V$.

So we can model information change over time by using a sheaf of $\mathcal{C}$-sets on the topological space $\mathbb{R}$. In other words, for every time interval, we give an $\mathcal{C}$-instance whose information is valid throughout that time interval. Definition 7.2.3.5 only defined sheaves with values in **Set**; we are now generalizing to sheaves in $\mathcal{C}$–**Set**. Namely we consider functors $\mathrm{Open}(\mathbb{R}) \to \mathcal{C}$–**Set** satisfying the same sheaf condition.

Example 7.2.3.14. Consider a hospital in which babies are born. In our scenario, mothers enter the hospital, babies are born, mothers and babies leave the hospital. Let $\mathcal{C}$ be the schema

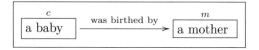

Consider the eight-hour intervals

$$\mathrm{Shift}_1 := (\mathrm{Jan}\ 1, \quad 00:00 - 08:00),$$
$$\mathrm{Shift}_2 := (\mathrm{Jan}\ 1, \quad 04:00 - 12:00),$$
$$\mathrm{Shift}_3 := (\mathrm{Jan}\ 1, \quad 08:00 - 16:00).$$

The nurses take shifts of eight hours, overlapping with their predecessors by four hours, and they record in the database only patients that were there *throughout* their shift or throughout any overlapping shift. Here is the schema:

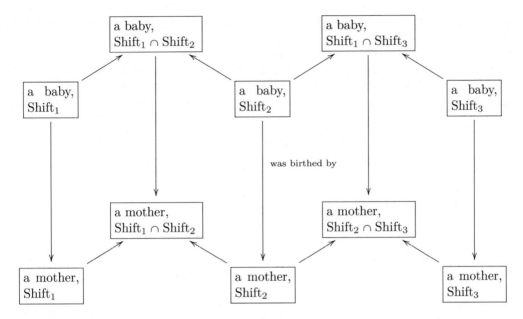

Whether or not this implementation of the sheaf semantics is most useful in practice is certainly debatable. But something like this could easily be useful as a semantics, i.e., a way of thinking about, the temporal nature of data.

7.3 Monads

Monads would probably not have been invented without category theory, but they have been useful in understanding algebraic theories, calculating invariants of topological spaces, and embedding nonfunctional operations into functional programming languages. We mainly discuss monads in terms of how they can help one make explicit a given modeling context and in so doing allow one to simplify the language used in such models. We use databases to give concrete examples.

Much of the following material on monads is taken from Spivak [40].

7.3.1 Monads formalize context

Monads can formalize assumptions about the way one does business throughout a domain. For example, suppose we want to consider functions that are not required to return a value for all inputs. These are not valid functions as defined in Section 2.1.2 (because they are not *total*), but in math classes one wants to speak of $f(x) = \frac{1}{x}$ and $g(x) = \tan(x)$ as *though* they were functions $\mathbb{R} \to \mathbb{R}$, so that they can be composed without constantly paying attention to domains.

Functions that are not required to be defined throughout their domain are called *partial functions*. We all know how they should work, so we need a way to make it mathematically legal. Monads, and the *Kleisli* categories to which they give rise, provide us with a way to do so. In particular, we will be able to formally discuss the composition

$$\mathbb{R} \xrightarrow{\frac{1}{x}} \mathbb{R} \xrightarrow{\tan(x)} \mathbb{R}.$$

Here we are drawing arrows between sets as though we were talking about total functions, but there is an implicit context in which we are actually talking about partial functions. Monads allow us to write maps between sets in the functional way while holding the underlying context. What makes them useful is that the notion of *context* we are using here is made formal.

Example 7.3.1.1 (Partial functions). Partial functions can be modeled by ordinary functions if we add a special "no answer" element to the codomain. That is, the set of partial functions $A \to B$ is in one-to-one correspondence with the set of ordinary functions $A \to B \sqcup \{\odot\}$. For example, suppose we want to model the partial function $f_p(x) := \frac{1}{x^2-1} : \mathbb{R} \to \mathbb{R}$ in this way; we would use the total function $f_t : \mathbb{R} \to \mathbb{R} \sqcup \{\odot\}$ defined as:

$$f(x) := \begin{cases} \frac{1}{x^2-1} & \text{if } x \neq -1 \text{ and } x \neq 1, \\ \odot & \text{if } x = -1, \\ \odot & \text{if } x = 1. \end{cases}$$

An ordinary function $g \colon A \to B$ can be considered a partial function because we can compose it with the inclusion

$$B \to B \sqcup \{\odot\}. \tag{7.11}$$

to get $A \to B \sqcup \{\odot\}$.

But how do we compose two partial functions written in this way? Suppose $f \colon A \to B \sqcup \{\odot\}$ and $g \colon B \to C \sqcup \{\odot\}$ are functions. First form a new function

$$g' := g \sqcup \{\odot\} \colon B \sqcup \{\odot\} \to C \sqcup \{\odot\} \sqcup \{\odot\},$$

then compose to get $(g' \circ f) \colon A \to C \sqcup \{\odot\} \sqcup \{\odot\}$, and finally send both $\odot$'s to the same element by composing with

$$C \sqcup \{\odot\} \sqcup \{\odot\} \to C \sqcup \{\odot\}. \tag{7.12}$$

How should one think about composing partial functions $g \circ f$? Every element $a \in A$ is sent by f either to an element $b \in B$ or to "no answer." If it has an answer $f(a) \in B$, then this again is sent by g either to an element $g(f(a)) \in C$ or to "no answer." We get a partial function $A \to C$ by sending a to $g(f(a))$ if possible or to "no answer" if it gets stopped along the way.

This monad is sometimes called the *maybe monad* in computer science, because a partial function $f \colon A \to B$ takes every element of A and may output just an element of B or may output nothing; more succinctly, it outputs a "maybe B."

Exercise 7.3.1.2.

a. Let $f \colon \mathbb{Z} \to \mathbb{Z} \sqcup \{\odot\}$ be the partial function given by $f(n) = \frac{1}{n^2 - n}$. Calculate the following: $f(-3), f(-2), f(-1), f(0), f(1)$, and $f(2)$.

b. Let $g \colon \mathbb{Z} \to \mathbb{Z} \sqcup \{\odot\}$ be the partial function given by

$$g(n) = \begin{cases} n^2 - 3 & \text{if } n \geq -1, \\ \odot & \text{if } n < -1 \end{cases}$$

Write $f \circ g(n)$ for $-3 \leq n \leq 2$.

◇

Solution 7.3.1.2.

a.)

f	
$\mathbb{Z}$	$\mathbb{Z} \sqcup \{\odot\}$
-3	$\frac{1}{12}$
-2	$\frac{1}{6}$
-1	$\frac{1}{2}$
0	$\odot$
1	$\odot$
2	$\frac{1}{2}$

b.)

$f \circ g$	
$\mathbb{Z}$	$\mathbb{Z} \sqcup \{\odot\}$
-3	$\odot$
-2	$\odot$
-1	$\frac{1}{6}$
0	$\frac{1}{12}$
1	$\frac{1}{6}$
2	$\odot$

◆

Application 7.3.1.3. Experiments are supposed to be performed objectively, but suppose we imagine that changing the person who performs the experiment, say, in psychology, may change the outcome. Let A be the set of experimenters, let X be the parameter space for the experimental variables (e.g., $X = \text{Age} \times \text{Income}$), and let Y be the observation space (e.g., $Y = \text{propensity for violence}$). We want to think of such an experiment as telling us about a function $f\colon X \to Y$ (how age and income affect propensity for violence). However, we may want to make some of the context explicit by including information about who performed the experiment. That is, we are really finding a function $f\colon X \times A \to Y$.

Given a set P of persons, the experimenter wants to know the age and income of each, i.e., a function $P \to X$. However, it may be the case that even ascertaining this basic information, which is achieved merely by asking each person these questions, is subject to which experimenter in A is doing the asking. Then we again want to consider the experimenter as part of the equation, replacing the function $P \to X$ with a function $P \times A \to X$. In such a case, we can use a monad to hide the fact that everything in sight is assumed to be influenced by A. In other words, we want to announce, once and for all, the modeling context—that every observable is possibly influenced by the observer—so that it can recede into the background.

We return to this in Examples 7.3.2.6 and 7.3.3.4.

$\diamond\diamond$

7.3.2   Definition and examples

What aspects of Example 7.3.1.1 are about monads, and what aspects are about partial functions in particular? Monads are structures involving a functor and a couple of natural transformations. Roughly speaking, the functor for partial functors was $B \mapsto B \sqcup \{\odot\}$, and the natural transformations were given in (7.11) and (7.12). This section gives the definition of monads and a few examples. We return to consider about how monads formalize context in Section 7.3.3.

Definition 7.3.2.1 (Monad). A *monad on* **Set** is defined as follows: One announces some constituents (A. functor, B. unit map, C. multiplication map) and shows that they conform to some laws (1. unit laws, 2. associativity law). Specifically, one announces

A. a functor $T\colon \mathbf{Set} \to \mathbf{Set}$,

B. a natural transformation $\eta\colon \mathrm{id}_{\mathbf{Set}} \to T$,

C. a natural transformation $\mu\colon T \circ T \to T$.

We sometimes refer to the functor T as though it were the whole monad; we call η the *unit map* and μ the *multiplication map*. One must then show that the following *monad laws* hold:

1. The following diagrams of functors **Set** → **Set** commute:

2. The following diagram of functors **Set** → **Set** commutes:

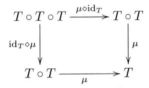

Example 7.3.2.2 (List monad). We now go through Definition 7.3.2.1 using the List monad. The first step is to give a functor List: **Set** → **Set**, which was done in Example 5.1.2.20. Recall that if $X = \{p, q, r\}$, then List(X) includes the empty list [ ], singleton lists such as $[p]$, and any other list of elements in X such as $[p, p, r, q, p]$. Given a function $f\colon X \to Y$, one obtains a function List(f): List(X) → List(Y) by entrywise application of f, as in Exercise 5.1.2.22.

As a monad, the functor List comes with two natural transformations, a unit map η and a multiplication map μ. Given a set X, the unit map $\eta_X\colon X \to$ List(X) returns singleton lists as follows:

$$X \xrightarrow{\quad \eta_X \quad} \text{List}(X)$$

$$p \mapsto [p]$$
$$q \mapsto [q]$$
$$r \mapsto [r]$$

Given a set X, the multiplication map $\mu_X\colon$ List(List(X)) → List(X) concatenates lists of lists as follows:

$$\text{List}(\text{List}(X)) \xrightarrow{\quad\quad\quad \mu_X \quad\quad\quad} \text{List}(X)$$

$$[[p],[q]] \longmapsto [p,q]$$

$$[[q,p,r],[\,],[q,r,p,r],[r]] \longmapsto [q,p,r,q,r,p,r,r]$$

The naturality of η and μ means that these maps work appropriately well under entrywise application of a function $f\colon X \to Y$. Finally, the three monad laws from Definition 7.3.2.1 can be exemplified as follows:

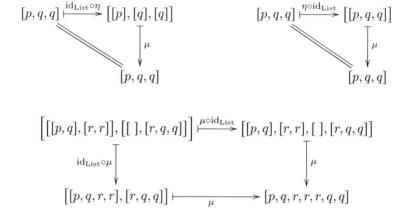

Exercise 7.3.2.3.

Let $\mathbb{P}\colon \mathbf{Set} \to \mathbf{Set}$ be the power-set functor, so that given a function $f\colon X \to Y$, the function $\mathbb{P}(f)\colon \mathbb{P}(X) \to \mathbb{P}(Y)$ is given by taking images.

a. Make sense of the statement, "With η defined by singleton subsets and with μ defined by union, $\top := (\mathbb{P}, \eta, \mu)$ is a monad."

b. With $X = \{a,b\}$, write the function η_X as a two-row, two-column table.

c. With $X = \{a,b\}$, write the function μ_X as a sixteen-row, two-column table (you can stop after five rows if you fully understand it).

d. Check that you believe the monad laws from Definition 7.3.2.1.

◇

Solution 7.3.2.3.

a. The statement suggests that the components of $\eta\colon \mathrm{id}_{\mathbf{Set}} \to \mathbb{P}$ can be defined using the concept of singleton subsets and that the components of $\mu\colon \mathbb{P}\circ\mathbb{P} \to \mathbb{P}$ can be defined using the concept of union. Given a set $X \in \mathrm{Ob}(\mathbf{Set})$, we need a function $\eta_X\colon X \to \mathbb{P}(X)$, meaning that for every element $x \in X$, we need a subset of X. The statement suggests we send x to the singleton subset $\{x\} \subseteq X$. The statement also suggests that we obtain $\mu_X\colon \mathbb{P}(\mathbb{P}(X)) \to \mathbb{P}(X)$ by sending a set of subsets to their union. For example, if $X = \{1,2,3,4,5\}$, then an element $T \in \mathbb{P}(\mathbb{P}(X))$ might look like $\{\{1,2\}, \varnothing, \{1,3,5\}\}$; the union of these subsets is $\mu_X(T) = \{1,2,3,5\}$, a subset of X. It is not hard to check that the given η and μ are natural transformations. The statement now asserts that the power-set functor $\mathbb{P}$, together with these natural transformations, forms a monad.

b.)

	η_X	
X		$\mathbb{P}(X)$
a		$\{a\}$
b		$\{b\}$

c.)

μ_X	
$\mathbb{P}(\mathbb{P}(X))$	$\mathbb{P}(X)$
$\varnothing$	$\varnothing$
$\{\varnothing\}$	$\varnothing$
$\{\{a\}\}$	$\{a\}$
$\{\{b\}\}$	$\{b\}$
$\{\{a,b\}\}$	$\{a,b\}$
$\{\varnothing,\{a\}\}$	$\{a\}$
$\{\varnothing,\{b\}\}$	$\{b\}$
$\{\varnothing,\{a,b\}\}$	$\{a,b\}$
$\{\{a\},\{b\}\}$	$\{a,b\}$
$\{\{a,\{a,b\}\}\}$	$\{a,b\}$
$\{\{b\},\{a,b\}\}$	$\{a,b\}$
$\{\varnothing,\{a\},\{b\}\}$	$\{a,b\}$
$\{\varnothing,\{a\},\{a,b\}\}$	$\{a,b\}$
$\{\varnothing,\{b\},\{a,b\}\}$	$\{a,b\}$
$\{\{a\},\{b\},\{a,b\}\}$	$\{a,b\}$
$\{\varnothing,\{a\},\{b\},\{a,b\}\}$	$\{a,b\}$

d. The monad laws hold. One says that if we take all the singleton subsets of X and union them, we get X. Another says that if we take the singleton set consisting of the whole set X and union it, we get X. The last says that the union of unions is a union.

◆

Example 7.3.2.4 (Partial functions as a monad). Here is the monad for partial functions, as discussed in Example 7.3.1.1. The functor $T\colon \mathbf{Set} \to \mathbf{Set}$ sends a set X to the set $X \sqcup \{\odot\}$. Clearly, given a function $f\colon X \to Y$, there is an induced function $(f \sqcup \{\odot\})\colon (X \sqcup \{\odot\}) \to$

$(Y \sqcup \{\odot\})$, so this is a functor. The natural transformation $\eta \colon \text{id} \to T$ is given on a set X by the component function

$$\eta_X \colon X \to X \sqcup \{\odot\}$$

that includes $X \hookrightarrow X \sqcup \{\odot\}$. Finally, the natural transformation $\mu \colon T \circ T \to T$ is given on a set X by the component function

$$\mu_X \colon X \sqcup \{\odot\} \sqcup \{\odot\} \longrightarrow X \sqcup \{\odot\}$$

that collapses both copies of $\odot$.

Exercise 7.3.2.5.

Let E be a set with elements refered to as *exceptions*. We imagine exceptions as warnings like "overflow!" or "division by zero!" and we imagine that a function $f \colon X \to Y$ outputs either a value or one of these exceptions. Let $T \colon \textbf{Set} \to \textbf{Set}$ be the functor $X \mapsto X \sqcup E$. Follow Example 7.3.2.4 and find a unit map η and a multiplication map μ for which (T, η, μ) is a monad. ◊

Solution 7.3.2.5.

Given a set X, we need $\eta_X \colon X \to X \sqcup E$ and $\mu_X \colon X \sqcup E \sqcup E \to X \sqcup E$. We let η_X be the inclusion, and we let μ_X be the map sending both copies of E in the domain to the codomain by identity on E. ♦

Example 7.3.2.6. Fix a set A. Let $T \colon \textbf{Set} \to \textbf{Set}$ be the functor given by $T(X) = X^A = \text{Hom}_{\textbf{Set}}(A, X)$; this is a functor. For a set X and an element $x \in X$, let $c_x \colon A \to X$ be the constant-x function, $c_x(a) = x$ for all $a \in A$. Define $\eta_X \colon X \to T(X)$ to be given by the constant-x function, $x \mapsto c_x$.

Now we have to specify a natural transformation $\mu \colon T \circ T \to T$, i.e., for each $X \in \text{Ob}(\textbf{Set})$, we need to provide an X-component function

$$\mu_X \colon (X^A)^A \to X^A.$$

By currying (see Example 7.1.1.8), this is equivalent to providing a function $(X^A)^A \times A \to X$. For any $Y \in \text{Ob}(\textbf{Set})$, we have an evaluation function (see Exercise 3.4.2.5) $ev \colon Y^A \times A \to Y$. We use it twice and find the desired function:

$$(X^A)^A \times A \xrightarrow{ev \times \text{id}_A} X^A \times A \xrightarrow{ev} X.$$

Remark 7.3.2.7. Monads can be defined on categories other than $\textbf{Set}$. In fact, for any category $\mathcal{C}$, one can take Definition 7.3.2.1 and replace every occurrence of $\textbf{Set}$ with $\mathcal{C}$ and obtain the definition for monads on $\mathcal{C}$. We have actually seen a monad $(\text{Paths}, \eta, \mu)$

on the category **Grph** of graphs before, namely, in Examples 5.3.1.15 and 5.3.1.16. That is, Paths: **Grph** → **Grph**, which sends a graph to its paths-graph is the functor part. The unit map η includes a graph into its paths-graph using the observation that every arrow is a path of length 1. And the multiplication map μ concatenates paths of paths. The Kleisli category of this monad (see Definition 7.3.3.1) is used, e.g., in (5.19), to define morphisms of database schemas.

7.3.3 Kleisli category of a monad

We are on our way to understanding how monads are used in computer science and how they may be useful for formalizing methodological context. There is only one more stop along the way, called the Kleisli category of a monad. For example, when we apply this Kleisli construction to the partial functions monad (Example 7.3.2.4), we obtain the category of partial functions (see Example 7.3.3.2). When we apply the Kleisli construction to the monad $X \mapsto X^A$ of Example 7.3.2.6 we get the psychological experiment example (Application 7.3.1.3) completed in Example 7.3.3.4.

Definition 7.3.3.1. Let $\top = (T, \eta, \mu)$ be a monad on **Set**. Form a new category, called the *Kleisli category for* $\top$, denoted **Kls**$(\top)$, with sets as objects, $\mathrm{Ob}(\mathbf{Kls}(\top)) := \mathrm{Ob}(\mathbf{Set})$, and with

$$\mathrm{Hom}_{\mathbf{Kls}(\top)}(X, Y) := \mathrm{Hom}_{\mathbf{Set}}(X, T(Y))$$

for sets X, Y. The identity morphism $\mathrm{id}_X \colon X \to X$ in **Kls**$(\top)$ is given by $\eta \colon X \to T(X)$ in **Set**. The composition of morphisms $f \colon X \to Y$ and $g \colon Y \to Z$ in **Kls**$(\top)$ is given as follows. Writing them as functions, we have $f \colon X \to T(Y)$ and $g \colon Y \to T(Z)$. The first step is to apply the functor T to g, giving $T(g) \colon T(Y) \to T(T(Z))$. Then compose with f to get $T(g) \circ f \colon X \to T(T(Z))$. Finally, compose with $\mu_Z \colon T(T(Z)) \to T(Z)$ to get the required function $X \to T(Z)$:

$$X \xrightarrow{\ f\ } TY \qquad\qquad\qquad (7.13)$$

$$Y \xrightarrow{\ g\ } TZ$$

$$\dotfill$$

$$X \xrightarrow{\ f\ } TY \xrightarrow{\ Tg\ } TTZ \xrightarrow{\ \mu_Z\ } TZ.$$

The associativity of this composition formula follows from the associativity law for monads.

Example 7.3.3.2. Recall the monad $\top$ for partial functions, $T(X) = X \sqcup \{\odot\}$, from Example 7.3.2.4. The Kleisli category **Kls**$(\top)$ has sets as objects, but a morphism $f \colon X \to Y$ means a function $X \to Y \sqcup \{\odot\}$, i.e., a partial function. Given another morphism

$g\colon Y \to Z$, the composition formula in $\mathbf{Kls}(\top)$ ensures that $g \circ f\colon X \to Z$ has the appropriate behavior.

Note how this monad allows us to make explicit a context in which all functions are assumed partial and then hide this context from our notation.

Remark 7.3.3.3. For any monad $\top = (T, \eta, \mu)$ on $\mathbf{Set}$, there is a functor $i\colon \mathbf{Set} \to \mathbf{Kls}(\top)$, given as follows. On objects we have $\mathrm{Ob}(\mathbf{Kls}(\top)) = \mathrm{Ob}(\mathbf{Set})$, so take $i = \mathrm{id}_{\mathrm{Ob}(\mathbf{Set})}$. Given a morphism $f\colon X \to Y$ in $\mathbf{Set}$, we need a morphism $i(f)\colon X \to Y$ in $\mathbf{Kls}(\top)$, i.e., a function $i(f)\colon X \to T(Y)$. We assign $i(f)$ to be the composite $X \xrightarrow{f} Y \xrightarrow{\eta} T(Y)$. The functoriality of this mapping follows from the unit law for monads.

Example 7.3.3.4. In this example we return to the setting laid out in Application 7.3.1.3, where we had a set A of experimenters and assumed that the person doing the experiment might affect the outcome. We use the monad $\top = (T, \eta, \mu)$ from Example 7.3.2.6 and hope that $\mathbf{Kls}(\top)$ will conform to the understanding of how to manage the effect of the experimenter on data.

The objects of $\mathbf{Kls}(\top)$ are ordinary sets, but a map $f\colon X \to Y$ in $\mathbf{Kls}(\top)$ is a function $X \to Y^A$. By currying, this is the same as a function $X \times A \to Y$, as desired. To compose f with $g\colon Y \to Z$ in $\mathbf{Kls}(\top)$, we follow the formula from (7.13). It turns out to be equivalent to the following. We have a function $X \times A \to Y$ and a function $Y \times A \to Z$. Multiplying by id_A, we have a function $X \times A \to Y \times A$, and we can now compose to get $X \times A \to Z$.

What does this say in terms of experimenters affecting data gathering? It says that if we work within $\mathbf{Kls}(\top)$, then we may assume that the experimenter is being taken into account; all proposed functions $X \to Y$ are actually functions $A \times X \to Y$. The natural way to compose these experiments is that we only consider the data from one experiment to feed into another if the experimenter is the same in both experiments.[11]

Exercise 7.3.3.5.

Exercise 7.3.2.3 discussed the power-set monad $\top = (\mathbb{P}, \eta, \mu)$.

a. Can you find a way to relate the morphisms in $\mathbf{Kls}(\top)$ to relations? That is, given a morphism $f\colon A \to B$ in $\mathbf{Kls}(\top)$, is there a natural way to associate to it a relation $R \subseteq A \times B$?

---

[11]This requirement is somewhat stringent, but it can be mitigated in a variety of ways. One such way would be to model the ability to hand off the experimental results to another person, who would then carry them forward. This could be done by defining a preorder structure on A to model who can hand off to whom (see Example 7.3.3.8).

b. How does the composition formula in **Kls**($\top$) relate to the composition of relations given in Definition 3.2.2.3?[12]

$\diamond$

Solution 7.3.3.5.

a. A morphism $A \to B$ in **Kls**($\top$) is a function $f\colon A \to \mathbb{P}(B)$ in **Set**. From such a function we need to obtain a binary relation, i.e., a subset $R \subseteq A \times B$. Recall that for any set X (e.g., $X = B$ or $X = A \times B$), we can identify the subsets of X with the functions $X \to \Omega = \{True, False\}$, using the characteristic function as in Definition 3.4.4.12. In other words, we have a bijection

$$\mathbb{P}(X) \cong \mathrm{Hom}_{\mathbf{Set}}(X, \Omega).$$

By currying, we get an isomorphism

$$\mathrm{Hom}_{\mathbf{Set}}(A, \mathbb{P}(B)) \cong \mathrm{Hom}_{\mathbf{Set}}(A, \mathrm{Hom}_{\mathbf{Set}}(B, \Omega))$$
$$\cong \mathrm{Hom}_{\mathbf{Set}}(A \times B, \Omega) \cong \mathbb{P}(A \times B).$$

In other words, we can identify the function $f\colon A \to \mathbb{P}(B)$ with an element of $\mathbb{P}(A \times B)$, i.e., with a subset $R \subseteq A \times B$, i.e., with a relation.

A more down-to-earth way to specify how $f\colon A \to \mathbb{P}(B)$ gives rise to a binary relation $R \subseteq A \times B$ is as follows. We ask, given $(a, b) \in A \times B$, when is it in R? We see that $f(a) \in \mathbb{P}(B)$ is a subset, so the answer is that we put $(a, b) \in R$ if $b \in f(a)$. This gives the desired relation.

b. It is the same.

$\blacklozenge$

Exercise 7.3.3.6.

   (Challenge) Let $\top = (\mathbb{P}, \eta, \mu)$ be the power-set monad. The category **Kls**($\top$) is closed under binary products, i.e., every pair of objects $A, B \in \mathrm{Ob}(\mathbf{Kls}(\top))$ has a product in **Kls**($\top$). What is the product of $A = \{1, 2, 3\}$ and $B = \{a, b\}$, and what are the projections? $\diamond$

---

[12]Actually, Definition 3.2.2.3 is about composing spans, but a relation $R \subseteq A \times B$ is a kind of span, $R \to A \times B$.

Solution 7.3.3.6.

The product of A and B in $\mathbf{Kls}(\top)$ is $A \times B = \{1, 2, 3, a, b\}$, which coincidentally would be their coproduct in **Set**. The projection maps are functions $\mathbb{P}(A) \xleftarrow{\pi_1} \{1, 2, 3, a, b\} \xrightarrow{\pi_2} \mathbb{P}(B)$; we use the obvious maps, e.g., $\pi_1(3) = \{3\}$ and $\pi_1(a) = \varnothing$. The question did not ask for the universal property, but we specify it anyway. Given $f \colon X \to \mathbb{P}(A)$ and $g \colon X \to \mathbb{P}(B)$, we take $\langle f, g \rangle \colon X \to \mathbb{P}(A \sqcup B\}$ to be given by union.   ♦

Exercise 7.3.3.7.

(Challenge.) Let $\top = (\mathbb{P}, \eta, \mu)$ be the power-set monad. The category $\mathbf{Kls}(\top)$ is closed under binary coproducts, i.e., every pair of objects $A, B \in \mathrm{Ob}(\mathbf{Kls}(\top))$ has a coproduct in $\mathbf{Kls}(\top)$. What is the coproduct of $A = \{1, 2, 3\}$ and $B = \{a, b\}$?   ◊

Solution 7.3.3.7.

It is $A \sqcup B = \{1, 2, 3, a, b\}$, which coincidentally would be their coproduct in **Set**, as in Exercise 7.3.3.6. The inclusion maps are functions $A \xrightarrow{i_1} \mathbb{P}(\{1, 2, 3, a, b\}) \xleftarrow{i_2} B$; we use the inclusion of singleton subsets. The question did not ask for the universal property, but we specify it anyway. Given $f \colon A \to \mathbb{P}(X)$ and $g \colon B \to \mathbb{P}(X)$, we take their coproduct $A \sqcup B \to \mathbb{P}(X)$ to be given by union.   ♦

Example 7.3.3.8. Let A be any preorder. We speak of A throughout this example as though it were the linear order given by time; however, the mathematics works for any $A \in \mathrm{Ob}(\mathbf{PrO})$.

There is a monad $\top = (T, \eta, \mu)$ that captures the idea that a function $f \colon X \to Y$ occurs in the context of time in the following sense: The output of f is determined not only by the element $x \in X$ on which it is applied but also by the time at which it was applied to x; and the output of f occurs at another time, which is not before the time of input.

The functor part of the monad is given on $Y \in \mathrm{Ob}(\mathbf{Set})$ by

$$T(Y) = \{p \colon A \to A \times Y \mid \text{ if } p(a) = (a', y) \text{ then } a \leqslant a'\}.$$

The unit $\eta_Y \colon Y \to T(Y)$ sends y to the function $a \mapsto (a, y)$. The multiplication map $\mu_Y \colon T(T(Y)) \to T(Y)$ is as follows. Suppose given $p \colon A \to A \times T(Y)$ in $T(T(Y))$. Then $\mu_Y(p) \colon A \to A \times Y$ is given on $a \in A$ as follows. Suppose $p(a) = (a', p')$, where $p' \colon A \to A \times Y$. Then we assign $\mu_Y(p)(a) = p'(a') \in A \times Y$.

Given two sets X, Y, what is the meaning of a morphism $X \to Y$ in the Kleisli category $\mathbf{Kls}(\top)$, i.e., a function $f \colon X \to T(Y)$? Note that $T(Y) \subseteq \mathrm{Hom}_{\mathbf{Set}}(A, A \times Y)$, and composing with f, we have a function $X \to \mathrm{Hom}_{\mathbf{Set}}(A, A \times Y)$, which can be curried to a function $f\colon A \times X \to A \times Y$. So we have an isomorphism

$$\mathrm{Hom}_{\mathbf{Kls}(\mathsf{T})}(X,Y) \cong \{f \in \mathrm{Hom}_{\mathbf{Set}}(A \times X, A \times Y) \mid \text{ if } f(a,x) = (a',y) \text{ then } a \leqslant a'\}.$$

The right-hand set could be characterized as time-sensitive functions $f\colon X \to Y$ for which the output arrives after the input.

Remark 7.3.3.9. One of the most important monads in computer science is the *state monad*. It is used when one wants to allow a program to mutate state variables (e.g., in the program

if $x \leqslant 4$, then $x := x + 1$ else Print "done"

x is a state variable). The state monad is a special case of the monad discussed in Example 7.3.3.8. Given any set A, the usual *state monad of type A* is obtained by giving A the indiscrete preorder (see Example 4.4.4.5). More explicitly, it is a monad with functor part

$$X \mapsto (A \times X)^A$$

(see Example 7.3.5.3).

Example 7.3.3.10. We reconsider Figure 1.1 reproduced as Figure 7.3.

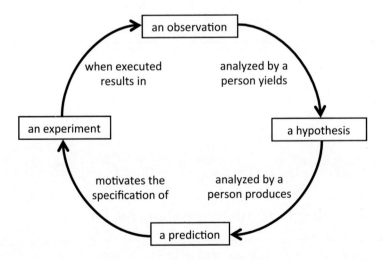

Figure 7.3 An olog whose arrows do not denote functions. It should be interpreted using a monad.

It looks like an olog, and all ologs are database schemas (see Section 4.5.2.15). But how is "analyzed by a person yields" a function? For it to be a function, there must be only one hypothesis corresponding to a given observation. The very name of this arrow belies the fact that it is an invalid aspect in the sense of Section 2.3.2.1, because given an observation, there may be more than one hypothesis yielded, corresponding to which person is doing the observing. In fact, all the arrows in this figure correspond to some hidden context involving people: the prediction is dependent on who analyzes the hypothesis, the specification of an experiment is dependent on who is motivated to specify it, and experiments may result in different observations by different observers.

Without monads, the model of science proposed by this olog would be difficult to believe in. But by choosing a monad we can make explicit (and then hide from discourse) the implicit assumption that "this is all dependent on which human is doing the science." The choice of monad is an additional modeling choice. Do we want to incorporate the partial order of time? Do we want the scientist to be modified by each function (i.e., the person is changed when analyzing an observation to yield a hypothesis)? These are all interesting possibilities.

One reasonable choice would be to use the state monad of type A, where A is the set of scientific models. This implies the following context. Every morphism $f\colon X \to Y$ in the Kleisli category of this monad is really a morphism $f\colon X \times A \to Y \times A$; while ostensibly giving a map from X to Y, it is influenced by the scientific model under which it is performed, and its outcome yields a new scientific model.

Reading the olog in this context might look like this:

> A hypothesis (in the presence of a scientific model) analyzed by a person produces a prediction (in the presence of a scientific model), which motivates the specification of an experiment (in the presence of a scientific model), which when executed results in an observation (in the presence of a scientific model), which analyzed by a person yields a hypothesis (in the presence of a scientific model).

The parenthetical statements can be removed if we assume them to be always there, which can be done using the preceding monad.

7.3.3.11   Relaxing functionality constraint for ologs

Section 2.3.2 said that every arrow in an olog has to be English-readable as a sentence, and it has to correspond to a function. For example, the arrow

$$\boxed{\text{a person}} \xrightarrow{\text{has}} \boxed{\text{a child}} \tag{7.14}$$

makes for a readable sentence, but it does not correspond to a function because a person may have no children or more than one child. We call an olog in which every arrow corresponds to a function (the only option proposed so far in this book) a *functional olog*. Requiring that ologs be functional comes with advantages and disadvantages. The main advantage is that creating a functional olog requires more conceptual clarity, and this has benefits for the olog creator as well as for anyone to whom he tries to explain the situation. The main disadvantage is that creating a functional olog takes more time, and the olog takes up more space on the page.

In the context of the power-set monad (see Exercise 7.3.2.3), a morphism $f\colon X \to Y$ between sets X and Y, as objects in $\mathbf{Kls}(\mathbb{P})$, becomes a binary relation on X and Y rather than a function (see Exercise 7.3.3.5). So in that context, the arrow in (7.14) becomes valid. An olog in which arrows correspond to mere binary relations rather than functions might be called a *relational olog*.

7.3.4 Monads in databases

This section discusses how to record data in the presence of a monad. The idea is quite simple. Given a schema (category) $\mathcal{C}$, an ordinary instance is a functor $I\colon \mathcal{C} \to \mathbf{Set}$. But if $\mathsf{T} = (T, \eta, \mu)$ is a monad, then a *Kleisli* T-*instance on* $\mathcal{C}$ is a functor $J\colon \mathcal{C} \to \mathbf{Kls}(\mathsf{T})$. Such a functor associates to every object $c \in \mathrm{Ob}(\mathcal{C})$ a set $J(c)$, and to every arrow $f\colon c \to c'$ in $\mathcal{C}$ a morphism $J(f)\colon J(c) \to J(c')$ in $\mathbf{Kls}(\mathsf{T})$. How does this look in terms of tables?

Recall that to represent an ordinary database instance $I\colon \mathcal{C} \to \mathbf{Set}$, we use a tabular format in which every object $c \in \mathrm{Ob}(\mathcal{C})$ is displayed as a table including one ID column and one additional column for each arrow $f\colon c \to c'$ emanating from c. The cells in the ID column of table c contain the elements of the set $I(c)$, and the cells in the f column contain elements of the set $I(c')$.

To represent a *Kleisli* database instance $J\colon \mathcal{C} \to \mathbf{Kls}(\mathsf{T})$ is similar; we again use a tabular format in which every object $c \in \mathrm{Ob}(\mathcal{C})$ is displayed as a table including one ID column and one additional column for each arrow $f\colon c \to c'$ emanating from c. The cells in the ID column of table c again contain the elements of the set $J(c)$; however the cells in the f column do not contain elements of $J(c')$, but T-values in $J(c')$, i.e., elements of $T(J(c'))$.

Example 7.3.4.1. Let $\mathsf{T} = (T, \eta, \mu)$ be the monad for partial functions (see Example 7.3.1.1). Given any schema $\mathcal{C}$, we can represent a Kleisli T-instance $I\colon \mathcal{C} \to \mathbf{Kls}(\mathsf{T})$ in tabular format. For every object $c \in \mathrm{Ob}(\mathcal{C})$ we have a set $I(c)$ of rows, and given a column $f\colon c \to c'$, applying f to a row either produces a value in $I(c')$ or fails to produce a value; this is the essence of partial functions. We might denote the absence of a value using ☺.

Consider the schema indexing graphs

$$\mathcal{C} := \boxed{\begin{array}{ccc} \texttt{Arrow} & \xrightarrow[tgt]{\ src\ } & \texttt{Vertex} \\ \bullet & \Rightarrow & \bullet \end{array}}$$

As discussed in Section 5.2.1.21, an ordinary instance on $\mathcal{C}$ represents a graph:

$$I := \quad$$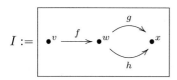

Arrow (I)		
ID	**src**	**tgt**
f	v	w
g	w	x
h	w	x

Vertex (I)
ID
v
w
x

A Kleisli $\top$-instance on $\mathcal{C}$ represents graphs in which edges can fail to have a source vertex, fail to have a target vertex, or both:

$$J := \quad$$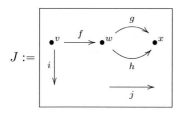

Arrow (J)		
ID	**src**	**tgt**
f	v	w
g	w	x
h	w	x
i	v	☺
j	☺	☺

Vertex (J)
ID
v
w
x

The context of these tables is that of partial functions, so we do not need a reference for ☺ in the vertex table. Mathematically, the morphism $J(src)\colon J(\texttt{Arrow}) \to J(\texttt{Vertex})$ in **Kls**($\top$) needs to be a function $J(\texttt{Arrow}) \to J(\texttt{Vertex}) \sqcup \{☺\}$, and it is.

7.3.4.2   Probability distributions

Let $[0,1] \subseteq \mathbb{R}$ denote the set of real numbers between 0 and 1. Let X be a set and $p\colon X \to [0,1]$ a function. We say that p is a *finitary probability distribution on X* if there exists a finite subset $W \subseteq X$ such that

$$\sum_{w \in W} p(w) = 1, \tag{7.15}$$

and such that $p(x) > 0$ if and only if $x \in W$. Note that the subset W is unique if it exists; we call it *the support of p* and denote it $\mathbf{Supp}(p)$.

For any set X, let $\mathbf{Dist}(X)$ denote the set of finitary probability distributions on X. It is easy to check that given a function $f: X \to Y$, one obtains a function $\mathbf{Dist}(f): \mathbf{Dist}(X) \to \mathbf{Dist}(Y)$ by $\mathbf{Dist}(f)(y) = \sum_{f(x)=y} p(x)$. Thus we can consider $\mathbf{Dist}: \mathbf{Set} \to \mathbf{Set}$ as a functor, and in fact the functor part of a monad. Its unit $\eta: X \to \mathbf{Dist}(X)$ is given by the Kronecker delta function $x \mapsto \delta_x$, where $\delta_x(x) = 1$ and $\delta_x(x') = 0$ for $x' \neq x$. Its multiplication $\mu: \mathbf{Dist}(\mathbf{Dist}(X)) \to \mathbf{Dist}(X)$ is given by weighted sum: given a finitary probability distribution $w: \mathbf{Dist}(X) \to [0,1]$ and $x \in X$, put $\mu(w)(x) = \sum_{p \in \mathbf{Supp}(w)} w(p)p(x)$.

Example 7.3.4.3 (Markov chains). Let $\mathcal{L}oop$ be the loop schema

$$\mathcal{L}oop := \boxed{\begin{array}{c} f \\ \circlearrowleft \, s \\ \bullet \end{array}}$$

as in Example 4.5.2.10. A $\mathbf{Dist}$-instance on $\mathcal{L}oop$ is equivalent to a time-homogeneous Markov chain. To be explicit, a functor $\delta: \mathcal{L}oop \to \mathbf{Kls}(\mathbf{Dist})$ assigns to the unique object $s \in \mathrm{Ob}(\mathcal{L}oop)$ a set $S = \delta(s)$, called the state space, and to $f: s \to s$ a function $\delta(f): S \to \mathbf{Dist}(S)$, which sends each element $x \in S$ to some probability distribution on elements of S. For example, the left-hand table δ (having states $\delta(s) = \{a, b, c, d\}$) corresponds to the right-hand Markov matrix M:

$$\delta := \begin{array}{|c||c|} \hline \multicolumn{2}{|c|}{\mathbf{s}} \\ \hline \mathbf{ID} & \mathbf{f} \\ \hline a & .5(a)+.5(b) \\ \hline b & 1(b) \\ \hline c & .7(a)+.3(c) \\ \hline d & .4(a)+.3(b)+.3(d) \\ \hline \end{array} \qquad M := \begin{pmatrix} 0.5 & 0.5 & 0 & 0 \\ 0 & 1 & 0 & 0 \\ 0.7 & 0 & 0.3 & 0 \\ 0.4 & 0.3 & 0 & 0.3 \end{pmatrix} \qquad (7.16)$$

As one might hope, for any natural number $n \in \mathbb{N}$, the map $f^n: S \to S$ in $\mathbf{Kls}(\mathbf{Dist})$ corresponds to the matrix M^n, which sends an element $s \in S$ to its probable location after n iterations of the transition map, $f^n(s) \in \mathbf{Dist}(S)$.

Application 7.3.4.4. Every star emits a spectrum of light, which can be understood as a distribution on the electromagnetic spectrum. Given an object B on earth, different parts of B will absorb radiation at different rates. Thus B produces a function from the electromagnetic spectrum to distributions of energy absorption. In the context of the probability distributions monad, we can record data on the schema

$$\underset{\bullet}{\overset{\mathtt{star}}{}} \xrightarrow{\mathtt{emits}} \underset{\bullet}{\overset{\mathtt{wavelengths}}{}} \xrightarrow{\mathtt{absorbed \ by} \ B} \underset{\bullet}{\overset{\mathtt{energies}}{}}$$

The composition formula for Kleisli categories is the desired one: to each star we associate the weighted sum of energy absorption rates over the set of wavelengths emitted by the star.

$$\Diamond\Diamond$$

7.3.5   Monads and adjunctions

There is a strong connection between monads and adjunctions: every adjunction creates a monad, and every monad comes from an adjunction. For example, the List monad (Example 7.3.2.2) comes from the free forgetful adjunction between sets and monoids

$$\mathbf{Set} \underset{U}{\overset{F}{\rightleftarrows}} \mathbf{Mon}$$

(see Proposition 7.1.1.2). That is, for any set X, the free monoid on X is

$$F(X) = (\mathrm{List}(X), [\,], +\!\!+\,),$$

and the underlying set of that monoid is $U(F(X)) = \mathrm{List}(X)$. So the List functor is given by $U \circ F \colon \mathbf{Set} \to \mathbf{Set}$. But a monad is more than a functor; it includes a unit map η and a multiplication map μ (see Definition 7.3.2.1). Luckily, the unit η and multiplication μ drop out of the adjunction too. First, we discuss the unit and counit of an adjunction.

Definition 7.3.5.1. Let $\mathcal{C}$ and $\mathcal{D}$ be categories, and let $L \colon \mathcal{C} \to \mathcal{D}$ and $R \colon \mathcal{D} \to \mathcal{C}$ be functors with adjunction isomorphism

$$\alpha_{c,d} \colon \mathrm{Hom}_{\mathcal{D}}(L(c), d) \overset{\cong}{\longrightarrow} \mathrm{Hom}_{\mathcal{C}}(c, R(d))$$

for any objects $c \in \mathrm{Ob}(\mathcal{C})$ and $d \in \mathrm{Ob}(\mathcal{D})$ (see Definition 7.1.1.1). The *unit* $\eta \colon \mathrm{id}_{\mathcal{C}} \to R \circ L$ (resp. the *counit* $\epsilon \colon L \circ R \to \mathrm{id}_{\mathcal{D}}$) of the adjunction is a natural transformation defined as follows.

Given an object $c \in \mathrm{Ob}(\mathcal{C})$, we apply α to $\mathrm{id}_{L(c)} \colon L(c) \to L(c)$ to get the c component

$$\eta_c \colon c \to R \circ L(c)$$

of η. Similarly given an object $d \in \mathrm{Ob}(\mathcal{D})$ we apply α^{-1} to $\mathrm{id}_{R(d)} \colon R(d) \to R(d)$ to get the d component

$$\epsilon_d \colon L \circ R(d) \to d.$$

One checks that these components are natural.

Later we see how to use the unit and counit of any adjunction to make a monad. We first walk through the process in Example 7.3.5.2.

Example 7.3.5.2. Consider the adjunction $\mathbf{Set} \underset{U}{\overset{F}{\rightleftarrows}} \mathbf{Mon}$ between sets and monoids. Let $T = U \circ F \colon \mathbf{Set} \to \mathbf{Set}$; this will be the functor part of the monad, and we have seen that $T = \mathrm{List}$. The unit of the adjunction, $\eta \colon \mathrm{id}_{\mathbf{Set}} \to U \circ F$ is precisely the unit of the monad: for any set $X \in \mathrm{Ob}(\mathbf{Set})$ the component $\eta_X \colon X \to \mathrm{List}(X)$ is the function that takes $x \in X$ to the singleton list $[x] \in \mathrm{List}(X)$. The monad also has a multiplication map $\mu_X \colon T(T(X)) \to T(X)$, which amounts to concatenating a list of lists. This function comes about using the counit ϵ, as follows

$$T \circ T = U \circ F \circ U \circ F \xrightarrow{\;\mathrm{id}_U \diamond \epsilon \diamond \mathrm{id}_F\;} U \circ F = T.$$

The general procedure for extracting a monad from an adjunction is analogous to the process shown in Example 7.3.5.2. Given any adjunction

$$\mathcal{C} \underset{R}{\overset{L}{\rightleftarrows}} \mathcal{D},$$

we define $T = R \circ L \colon \mathcal{C} \to \mathcal{C}$, we define $\eta \colon \mathrm{id}_{\mathcal{C}} \to T$ to be the unit of the adjunction (as in Definition 7.3.5.1), and we define $\mu \colon T \circ T \to T$ to be the natural transformation $\mathrm{id}_R \diamond \epsilon \diamond \mathrm{id}_L \colon RLRL \to RL$, obtained by applying the counit $\epsilon \colon LR \to \mathrm{id}_{\mathcal{D}}$.

This procedure produces monads on arbitrary categories $\mathcal{C}$, whereas the definition of monad (Definition 7.3.2.1) considers only the case $\mathcal{C} = \mathbf{Set}$. However, Definition 7.3.2.1 can be generalized to arbitrary categories $\mathcal{C}$ by simply replacing every occurrence of the string $\mathbf{Set}$ with the string $\mathcal{C}$. Similarly, the definition of Kleisli categories (Definition 7.3.3.1) considers only the case $\mathcal{C} = \mathbf{Set}$, but again the generalization to arbitrary categories $\mathcal{C}$ is straightforward.

Example 7.3.5.3. Let $A \in \mathrm{Ob}(\mathbf{Set})$ be a set, and recall the currying adjunction

$$\mathbf{Set} \underset{Y \mapsto Y^A}{\overset{X \mapsto X \times A}{\rightleftarrows}} \mathbf{Set},$$

discussed briefly in Example 7.1.1.8. The corresponding monad St_A is typically called the *state monad of type A* in programming language theory. Given a set X, we have

$$St_A(X) = (A \times X)^A.$$

In the Kleisli category $\mathbf{Kls}(St_A)$ a morphism from X to Y is a function of the form $X \to (A \times Y)^A$, but this can be curried to a function $A \times X \to A \times Y$.

As discussed in Remark 7.3.3.9, this monad is related to holding onto an internal state variable of type A. Under the state monad St_A, every morphism written $X \to Y$, when viewed as a function, takes as input not only an element of X, but also the current state $a \in A$, and it produces as output not only an element of Y, but also an updated state.

Computer scientists in programming language theory have found monads very useful (Moggi [33]). In much the same way, monads on **Set** might be useful in databases (see Section 7.3.4). Another, totally different way to use monads in databases is by using a mapping between schemas to produce in each one an internal model of the other. That is, for any functor $F\colon \mathcal{C} \to \mathcal{D}$, i.e., mapping of database schemas, the adjunction (Σ_F, Δ_F) produces a monad on $\mathcal{C}$–**Set**, and the adjunction (Δ_F, Π_F) produces a monad on $\mathcal{D}$–**Set**. If one interprets the List monad as producing in **Set** an internal model of the category **Mon** of monoids, one can similarly interpret these monads on $\mathcal{C}$–**Set** and $\mathcal{D}$–**Set** as producing internal models of each within the other.

7.4   Operads

This section briefly introduces operads, which are generalizations of categories. They often are useful for speaking about self-similarity of structure. For example, we use operads to model agents made up of smaller agents, or materials made up of smaller materials. This association with self-similarity is not really inherent in the definition, but it tends to emerge in thinking about many operads used in practice.

Let me begin with a warning.

Warning 7.4.0.4. My use of the term *operad* is not entirely standard and conflicts with widespread usage. The more common term for what I am calling an operad is *colored operad* or *symmetric multicategory*. An operad classically is a multicategory with one object, and a colored operad is a multicategory with possibly many objects (one for each "color"). The term *multicategory* stems from the fact that the morphisms in a multicategory have many, rather than one, domain object. One reason I prefer not to use the term *multicategory* is that there is nothing really "multi" about the multicategory itself, only its morphisms. Further, I do not see enough reason to differentiate, given that the term *multicategory* seems rather clunky and the term *operad* seems rather sleek. I hope my break with standard terminology does not cause confusion.

This introduction to operads is quite short; see Leinster [25] for an excellent treatment. Operads are also related to monoidal categories, a subject that is not elaborated in this book to discuss, but which was briefly mentioned when discussing topological enrichment in Example 5.2.3.3. Many of the following operads are actually monoidal categories in disguise.

7.4.1   Definition and classical examples

An operad is like a category in that it has objects, morphisms, and a composition formula, and it obeys an identity law and an associativity law. The difference is that each morphism f in an operad can have many inputs (and one output):

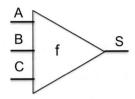

The description of composition in an operad is a bit more complicated than for a category, because it involves much more variable indexing; however, the idea is straightforward. Figure **??** shows morphisms being composed. Note that S and T disappear from the composition, but this is analogous to the way the middle object disappears from the composition of morphisms in a category

$$A \xrightarrow{f} S \xrightarrow{g} X \qquad \text{the morphisms to the left compose to give} \qquad A \xrightarrow{g \circ f} X$$

 Here is the definition, taken from Spivak [41]. Skip to Example 7.4.1.3 if the definition gets too difficult.

Definition 7.4.1.1. An *operad* $\mathcal{O}$ is defined as follows: One announces some constituents (A. objects, B. morphisms, C. identities, D. compositions) and shows that they conform to some laws (1. identity law, 2. associativity law). Specifically, one announces

A. a collection $\mathrm{Ob}(\mathcal{O})$, each element of which is called an *object* of $\mathcal{O}$;

B. for each object $y \in \mathrm{Ob}(\mathcal{O})$, finite set $n \in \mathrm{Ob}(\mathbf{Fin})$, and n-indexed set of objects $x\colon n \to \mathrm{Ob}(\mathcal{O})$, a set $\mathcal{O}_n(x; y) \in \mathrm{Ob}(\mathbf{Set})$; its elements are called *morphisms from x to y* in $\mathcal{O}$;

C. for every object $x \in \mathrm{Ob}(\mathcal{O})$, a specified morphism, denoted $\mathrm{id}_x \in \mathcal{O}_1(x; x)$ and called *the identity morphism on x.*

D. Let $s\colon m \to n$ be a morphism in $\mathbf{Fin}$. Let $z \in \mathrm{Ob}(\mathcal{O})$ be an object, let $y\colon n \to \mathrm{Ob}(\mathcal{O})$ be an n-indexed set of objects, and let $x\colon m \to \mathrm{Ob}(\mathcal{O})$ be an m-indexed set of objects. For each element $i \in n$, write $m_i := s^{-1}(i)$ for the pre-image of s under i, and write $x_i = x|_{m_i}\colon m_i \to \mathrm{Ob}(\mathcal{O})$ for the restriction of x to m_i. Then one announces a function

$$\circ\colon \mathcal{O}_n(y; z) \times \prod_{i \in n} \mathcal{O}_{m_i}(x_i; y(i)) \longrightarrow \mathcal{O}_m(x; z), \qquad (7.17)$$

called *the composition formula.*

Given an n-indexed set of objects $x \colon n \to \mathrm{Ob}(\mathcal{O})$ and an object $y \in \mathrm{Ob}(\mathcal{O})$, we sometimes abuse notation and denote the set of morphisms from x to y by $\mathcal{O}(x_1, \ldots, x_n; y)$.[13] We may write $\mathrm{Hom}_{\mathcal{O}}(x_1, \ldots, x_n; y)$, in place of $\mathcal{O}(x_1, \ldots, x_n; y)$, when convenient. We can denote a morphism $\phi \in \mathcal{O}_n(x; y)$ by $\phi \colon x \to y$ or by $\phi \colon (x_1, \ldots, x_n) \to y$; we say that each x_i is a *domain object* of ϕ and that y is the *codomain object* of ϕ. We use infix notation for the composition formula, e.g., $\psi \circ (\phi_1, \ldots, \phi_n)$.

One must then show that the following *operad laws* hold:

1. For every $x_1, \ldots, x_n, y \in \mathrm{Ob}(\mathcal{O})$ and every morphism $\phi \colon (x_1, \ldots, x_n) \to y$, we have
$$\phi \circ (\mathrm{id}_{x_1}, \ldots, \mathrm{id}_{x_n}) = \phi \qquad \text{and} \qquad \mathrm{id}_y \circ \phi = \phi.$$

2. Let $m \xrightarrow{s} n \xrightarrow{t} p$ be composable morphisms in **Fin**. Let $z \in \mathrm{Ob}(\mathcal{O})$ be an object, let $y \colon p \to \mathrm{Ob}(\mathcal{O})$, $x \colon n \to \mathrm{Ob}(\mathcal{O})$, and $w \colon m \to \mathrm{Ob}(\mathcal{O})$ respectively be a p-indexed, n-indexed, and m-indexed set of objects. For each $i \in p$, write $n_i = t^{-1}(i)$ for the pre-image and $x_i \colon n_i \to \mathrm{Ob}(\mathcal{O})$ for the restriction. Similarly, for each $k \in n$, write $m_k = s^{-1}(k)$ and $w_k \colon m_k \to \mathrm{Ob}(\mathcal{O})$; for each $i \in p$, write $m_{i,-} = (t \circ s)^{-1}(i)$ and $w_{i,-} \colon m_{i,-} \to \mathrm{Ob}(\mathcal{O})$; for each $j \in n_i$, write $m_{i,j} := s^{-1}(j)$ and $w_{i,j} \colon m_{i,j} \to \mathrm{Ob}(\mathcal{O})$. Then the following diagram commutes:

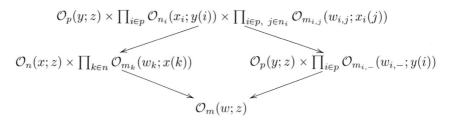

Remark 7.4.1.2. This remark considers the abuse of notation in Definition 7.4.1.1 and how it relates to an action of a symmetric group on each morphism set in the definition of operad. We follow the notation of Definition 7.4.1.1, especially the use of subscripts in the composition formula.

Suppose that $\mathcal{O}$ is an operad, $z \in \mathrm{Ob}(\mathcal{O})$ is an object, $y \colon n \to \mathrm{Ob}(\mathcal{O})$ is an n-indexed set of objects, and $\phi \colon y \to z$ is a morphism. If we linearly order n, enabling us to write $\phi \colon (y(1), \ldots, y(|n|)) \to z$, then changing the linear ordering amounts to finding an

---

[13]There are three abuses of notation when writing $\mathcal{O}(x_1, \ldots, x_n; y)$. First, it confuses the set $n \in \mathrm{Ob}(\mathbf{Fin})$ with its cardinality $|n| \in \mathbb{N}$. But rather than writing $\mathcal{O}(x_1, \ldots, x_{|n|}; y)$, it would be more consistent to write $\mathcal{O}(x(1), \ldots, x(|n|); y)$ because we have assigned subscripts another meaning in part D. But even this notation unfoundedly suggests that the set n has been endowed with a linear ordering, which it has not. This may be seen as a more serious abuse, but see Remark 7.4.1.2.

isomorphism of finite sets $\sigma\colon m \xrightarrow{\cong} n$, where $|m| = |n|$. Let $x = y \circ \sigma$, and for each $i \in n$, note that $m_i = \sigma^{-1}(\{i\}) = \{\sigma^{-1}(i)\}$, so $x_i = x|_{\sigma^{-1}(i)} = y(i)$. Taking $\mathrm{id}_{x_i} \in \mathcal{O}_{m_i}(x_i; y(i))$ for each $i \in n$, and using the identity law, we find that the composition formula induces a bijection $\mathcal{O}_n(y; z) \xrightarrow{\cong} \mathcal{O}_m(x; z)$, which we might denote

$$\sigma\colon \mathcal{O}\big(y(1), y(2), \ldots, y(n); z\big) \cong \mathcal{O}\big(y(\sigma(1)), y(\sigma(2)), \ldots, y(\sigma(n)); z\big). \tag{7.18}$$

In other words, the permutation group $\mathrm{Aut}(n)$ acts on the set $\mathcal{O}_n$ of n-ary morphisms by permuting the order of the domain objects $\mathrm{Ob}(\mathcal{O})^n$.

Throughout this book, we allow this abuse of notation and speak of morphisms $\phi\colon (y_1, y_2, \ldots, y_n) \to z$ for a natural number $n \in \mathbb{N}$, without mentioning the abuse inherent in choosing an order, as long as it is clear that permuting the order of indices would not change anything up to the canonical isomorphism of (7.18).

Example 7.4.1.3 (Little squares operad). An operad commonly used in mathematics is called the *little n-cubes operad*. We will focus on $n = 2$ and talk about the little squares operad $\mathcal{O}$. Here the set of objects has only one element, denoted by a square, $\mathrm{Ob}(\mathcal{O}) = \{\square\}$. For a natural number $n \in \mathbb{N}$, a morphism $f\colon (\square, \square, \ldots, \square) \longrightarrow \square$ is a positioning of n nonoverlapping squares inside of a square. Figure 7.5 shows a morphism $(X_1, X_2, X_3) \to Y$, where $X_1 = X_2 = X_3 = Y = \square$.

The composition formula says that given a positioning of small squares inside a large square, and given a positioning of tiny squares inside each of those small squares, we get a positioning of tiny squares inside a large square. See Figure 7.6.

Example 7.4.1.3 exemplifies the kind of self-similarity mentioned on page 452.

Exercise 7.4.1.4.

Consider an operad $\mathcal{O}$ like the little squares operad from Example 7.4.1.3, except with three objects: square, circle, equilateral triangle. A morphism is again a nonoverlapping positioning of shapes inside a shape.

a. Draw an example of a morphism f from two circles and a square to a triangle.

b. Find three other morphisms that compose into f, and draw the composite.

$\diamond$

Solution 7.4.1.4.

a.

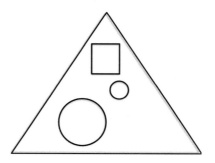

b.

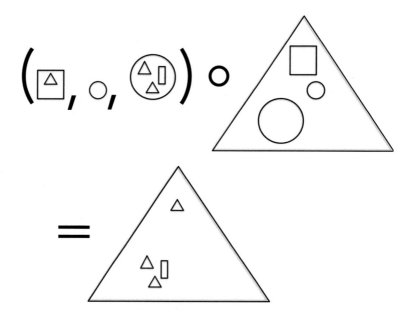

♦

Example 7.4.1.5. Let **Sets** denote the operad defined as follows. As objects we put
$\mathrm{Ob}(\mathbf{Sets}) = \mathrm{Ob}(\mathbf{Set})$. For a natural number $n \in \mathbb{N}$ and sets $X_1, \ldots, X_n, Y$, put

$$\mathrm{Hom}_{\mathbf{Sets}}(X_1, \ldots, X_n; Y) := \mathrm{Hom}_{\mathbf{Set}}(X_1 \times \cdots \times X_n, Y).$$

Given functions $f_1 \colon (X_{1,1} \times \cdots \times X_{1,m_1}) \to Y_1$ through $f_n \colon (X_{n,1} \times \cdots \times X_{n,m_n}) \to Y_n$ and a function $Y_1 \times \cdots \times Y_n \to Z$, the universal property provides a unique function of the form $(X_{1,1} \times \cdots \times X_{n,m_n}) \longrightarrow Z$, giving rise to the composition formula in **Sets**.

7.4.1.6 Operads: functors and algebras

If operads are like categories, then we can define things like functors and call them *operad functors*.

Warning 7.4.1.7. What is called an operad functor in Definition 7.4.1.8 is usually called an *operad morphism*. I think the terminology clash between morphisms *of* operads and morphisms *in* an operad is confusing. It is similar to what would occur in regular category theory (see Chapter 5) if we replaced the term *functor* with the term *category morphism*.

Definition 7.4.1.8. Let $\mathcal{O}$ and $\mathcal{O}'$ be operads. An *operad functor from $\mathcal{O}$ to $\mathcal{O}'$*, denoted $F \colon \mathcal{O} \to \mathcal{O}'$, is defined as follows. One announces some constituents (A. on-objects part, B. on-morphisms part) and shows that they conform to some laws (1. preservation of identities, 2. preservation of composition). Specifically, one announces

    A. a function $\mathrm{Ob}(F) \colon \mathrm{Ob}(\mathcal{O}) \to \mathrm{Ob}(\mathcal{O}')$, sometimes denoted simply $F \colon \mathrm{Ob}(\mathcal{O}) \to \mathrm{Ob}(\mathcal{O}')$;

    B. for each object $y \in \mathrm{Ob}(\mathcal{O})$, finite set $n \in \mathrm{Ob}(\mathbf{Fin})$, and n-indexed set of objects $x \colon n \to \mathrm{Ob}(\mathcal{O})$, a function

$$F_n \colon \mathcal{O}_n(x; y) \to \mathcal{O}'_n(Fx; Fy).$$

One must then show that the following *operad functor laws* hold:

    1. For each object $x \in \mathrm{Ob}(\mathcal{O})$, the equation $F(\mathrm{id}_x) = \mathrm{id}_{Fx}$ holds.

    2. Let $s \colon m \to n$ be a morphism in **Fin**. Let $z \in \mathrm{Ob}(\mathcal{O})$ be an object, let $y \colon n \to \mathrm{Ob}(\mathcal{O})$ be an n-indexed set of objects, and let $x \colon m \to \mathrm{Ob}(\mathcal{O})$ be an m-indexed set of objects. Then, with notation as in Definition 7.4.1.1, the following diagram of sets commutes:

$$
\begin{array}{ccc}
\mathcal{O}_n(y; z) \times \prod_{i \in n} \mathcal{O}_{m_i}(x_i; y(i)) & \xrightarrow{\ F\ } & \mathcal{O}'_n(Fy; Fz) \times \prod_{i \in n} \mathcal{O}'_{m_i}(Fx_i; Fy(i)) \\
{\scriptstyle\circ}\downarrow & & \downarrow{\scriptstyle\circ} \\
\mathcal{O}_m(x; z) & \xrightarrow[\ F\ ]{} & \mathcal{O}'_m(Fx; Fz)
\end{array}
$$

$$(7.19)$$

We denote the category of operads and operad functors **Oprd**.

Exercise 7.4.1.9.

Let $\mathcal{O}$ denote the little squares operad from Example 7.4.1.3, and let $\mathcal{O}'$ denote the little shapes operad you constructed in Exercise 7.4.1.4.

a. Can you find an operad functor $F\colon \mathcal{O} \to \mathcal{O}'$?

b. Is it possible to find an operad functor $G\colon \mathcal{O}' \to \mathcal{O}$?

$\diamond$

Solution 7.4.1.9.

a. Yes. One of the shapes in $\mathrm{Ob}(\mathcal{O}')$ was a square, so we know F on objects. And a morphism in $\mathcal{O}$ is a way to draw squares in a square, which is in particular, a way to draw shapes in a shape, meaning that it can be assigned to a morphism in $\mathcal{O}'$. The composition formula works correctly, so we have defined the operad functor.

b. I cannot think of one. If you think you have one, beware if you are using any kind of resizing or scaling operation. Think about how that resizing interacts with the fact that your functor needs to preserve identity morphisms.

♦

Definition 7.4.1.10 (Operad algebra). Let $\mathcal{O}$ be an operad, and let **Sets** be the operad from Example 7.4.1.5. An *algebra on* $\mathcal{O}$ is an operad functor $A\colon \mathcal{O} \to \mathbf{Sets}$.

Remark 7.4.1.11. Every category can be construed as an operad (there is a functor **Cat** → **Oprd**), one in which every morphism is unary. That is, given a category $\mathcal{C}$, one makes an operad $\mathcal{O}$ with $\mathrm{Ob}(\mathcal{O}) := \mathrm{Ob}(\mathcal{C})$ and with

$$\mathrm{Hom}_{\mathcal{O}}(x_1, \ldots, x_n; y) = \begin{cases} \mathrm{Hom}_{\mathcal{C}}(x_1, y) & \text{if } n = 1, \\ \varnothing & \text{if } n \neq 1. \end{cases}$$

Throughout the book a connection is made between database schemas and categories (see Section 5.2.2), under which a schema $\mathcal{C}$ is construed as a category presentation, i.e., by generators and relations. Similarly, it is possible to discuss operad presentations $\mathcal{O}$, again by generators and relations. Under this analogy, an instance $\mathcal{C} \to$ **Set** of the database (see Section 5.2.2.6) corresponds to an algebra $\mathcal{O} \to$ **Sets** of the operad.

7.4.2   Applications of operads and their algebras

Hierarchical structures seem to be well modeled by operads. A hierarchical structure often has basic building blocks and instructions for how they can be put together into larger building blocks. Describing such structures using operads and their algebras allows one to make appropriate distinctions between different types of thinking, which may otherwise be blurred. For example, the abstract building instructions should be encoded in the operad, whereas the concrete building blocks should be encoded in the algebra. Morphisms of algebras are high-level understandings of how building blocks of very different types (such as materials versus numbers) can occupy the same place in the structure and be compared.

We get a general flavor of these ideas in the following examples.

Application 7.4.2.1. Every material is composed of constituent materials, arranged in certain patterns. (In case the material is pure, we consider the material to consist of itself as the sole constituent.) Each of these constituent materials is itself an arrangement of constituent materials. Thus a kind of self-similarity can be modeled with operads.

For example, a tendon is made of collagen fibers that are assembled in series and then in parallel, in a specific way. Each collagen fiber is made of collagen fibrils that are again assembled in series and then in parallel, with slightly different specifications. We can continue, perhaps indefinitely. Going a bit further, each collagen fibril is made up of tropocollagen collagen molecules, which are twisted ropes of collagen molecules, and so on.[14]

Here is how operads might be employed. We want the same operad to model all three of the following: actual materials, theoretical materials, and functional properties. That is, we want more than one algebra on the same operad.

The operad $\mathcal{O}$ should abstractly model the structure but not the substance being structured. Imagine that each of the shapes, say a triangle, in Figure (7.7) is a placeholder that indicates "your triangular material here." Each morphism represents a construction of a material out of parts.

◊◊

Application 7.4.2.2. Suppose we have chosen an operad $\mathcal{O}$ to model the structure of materials. Say each object of $\mathcal{O}$ corresponds to a certain quality of material, and each morphism corresponds to an arrangement of various qualities to form a new quality. An algebra $A\colon \mathcal{O} \to \mathbf{Sets}$ on $\mathcal{O}$ requires us to choose what substances will fill in for these qualities. For every object $x \in \mathrm{Ob}(\mathcal{O})$, we want a set $A(x)$ that will be the set of materials with that quality. For every arrangement, i.e., morphism, $f\colon (x_1, \ldots, x_n) \to y$, and every choice $a_1 \in A(x_1), \ldots, a_n \in A(x_n)$ of materials, we need to understand what

---

[14]Thanks to Professor Sandra Shefelbine for explaining the hierarchical nature of collagen to me. Any errors are my own.

material $a' = A(f)(a_1, \ldots, a_n) \in A(y)$ will emerge when materials $a_1, \ldots, a_n$ are arranged in accordance with f.

There may be more than one interesting algebra on $\mathcal{O}$. Suppose that $B \colon \mathcal{O} \to \mathbf{Sets}$ is an algebra of strengths rather than of materials. For each object $x \in \mathrm{Ob}(\mathcal{O})$, which represents some quality, we let $B(x)$ be the set of possible strengths that something of quality x can have. Then for each arrangement, i.e., morphism, $f \colon (x_1, \ldots, x_n) \to y$, and every choice $b_1 \in B(x_1), \ldots, b_n \in B(x_n)$ of strengths, we need to understand what strength $b' = B(f)(b_1, \ldots, b_n) \in B(y)$ will emerge when strengths $b_1, \ldots, b_n$ are arranged in accordance with f.

Finally, a morphism of algebras $S \colon A \to B$ would consist of a coherent system for assigning to each material $a \in A(X)$ of a given quality x a specific strength $S(a) \in B(X)$, in such a way that morphisms behave appropriately. One can use the language of operads and algebras to state a very precise goal for the field of material mechanics.

$\diamond\diamond$

Exercise 7.4.2.3.

Consider again the little squares operad $\mathcal{O}$ from Example 7.4.1.3. Suppose we want to use this operad to describe photographic mosaics.

a. Devise an algebra $P \colon \mathcal{O} \to \mathbf{Sets}$ that sends the square to the set M of all photos that can be pasted into that square. What does P do on morphisms in $\mathcal{O}$?

b. Devise an algebra $C \colon \mathcal{O} \to \mathbf{Sets}$ that sends each square to the set of all colors (visible frequencies of light). In other words, $C(\square)$ is the set of colors, not the set of ways to color the square. What does C do on morphisms in $\mathcal{O}$. Hint: Use some kind of averaging scheme for the morphisms.

c. Guess: If someone were to appropriately define morphisms of $\mathcal{O}$-algebras (something akin to natural transformations between functors $\mathcal{O} \to \mathbf{Sets}$), do you think there would be some morphism of algebras $P \to C$?

$\diamond$

Solution 7.4.2.3.

a. Suppose given a morphism $f \colon (\square_1, \square_2, \ldots \square_n) \longrightarrow \square$ in $\mathcal{O}$, i.e., an arrangement of n little squares in a square. We need a function $P(f) \colon M^n \to M$, i.e., for every n-tuple $(m_1, \ldots, m_n)$ of mosaics, we need a new mosaic $P(f)(m_1, \ldots, m_n) \in M$. One solution would be to put the mosaics $m_1, \ldots, m_n$ in their respective places (as dictated by f) and then to fill in the rest of the big square with white. This would indeed give an algebra.

b. Think of the set of colors, i.e., the set $C(\square)$, as the set of points in the color cube, $L = [0,1]^{\{r,g,b\}}$, whose coordinate axes correspond to the amounts of red, green, and blue light that are in the color. Given an arrangement $f\colon (\square_1, \square_2, \ldots \square_n) \longrightarrow \square$ of squares in the square, we need a function $C(f)\colon L^n \to L$, i.e., for every n-tuple $(\ell_1, \ldots, \ell_n)$ of color choices, we need a new color choice $C(f)(\ell_1, \ldots, \ell_n) \in L$. To produce this color, we begin by drawing the little squares inside the big one and coloring them according to $\ell_1, \ldots, \ell_n$. We then sum up the quantities of red, green, and blue light in all of them, perhaps as some sort of integral over the outer square. We divide this quantity by the total area of the outer square. Then $C(f)(\ell_1, \ldots, \ell_n)$ assigns the outer square this color.

c. Yes, this should work by taking the average color of the mosaic.

$\blacklozenge$

7.4.2.4   Relations and wiring diagrams

Example 7.4.2.5. Here we describe an *operad of relations*, denoted $\mathcal{R}$. The objects are sets, $\mathrm{Ob}(\mathcal{R}) = \mathrm{Ob}(\mathbf{Set})$. A morphism $f\colon (X_1, X_2, \ldots, X_n) \longrightarrow Y$ in $\mathcal{R}$ is a relation

$$R \subseteq X_1 \times X_2 \times \cdots \times X_n \times Y. \tag{7.20}$$

We use a composition formula similar to that in Definition 3.2.2.3. Namely, to compose relations $R_1, \ldots, R_n$ with S, we first form a fiber product, denoted FP:

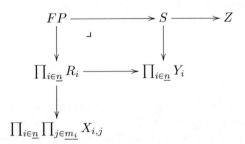

We have an induced function $FP \longrightarrow \left(\prod_{i\in\underline{n}} \prod_{j\in\underline{m_i}} X_{i,j}\right) \times Z$, and its image is the subset we take to be the composite: $S \circ (R_1, \ldots, R_n) \subseteq \left(\prod_{i\in\underline{n}} \prod_{j\in\underline{m_i}} X_{i,j}\right) \times Z$. This gives a composition formula, for which the associativity and identity laws hold, so we indeed have an operad $\mathcal{R}$.

Application 7.4.2.6. Suppose we are trying to model life in the following way. We define an entity as a set of *available experiences*. We also want to be able to put entities together to form a superentity, so we have a notion of morphism $f\colon (X_1, \ldots, X_n) \longrightarrow Y$ defined as a relation, as in (7.20).

The idea is that the morphism f is a way of translating between the experiences available to the subentities and the experiences available to the superentity. The superentity Y consists of some available experiences, like "hunger" $\in Y$. The subentities X_i each have their own set of available experiences, like "U88fh" $\in X_2$. The relation $R \subseteq X_1 \times \ldots \times X_n \times Y$ provides a way to translate between them. It says that when X_1 is experiencing "acidic" and X_2 is experiencing "U88fh," and so on, this is the same as Y experiencing "hunger."

The operad $\mathcal{R}$ from Example 7.4.2.5 becomes useful as a language for discussing issues in this domain. $\diamond\diamond$

Example 7.4.2.7. Let $\mathcal{R}$ be the operad of relations from Example 7.4.2.5, and recall that $\mathrm{Ob}(\mathcal{R}) = \mathrm{Ob}(\mathbf{Set})$. Consider the algebra $S\colon \mathcal{R} \to \mathbf{Sets}$ given by $S(X) = \mathbb{P}(X)$ for $X \in \mathrm{Ob}(\mathcal{R})$. Given a morphism $R \subseteq \prod_i X_i \times Y$ and subsets $X_i' \subseteq X_i$, we have a subset $\prod_i X_i' \subseteq \prod_i X_i$. We take the fiber product

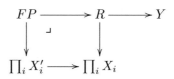

and the image of $FP \to Y$ is a subset of Y, as needed. We will continue with Application 7.4.2.8 using this algebra.

Application 7.4.2.8. Following Application 7.4.2.6 we can use Example 7.4.2.7 as a model of survival. Each entity Y survives only for a subset of the phenomena that it can experience. Under this interpretation, the algebra from Example 7.4.2.7 defines survival of an entity as the survival of all parts.

Suppose that we understand how the experiences of a superentity Y relate to those of subentities $X_1, \ldots, X_n$ in the sense that we have a morphism $f\colon (X_1, \ldots, X_n) \to Y$ in $\mathcal{R}$. In the language of Application 7.4.2.6, we have a translation between the set of experiences available across the sub-entities and the set of experiences available to the superentity. Our algebra postulates that the superentity will survive exactly those experiences for which each subentity survives.

Another way to phrase this, rather than in terms of survival, would be in terms of allowance. A bureaucracy consists of a set of smaller bureaucracies, each of which allows certain requests to pass; the whole bureaucracy allows a request to pass if and only if, when the request is translated into the perspective of each subbureaucracy, it is allowed to pass there.

                                                                                           $\diamond\diamond$

Exercise 7.4.2.9.

Define the following six sets, $A = B = M = C = N = Z = \mathbb{Z}$, and consider them as objects $A, B, M, C, N, Z \in \mathrm{Ob}(\mathcal{R})$.

a. How would you encode the relations

$$ab = m^2, \quad c^2 = n^3, \quad m + n = z$$

as a 2-ary morphism $R_1 \colon (A, B) \to M$, a 1-ary morphism $R_2 \colon (C) \to N$, and a 2-ary morphism $S \colon (M, N) \to Z$ in the operad $\mathcal{R}$?

b. What is the domain and codomain of the composite $S \circ (R_1, R_2)$?

c. Write the composite $S \circ (R_1, R_2)$ as a relation.

◇

Solution 7.4.2.9.

a. These are the relations:

$$R_1 = \{(a, b, m) \mid ab = m^2\} \subseteq A \times B \times M.$$
$$R_2 = \{(c, n) \mid c^2 = n^3\} \subseteq C \times N.$$
$$S = \{(m, n, z) \mid m + n = z\} \subseteq M \times N \times Z.$$

b. The composition is a 3-ary morphism $S \circ (R_1, R_2) \colon (A, B, C) \to Z$.

c. The composition is given by the relation

$$
\begin{aligned}
S \circ (R_1, R_2) = \{(a, b, c, z) \subseteq & A \times B \times C \times Z \mid \\
& \exists m \in M, \exists n \in N \text{ such that} \\
& ab = m^2, c^2 = n^3, m + n = z\}.
\end{aligned}
$$

◆

Example 7.4.2.10. This example discusses wiring diagrams. This operad is denoted $\mathcal{W}$ (see [41]). An object of $\mathcal{W}$ is just a finite set, $\mathrm{Ob}(\mathcal{W}) = \mathrm{Ob}(\mathbf{Fin})$, elements of which are called *wires*. A morphism in $\mathcal{W}$ is shown in Figure 7.8 (see page 474) and is formalized as follows. Given objects $C_1, \ldots, C_n$, and D, a morphism $(C_1, \ldots, C_n) \to D$ is a commutative diagram of sets

$$
\begin{array}{ccc}
 & & D \\
 & & \downarrow{\scriptstyle q} \\
\bigsqcup_{i \in \underline{n}} C_i & \xrightarrow{\ p\ } & G
\end{array}
\tag{7.21}
$$

such that p and q are jointly surjective.

Composition of morphisms is easily understood in graphic form: Given wiring diagrams inside of wiring diagrams, we can throw away the intermediary circles. In terms of sets, we first take the pushout PO:

$$
\begin{array}{ccc}
 & & E \\
 & & \downarrow \\
\bigsqcup_{i \in \underline{n}} D_i & \longrightarrow & H \\
\downarrow & {\scriptstyle \ulcorner} & \downarrow \\
\bigsqcup_{i \in \underline{n}} \bigsqcup_{j \in \underline{m_i}} C_{i,j} & \longrightarrow & \bigsqcup_{i \in \underline{n}} G_i \longrightarrow PO
\end{array}
$$

and then take the composition to be the image of $\left(\bigsqcup_{i \in \underline{n}} \bigsqcup_{j \in \underline{m_i}} C_{i,j}\right) \sqcup E \longrightarrow PO$.

Exercise 7.4.2.11.

Let $C_1 = \{a, b, m\}, C_2 = \{c, n\}, C_3 = \{m, n, z\}$, let $C = C_1 \sqcup C_2 \sqcup C_3$, and let $D = \{a, c, z\}$.

a. Suppose we draw C_1, C_2, and C_3 as follows:

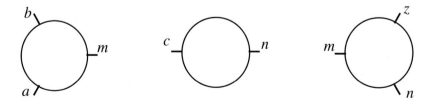

Follow those examples to draw D.

b. What set G and functions $C \xrightarrow{p} G \xleftarrow{q} D$ in (7.21) correspond to this picture?

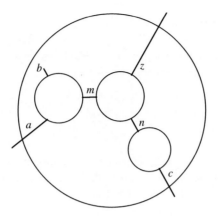

◇

Solution 7.4.2.11.

a. We can draw $D = \{a, c, z\}$ as follows:

b. Here $G = \{a, b, m, c, n, z\}$. The functions $C \xrightarrow{p} G \xleftarrow{q} D$ are given in the following tables:

$p: C_1 \sqcup C_2 \sqcup C_3 \to G$		
ID	(From)	**G**
a	C_1	a
b	C_1	b
m	C_1	m
c	C_2	c
n	C_2	n
m	C_3	m
n	C_3	n
z	C_3	z

$q: D \to G$		
ID	(From)	**G**
a	D	a
c	D	c
z	D	z

♦

Example 7.4.2.12. Let's continue with the operad $\mathcal{W}$ of wiring diagrams, and try to form an algebra on it. Taking $\mathcal{R}$ to be the operad of relations as described in Example 7.4.2.5, there is an operad functor $Q\colon \mathcal{W} \to \mathcal{R}$. It assigns to each $C \in \mathrm{Ob}(\mathcal{W})$ the set $\mathbb{Z}^C \in \mathrm{Ob}(\mathcal{R}) = \mathrm{Ob}(\mathbf{Set})$. To a morphism $G\colon (C_1, \dots, C_n) \longrightarrow D$ as in (7.21) it assigns the relation

$$\mathbb{Z}^G \subseteq \left(\prod_{i \in \underline{n}} \mathbb{Z}^{C_i}\right) \times \mathbb{Z}^D.$$

The idea is that to an entity defined as having a bunch of cables carrying integers, a phenomenon is the same thing as a choice of integer on each cable. A wiring diagram translates between phenomena experienced locally and phenomena experienced globally.

Now recall the algebra $S\colon \mathcal{R} \to \mathbf{Set}$ from Example 7.4.2.7. We can compose with Q to get $Q' := S \circ Q\colon \mathcal{W} \to \mathbf{Set}$.

Exercise 7.4.2.13.

Consider the wiring diagrams operad $\mathcal{W}$ from Example 7.4.2.10. Let's continue with Exercise 7.4.2.11 so that "everything," i.e., C_1, C_2, C_3, D, G, i, and j, are as in that exercise. By Example 7.4.2.12 we have an algebra $Q'\colon \mathcal{W} \to \mathbf{Set}$.

a. What might we mean by saying that the following picture represents an element $q_1 \in Q'(C_1)$?

b. Suppose we have the following elements $q_1 \in Q'(C_1), q_2 \in Q'(C_2)$, and $q_3 \in Q'(C_3)$:

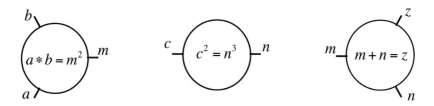

Given the wiring diagram $G\colon (C_1, C_2, C_3) \to D$ pictured here,

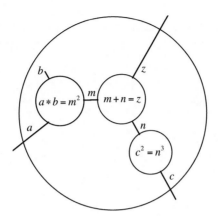

what is $G(q_1, q_2, q_3) \in Q'(D)$?

◇

Solution 7.4.2.13.

a. First, what is $Q'(C_1)$? Recall from Example 7.4.2.7 that $S \colon \mathcal{R} \to \mathbf{Set}$ is the algebra of subsets. We have

$$Q'(C_1) = S \circ Q(C_1) = S(\mathbb{Z}^C) \cong \mathbb{P}(\mathbb{Z}^3).$$

In other words, $q_1 \in Q'(C_1)$ should be a subset of $\mathbb{Z}^3$. The picture indicates that the desired subset should be

$$q_1 = \{(a, b, m) \in \mathbb{Z}^3 \mid a * b = m\}.$$

b. Following the mathematics, we find a situation similar to that of Exercise 7.4.2.11, and the answer is similar. Namely, $G(q_1, q_2, q_3)$ is the following subset:

$$G(q_1, q_2, q_3) = \{(a, c, z) \mid \exists b, m, n \text{ such that } a * b = m^2, m + n = z, c^2 = n^3\}.$$

♦

Application 7.4.2.14. In cognitive neuroscience or in industrial economics, it may be that we want to understand the behavior of an entity such as a mind, a society, or a business in terms of its structure. Knowing the connection pattern (connectome, supply chain) of subentities should help us understand how big changes are generated from small ones.

◇◇

Application 7.4.2.15. In [36], Radul and Sussman discuss propagator networks. Their implementation can presumably be understood in terms of wiring diagrams and their algebra of relations.

                                                                                    ◇◇

Figure 7.1 Three overlapping views of the night sky. Source: NASA, ESA, Digitized Sky Survey Consortium.

Figure 7.2 The three overlapping views have been glued together into one coherent view.

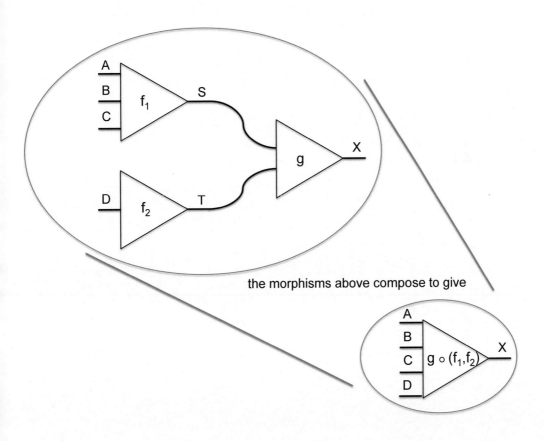

the morphisms above compose to give

Figure 7.4 The composition of morphisms f_1 and f_2 with g.

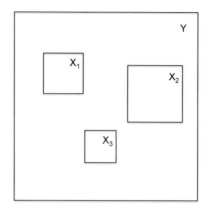

Figure 7.5 A morphism $(X_1, X_2, X_3) \longrightarrow Y$ in an operad with only one object, $\square$.

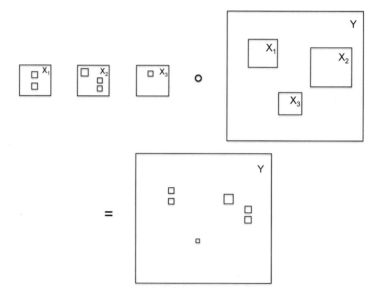

Figure 7.6 A morphism $(X_1, X_2, X_3) \rightarrow Y$ and morphisms $(W_{1,1}, W_{1,2}) \rightarrow X_1$, $(W_{2,1}, W_{2,2}, W_{2,3}) \rightarrow X_2$, and $(W_{3,1}) \rightarrow X_3$, each of which is a positioning of squares inside a square. The composition formula is given by scaling and positioning the squares to give $(W_{1,1}, W_{1,2}, W_{2,1}, W_{2,2}, W_{2,3}, W_{3,1}) \longrightarrow Y$.

Figure 7.7 A morphism expressing the construction of a material from smaller materials.

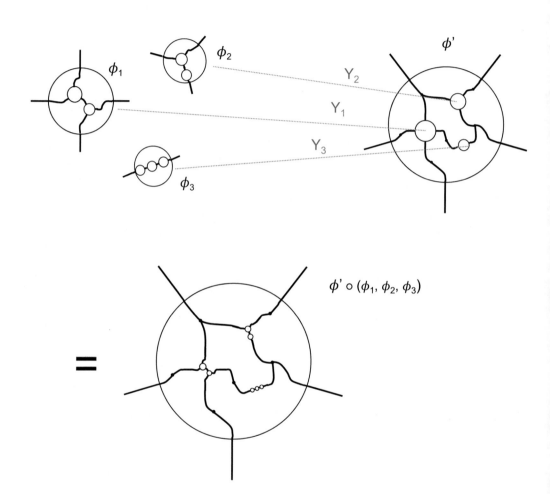

Figure 7.8 Morphisms in a wiring diagram operad $\mathcal{W}$. Composition of wiring diagrams is given by substitution.

References

[1] Abramsky, S. (2012) Relational databases and Bell's Theorem. Available at http://arxiv.org/abs/1208.6416

[2] Atiyah, M. (1989) Topological quantum field theories. *Publications Mathématiques de l'IHÉS* 68(1), 175–186.

[3] Axler, S. (1997) *Linear Algebra Done Right.* 2d ed. New York: Springer.

[4] Awodey, S. (2010) *Category Theory.* 2d ed. Oxford: Oxford University Press.

[5] Bralow, H. (1961) Possible principles underlying the transformation of sensory messages. In *Sensory Communication*, ed. W. Rosenblaith, 217–234. Cambridge, MA: MIT Press.

[6] Baez, J.C.; Dolan, J. (1995) Higher-dimensional algebra and topological quantum field theory. *Journal of Mathematical Physics* 36: 6073–6105.

[7] Baez, J.C.; Fritz, T.; Leinster, T. (2011) A characterization of entropy in terms of information loss. *Entropy* 13(11): 1945–1957.

[8] Baez, J.C.; Stay, M. (2011) Physics, topology, logic and computation: a Rosetta Stone. In *New Structures for Physics*, ed. B. Coecke, 95Ð172. Lecture Notes in Physics 813. Heidelberg: Springer.

[9] Brown, R.; Porter, T. (2006) Category Theory: An abstract setting for analogy and comparison. In: *What Is Category Theory?* ed. G. Sica, 257–274. Advanced Studies in Mathematics and Logic. Monza Italy: Polimetrica.

[10] Brown, R.; Porter, T. (2003) Category theory and higher dimensional algebra: potential descriptive tools in neuroscience. In *Proceedings of the International Conference on Theoretical Neurobiology*, vol. 1, 80–92.

475

[11] Barr, M.; Wells, C. (1990) *Category Theory for Computing Science*. New York: Prentice Hall.

[12] Biggs, N.M. (2004) *Discrete Mathematics*. New York: Oxford University Press.

[13] Diaconescu, R. (2008) *Institution-Independent Model Theory* Boston: Birkhäuser.

[14] Döring, A.; Isham, C. J. (2008) A topos foundation for theories of physics. I. Formal languages for physics. *Journal of Mathematical Physics* 49(5): 053515.

[15] Ehresmann, A.C.; Vanbremeersch, J-P. (2007) *Memory Evolutive Systems: Hierarchy, Emergence, Cognition*. Amsterdam: Elsevier.

[16] Everett III, H. (1973). The theory of the universal wave function. In *The Many-Worlds Interpretation of Quantum Mechanics*, ed. B.S. DeWitt and N. Graham, 3–140. Princeton, NJ: Princeton University Press.

[17] Goguen, J. (1992) Sheaf semantics for concurrent interacting objects *Mathematical Structures in Computer Science* 2(2): 159–191.

[18] Grothendieck, A.; Raynaud, M. (1971) *Revêtements étales et groupe fondamental* Séminaire de Géométrie Algébrique du Bois Marie, 1960/61 (SGA 1) Lecture Notes in Mathematics 224. In French. New York: Springer.

[19] Krömer, R. (2007) *Tool and Object: A History and Philosophy of Category Theory*. Boston: Birkhäuser.

[20] Lambek, J. (1980) From λ-calculus to Cartesian closed categories. In *To H. B. Curry: Essays on Combinatory Logic, Lambda Calculus and Formalism*, ed. J.P. Seldin and J. Hindley, 376–402. London: Academic Press.

[21] Khovanov, M. (2000) A categorificiation of the Jones polynomial. *Duke Mathematical Journal* 101(3):359–426.

[22] Landry, E.; Marquis, J.-P. (2005) Categories in contexts: Historical, foundational, and philosophical. *Philosophia Mathematica* 13(1): 1–43.

[23] Lawvere, F.W. (2005) An elementary theory of the category of sets (long version) with commentary. *Reprints in Theory and Applications of Categories*. no. 11, 1–35. Expanded from *Procedings of the National Academy of Sciences* 1964; 52(6):1506–1511.

[24] Lawvere, F.W.; Schanuel, S.H. (2009) *Conceptual Mathematics. A First Introduction to Categories*. 2d ed. Cambridge: Cambridge University Press.

[25] Leinster, T. (2004) *Higher Operads, Higher Categories*. London Mathematical Society Lecture Note Series 298. New York: Cambridge University Press.

[26] Leinster, T. (2012) Rethinking set theory. Available at http://arxiv.org/abs/1212.6543.

[27] Linsker, R. (1988) Self-organization in a perceptual network. *Computer* 21(3): 105–117.

[28] MacKay, D.J. (2003) *Information Theory, Inference and Learning Algorithms*. Cambridge: Cambridge University Press.

[29] Mac Lane, S. (1998) *Categories for the Working Mathematician*. 2d ed. New York: Springer.

[30] Marquis, J.-P. (2009) *From a Geometrical Point of View: A Study in the History and Philosophy of Category Theory*. New York: Springer.

[31] Marquis, J.-P. (2013) Category theory. In *Stanford Encyclopedia of Philosophy* (summer ed.), ed. E.N. Zalta, Available at http://plato.stanford.edu/archives/spr2011/entries/category-theory.

[32] Minsky, M. (1985) *The Society of Mind*. New York: Simon and Schuster.

[33] Moggi, E. (1991) Notions of computation and monads. *Information and Computation* 93(1): 52–92.

[34] nLab. http://ncatlab.org/nlab/show/HomePage.

[35] Penrose, R. (2005) *The Road to Reality*. New York: Knopf.

[36] Radul, A.; Sussman, G.J. (2009). The Art of the Propagator. MIT Computer Science and Artificial Intelligence Laboratory Technical Report.

[37] Simmons, H. (2011) *An Introduction to Category Theory*. New York: Cambridge University Press.

[38] Spivak, D.I. (2012) Functorial data migration. *Information and Computation* 217 (August): 31–51.

[39] Spivak, D.I. (2013) Database queries and constraints via lifting problems. *Mathematical structures in computer science* 1–55. Available at http://arxiv.org/abs/1202.2591.

[40] Spivak, D.I. (2012) Kleisli database instances. Available at http://arxiv.org/abs/1209.1011.

[41] Spivak, D.I. (2013) The operad of wiring diagrams: Formalizing a graphical language for databases, recursion, and plug-and-play circuits. Available at: http://arxiv.org/abs/1305.0297.

[42] Spivak, D.I.; Giesa, T.; Wood, E.; Buehler, M.J. (2011) Category-theoretic analysis of hierarchical protein materials and social networks. *PLoS ONE* 6(9): e23911.

[43] Spivak, D.I.; Kent, R.E. (2012) Ologs: A categorical framework for knowledge representation. *PLoS ONE* 7(1): e24274.

[44] Weinberger, S. (2011) What is … persistent homology? *Notices of the AMS* 58(1): 36–39.

[45] Weinstein, A. (1996) Groupoids: Unifying internal and external symmetry. *Notices of the AMS* 43(7): 744–752.

[46] Wikipedia. Accessed between December 6, 2012 and December 31, 2013.

Index